Ultrafine-Grain Ceramics

SAGAMORE ARMY MATERIALS RESEARCH
CONFERENCE PROCEEDINGS
Published by Syracuse University Press

Fundamentals of Deformation Processing
eds., Walter A. Backofen and others
(9th Proceeding)

Fatigue—An Interdisciplinary Approach
eds., John J. Burke, Norman L. Reed, and Volker Weiss
(10th Proceeding)

Strengthening Mechanisms—Metals and Ceramics
eds., John J. Burke, Norman L. Reed, and Volker Weiss
(12th Proceeding)

Surfaces and Interfaces I
Chemical and Physical Characteristics
eds., John J. Burke, Norman L. Reed, and Volker Weiss
(13th Proceeding)

Surfaces and Interfaces II
Physical and Mechanical Properties
eds., John J. Burke, Norman L. Reed, and Volker Weiss
(14th Proceeding)

Ultrafine-Grain Ceramics

• • • • • • •

EDITORS

JOHN J. BURKE

*Staff Scientist, Army Materials and
Mechanics Research Center*

NORMAN L. REED

*Associate Director, Army Materials and
Mechanics Research Center*

VOLKER WEISS

Professor, Syracuse University

Proceedings of the 15th Sagamore Army Materials
Research Conference. Held at Sagamore Conference
Center, Raquette Lake, New York, August 20–23.
1968. Sponsored by Army Materials and Mechanics
Research Center, Watertown, Mass., in cooperation
with Syracuse University. Organized and directed by
Army Materials and Mechanics Research Center in
cooperation with Syracuse University.

SYRACUSE UNIVERSITY PRESS

Standard Book Number
8156-5026-4

Library of Congress
Catalog Card Number: 64–12568

Printed in United States of America
Composed and printed by The Science Press, Inc., Ephrata, Pa.
Bound by Vail-Ballou Press, Inc., Binghamton, N. Y.

Sagamore Conference Committee

Chairman
Norman L. Reed, Army Materials and Mechanics
Research Center

Vice Chairman
John J. Burke, Army Materials and Mechanics
Research Center

Program Director
Volker Weiss, Syracuse University

Program Coordinator
Robert J. Sell, Syracuse University

Program Committee
Irving Berman, Army Materials and Mechanics
Research Center
Capt. Stuart L. Blank, Army Materials and Mechanics
Research Center
John J. Burke, Army Materials and Mechanics
Research Center
H. M. Davis, U. S. Army Research Office
Alexis G. Pincus, IIT Research Institute
Norman L. Reed, Army Materials and Mechanics
Research Center
Richard M. Spriggs, Lehigh University

Arrangements at
Sagamore Conference Center
Richard A. Jones, Syracuse University

Foreword

The Army Materials and Mechanics Research Center has conducted the Sagamore Army Materials Research Conferences, in cooperation with the Metallurgical Research Laboratories of the Department of Chemical Engineering and Metallurgy of Syracuse University, since 1954. The purpose of the conferences has been to gather together scientists and engineers from academic institutions, industry, and government who are uniquely qualified to explore in depth a subject of importance to the Army, the Department of Defense and the scientific community.

Grain size and its control is a major factor governing the behavior of materials. Knowledge that one can obtain grain sizes in the 10 micron down to submicron range has established the basis for a new technology.

This volume, Ultrafine-Grain Ceramics, addresses itself to the broad areas of: realization of fine-grain ceramics; preparation of ultrafine particle size materials; characterization of ultrafine particles and ultrafine-grain ceramics; processing and behavior of ultrafine-grain ceramics.

The technical advice provided by Dr. S. K. Dutta of the Army Materials and Mechanics Research Center during the editing of this book is acknowledged.

The continued active interest and support of these conferences by Dr. E. Scala, Director, Lt. Col. Joseph B. Mason, Commanding Officer, and J. F. Sullivan, Deputy Technical Director, of the Army Materials and Mechanics Research Center is appreciated.

Sagamore Conference Committee   The Editors
Raquette Lake, New York
August, 1968

Contents

ix

SESSION V
BEHAVIOR OF ULTRAFINE-GRAIN CERAMICS
S. M. Copley, *Moderator*

SESSION I

INTRODUCTION

MODERATOR: NORMAN L. REED
Army Materials and Mechanics Research Center
Watertown, Massachusetts

1. Realization of Fine-Grain Ceramics

ALEXIS G. PINCUS
IIT Research Institute
Chicago, Illinois

' *ABSTRACT*

Grain size is but one factor in the complex microstructures of ceramics. Ability to control or minimize unwanted phases has made feasible evaluation of the effect of grain size on physical properties, but interpretations must also account for other influences such as grain bonding and internal strains, surface condition, and extrinsic flaws. Recently, attention has been focused on processing as the key to grain refinement. The diverse approaches within particulate, melt, and vapor-processing will be highlighted in terms of potential for control over grain size.

' Introduction

This conference is concerned with ultrafine-grain ceramics and how to prepare them.* The reason for our interest is that some remarkable improvements in various properties of practical significance appear to take place as grain sizes grow smaller. For example, we are at a stage where tantalizing promise exists that grain refinement will provide dramatic improvements in the strength of monolithic ceramics. A ceiling which has existed for more than a decade in the strength of alumina ceramics at 50,000 psi flexural is now being upgraded to 100,000 psi reproducibly, with promise of an even higher plateau. Similar dramatic property improvements with grain refinement have been propounded for one after another of the large variety of compositions of oxide ceramics which have been finding unique uses: barium titanate for capacitors, ferrites for recording heads or memory cores, zirconia for refractoriness in oxidizing atmospheres, lithium aluminosilicates for thermal shock resistant materials, and so on.

But the promises have remained tantalizing because of lack of assurance that the higher property levels can be achieved outside of the laboratory, reproducibly and economically. This lack of reproducibility has been ascribed to two factors: (1) inadequate knowledge about all the other interrelated features of

---

*Ultrafine grain is being arbitrarily defined as less than 10 μ in diameter.

microstructure in addition to grain size; and (2) inadequate capability in the processing of ceramics.

The first factor lies in the domain of "Physical Ceramics." Great as have been the advances in this discipline, it is being recognized that knowledge about character–property relationships is still insufficient to satisfy the ever more stringent demands for reliability and enhancement of properties. A further handicap has been the communication gap between the physical ceramist and the process ceramist. This gap has slowed utilization of the new knowledge, and conversely has hindered the feedback of problems in processing which indicate needed areas of research in physical ceramics.

The deficiencies of ceramic processing have been a favorite complaint for a decade now. The process ceramist has been made a whipping boy for all the cries of "non-uniformity," "non-reproducibility," and "unreliability" which have issued from the customers for ceramics in government and industry. A series of surveys have been published which have attempted to identify the shortcomings and how they might be cured by liberal funding of research and development.

Meanwhile technologists, inventors, engineers, and entrepreneurs have been active in advancing processing by developing new methods and equipment, improved starting materials, and more sophisticated controls. It is my belief that processing has been unfairly made a scapegoat. Again the problem may be in communications, this time between processors and users.

Conferences like this, then, serve an important purpose in bringing users together with scientists from the research laboratories and technologists from development and manufacturing to jointly appraise their achievements, their needs, and their unsolved problems.

Some Definitions And Relationships

When groups of people meet over a period of time to debate an area of specialization, they develop a jargon which renders the subject esoteric to outsiders. This has been a consequence of the three-year effort of the Materials Advisory Board Ad Hoc Committee on Ceramic Processing and its panels [1]. Therefore, let me begin with a review of the terms and relationships evolved by that Committee.

Figure 1 reproduces the chart which describes the sequence in the processing of raw materials to useful products. *Starting materials*—almost always particulates—are converted by *controlled processing* into a *ceramic product*, which has to be characterized by observations and measurements of its *character* and related *properties and behavior* in order that the prospective *users* know how to design it into useful devices and systems. The coupling of character to properties and behavior is called *evaluation*.

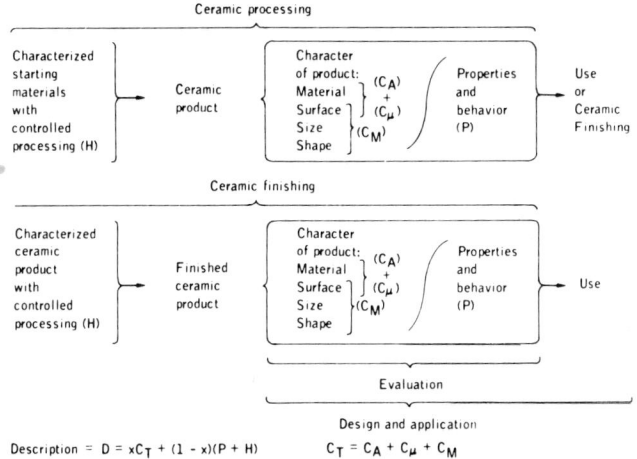

Figure 1. Ceramic Processing and Finishing

Processing may follow the route of *solids processing* based on solid particulates as the precursor of the consolidated product, or it may pass through a *fluid* stage where the immediate precursor of the solid product is largely or completely in the liquid or the vapor state. In either case the product is often and increasingly handed over to finishing operations which further modify the surface, and sometimes the bulk character of the material, and help in meeting dimensional specifications. The finishing operations and their products have to be subjected to the same evaluation as the initial ceramic, though usually with more emphasis on surfaces.

Character was originally defined as the sum of the compositional and structural descriptions which identify a specimen uniquely and may be correlated with its properties. *Characterization* is, then, the determination of the character of a specimen. More explicitly, and following the definition of a Materials Advisory Board committee on "Characterization of Materials" [2], the definition becomes: Characterization describes those features of the composition and structure (including defects) of a material that are significant for a particular preparation, study of properties, or use, and suffice for the reproduction of a material.

It was also recognized that we are not yet knowledgeable enough to avoid introduction of processing history and property measurements into attempts to characterize materials. But to retain the desirable limits to the definition of character, this added information should be called "Description":

$$D = xC_T + (1 - x)(P + H)$$

where

D = description
C_T = total character

P = property measurements not presently capable
     of being related to character features

H = processing history (pressure, temperature, time,
     environment, etc.)

x = a factor with values between 0 and 1

Ideally, the use of processing history to describe a material should be abandoned as soon as possible. Use of property measurements will also become less necessary as the ability improves in measuring character features and relating them to behavior in testing and service. Ultimately, a stage may be reached where x will become unity, and a material will then be completely described in terms of character.

By total character is meant the sum of the characters on the atomic, micro, and macro levels.

$$C_T = C_A + C_\mu + C_M$$

C_A = atomic character

C_μ = microcharacter

C_M = macrocharacter (including size and shape)

At each level of size resolution, character needs to include both the bulk and the surface of the material.

Atomic characterization is concerned with the chemical identity and location of all the atoms in a solid, and includes their chemical bonding. Microcharacterization deals with the scale that can be seen directly by microscope techniques. Macrocharacterization is concerned with size and shape, density, bulk chemical composition, homogeneity, surface topography, and gross stress distribution.

Microstructure was defined as those elements of character referring to identifiable grains and phases, their geometric features, and their distributions.

Microstructures of Ceramics

The factors of a microstructure that are recognized at present include those listed in Table I [3]. It is apparent that grain size is only one factor among at least three dozen. And remember that microstructure is only part of total character.

Until process ceramists, guided by physical ceramists, learned how to eliminate or minimize certain extraneous features of microstructure—pores, unwanted crystalline and glassy phases, and wide grain-size distributions [4]—it was not at all possible to establish the significance of grain sizes in determination of a property level such as strength. Even now with the adoption of "ultra-pure" raw materials, processed to avoid contaminants, and matured by solid-state

TABLE I

Microstructural Features in Ceramic Products

1. Predominating Crystal Phase
    1.1 Phase identification
    1.2 Quantity
    1.3 Size and distribution
    1.4 Shape and shape distribution (including connectivity)
    1.5 Preferred orientation
    1.6 Internal strain
    1.7 Composition
    1.8 Dislocations

2. Additional Phases (Glassy and Crystalline)
    2.1 to 2.8 as above

3. Grain Boundaries
    3.1 Composition and composition distribution
    3.2 Structure
    3.3 Relative orientation of grains
    3.4 Grain boundary strain

4. Pores, Microcracks, Cracks
    4.1 Quantity
    4.2 Distribution in structure
    4.3 Size and size distribution
    4.4 Shape and shape distribution (including connectivity)
    4.5 Relative orientation
    4.6 Gas composition

5. Surfaces
    5.1 Topography
    5.2 State of stress (strain)
    5.3 Adsorbed gases
    5.4 Chemistry

6. Extrinsic Defects
    6.1 Type
    6.2 Position

Adapted from "Ceramic Processing" [3], page 184–85.

sintering to full density without exaggerated grain growth, it is not possible to avoid significant concentrations of secondary phases at grain boundaries, as de Wys and Leipold have shown [5].

At best, the interactions between the various character features demand care and judgment in interpretation [6]; Until surface finish is eliminated as the major controller of strength measurements, the effects of porosity or grain size are masked. Even when the quality of the surface finish is improved and the strength increased until a grain-size effect is observable, some other factor may become limiting. For example, as the grain size is varied by controlling thermal history, the extent of segregation at the grain boundaries may cause variations in strength not related to the grain size. When one factor is being appraised, the state of several other potentially limiting factors must be specified. In the case of ultimate tensile strength used as a descriptive property, 29 character features in addition to grain size must be maintained constant!

With such complexities it is not surprising that the contribution of grain size to strength is barely beginning to be sorted out from other factors. The papers comprising this book will help us to appraise the state-of-the-science and point to future directions for research and development.

In the absence of definitive proof that it is fine grain size alone which can bring about higher strength and toughness, a surprising amount of activity already exists in pursuit of fine-grained microstructures for many compositions and properties and resulting applications. Concerning microstructure, these take the direction both of essentially single-phase, pure oxide ceramics and of two-or-more-phase, mixed grain materials. The two-phase materials have to be -designed in terms of composite theory, whether the second phase is glassy or crystallized, ceramic or metal, oxide or other -ide, euhedral or fibrous, continuous or enclosed by grains of the predominant crystal phase. With single-phase ceramics, grain boundaries, grain bonding, and internal strains become the critical features supplementary to grain size.

Determination of grain size and size distribution is commonly made by light-microscopy, using reflection on polished sections or transmission on thin sections. These reach their limits of resolution at approximately 1μ. For smaller sizes, recourse is had to electron-microscopy by replica or direct transmission, or to scanning electron-microscopy. Line-broadening from X-ray or electron diffraction can be used to assess grain size and also for a measurement of internal strain. But all of these techniques give only a limited picture of size distribution, and there has been no standard accepted for qualitative or quantitative presentation of size data. This book is fortunate in that it contains two authoritative chapters on this subject.

Processing to Achieve Fine Grain

Paralleling the activity on fine-grain ceramics has been extensive and intensive development of improved and novel processing techniques. Many of these have been pursued because of their promise for assisting the realization of full-density, fine-grained materials. Among them may be mentioned controlled comminution and compaction; sintering with grain growth inhibitors or with ultrafine particles of controlled agglomeration; high energy rate forming; hot pressing and other types of hot working; isostatic hot pressing; melting by either the method of immediate crystallization or by crystallization from a glass; particle impact; and vapor deposition.

These developments have been critically evaluated in several current state-of-the-art reviews [1,7,8,9,10] and are the focus of continuing discussions at every gathering of ceramists.

Fluid processing has an advantage in that it bypasses the problems associated with preparing, storing, characterizing, and using ultrafine particles [11]. However, it has its own limitations and problems. Considering processing by crystallization from glasses, for example, major successes have been the achievement of extremely fine grain (hundreths of a micron) and outstanding properties, but the limitations include the inability to produce all the desired mineral compositions, the presence of residual glassy phase, and the development of microcracks when crystallization proceeds too rapidly or too completely. Vapor deposition (or incremental gas-phase molecular transport) processes are still slow for forming massive ceramics, and there is insufficient understanding of fundamentals or experience with the technologies involved.

Nevertheless, a strong case can be made for large-scale efforts on improving the capability of fluid processing for producing fine-grained ceramics [11].

Despite all the attention given to more novel processing approaches, solids processing and its precursor, particle preparation, remain the sources of most effort in the attempt to prepare ultrafine-grain ceramics. Paths to finer grain which are being pursued by the route of particulate processing include:

1. Preparation of ultrafine ("reactive") particles by a variety of approaches.
2. Addition of grain growth inhibitors to the batch or in the sintering atmosphere.
3. Milling to break up agglomerates, while avoiding excessive reaction with water in the atmosphere.
4. Compaction by fragmentation whether cold, hot, or explosive (high energy rate).
5. Control over sintering kinetics.
6. Understanding of the role of surface chemistry, of adsorbed gases, in both compaction and sintering.
7. Imposition of pressure simultaneously with heat to lower maturing temperature and promote high density and fine grain.
8. Hot working after sintering (extrusion, forging, strain annealing).

A select few of these approaches to fine-grain ceramics are presented in this volume. Consideration of all of these approaches in the depth which they merit, as well as in their inter-relationships and competitive advantages, would take a semester course at least, featuring many guest lecturers on their specialties.

Particle Preparation and Characterization

The preparation of ultrafine-particle materials is the subject of three chapters in this book, as well as inherent in other chapters. An impressive fact is the variety of approaches which are being pursued to produce fine particles as starting materials for ultrafine-grain ceramics. A classification of the methods parallels every one of the techniques being explored for ceramic processing and adds a few others. Fine particles are being provided by:

1. Geology—natural mineral deposits which need less or more beneficiation.
2. Mechanical size reduction (comminution).
3. Solid-state reactions.
4. Hydrothermal formation.
5. Melts in bulk, followed by comminution with or without accompanying chemical purification.
6. Particulate melting (atomization).
7. Precipitation or co-precipitation from solutions.
8. The sol-gel process.
9. The cryochemical (freeze-drying) approach.
10. Chemical vapor phase reactions ("snowing out").
11. Physical vaporization-condensation.
12. Oxidation of metals.
13. Electrolytic anodization or deposition.

TABLE II

List of Characteristics, Properties,
and Performance of Particulate Materials

Characteristics†	Properties of a Particulate System	Performance of a Particulate System
elemental composition	capillarity	green strength
(impurity content)	permeability	angle of repose
(stoichiometry)	electrical conductivity	grindability
crystal structure	magnetic susceptibility	
surface energy	heat capacity	rheology
defects	thermal expansion	sinterability
size*	thermal conductivity	compactibility
shape*	adsorption of molecules	flowability
density, porosity, pore	zero point of charge	flocculation and
distribution*		dispersion
size distribution**	reactivity of solids	
specific surface**	heat of immersion	
mixedness**		
bulk density**		

†Those characteristics marked (*) apply to a particle, whereas those marked (**) apply to an assembly of particles. The remainder apply to both.

E. R. Stover (U. Calif., Berkeley) "A Critical Survey of Characterization of Particulate Ceramic Raw Materials" [14]. AFML-TR-67-56 (May 1967).

Characterization of ultrafine particles is still considered so difficult that the MAB Committee on Characterization recommended a major mission in this area [12]. However, a start has been made in various symposia [13], critical surveys [14, 15], and research in progress. Measurements of true particle sizes and distributions are being established on a rational basis. Surface area measurements have become a routine control tool. New instruments and techniques, such as scanning electron-microscopy, combined with infra-red and dielectric measurements, are producing useful information. Much can be learned from researches on catalysts. A feel is developing for those character features from the long lists (Table II) which must be appraised if ultrafine particles are to be properly processed into reproducible ceramics.

Ability to characterize systems of ultrafine particles is ovbiously not a hopeless task. Instead, the need is to get on with the hard work which will lead to mastery over the materials. Just as ceramists have not feared to deal with complex, polycrystalline materials, while the solid-state physicists called for working only on "perfect single crystals," ways will be found to adequately characterize ultrafine particles.

Consequences of Fine-Grain Ceramics on Usages

From correspondence with colleagues active in research and production of fine-grain ceramics, concern emerges over so much work for so little apparent gain. Yet a more realistic assessment affirms how much learning has been accumulated during the recent period, and how securely the foundations have been laid for the jump into the next levels of improved properties.

The usefulness of fine grain is being demonstrated in one after another of the compositions and functional applications of ceramics. The material class which has the most revealing history is aluminum oxide ceramics [16]. First developed for spark plugs [17] because of their resistance to corrosion by lead oxide coupled with thermal shock resistance and electric strength, the materials were then adapted for electron-tube uses as electrical insulators, envelopes, and spacers [18], and later as radomes and substrates. Important user requirements were low dielectric losses at ultrahigh frequencies, strength superior to then-available glasses and porcelains, thermal shock resistance, and sealability to metals. The knowledge about compositions and processing developed for these areas led to increased utilization where mechanical wear is important: bearings, mill linings and balls [19], culminating in the intensive development of alumina cutting tools.

Here the significance of fine grain was recognized, and both hot pressing and unique additives were adopted commercially to achieve grain sizes around 2 μ. Table III, adapted from King and Wheildon [20], summarizes the average grain

TABLE III

Properties of Alumina Ceramic Tools

Tool	Grain Size Average (microns)	Transverse Rupture Strength (10^3 psi)	Knoop Microhardness K100
0–30	2.	85	1665
CCT–707	3.	85	1570
VR97	4.	87	1655
France	3.5	70	1580
Germany	3.	66	1595
Diamonite	10.	49	1405
U.S.S.R.	5.	51	1525
Japan	5.5	57	1515

From King and Wheildon (1966) [20].

size and strengths of commercial tool bit materials as they were available in 1966. Recent developments appear to be leading to tool bits at around 1 μ grain size achieved at 99.9% Al_2O_3 by conventional sintering, and in some respects better than the hot pressed products [21]. This further illustrates the recurrent history that non-conventional processing leads the way in demonstrating improved microstructure and properties, but that more conventional processing then learns how to achieve the same results more economically.

Research on sintering of alumina led to the translucent, full-density but coarse-grained alumina ceramics best known as Lucalox [22]. These are useful in lamps and other applications, not only for their optical properties but for chemical properties such as resistance to alkali metal vapors, as well as for good mechanical and electrical properties above 1000°C, and ability to be sealed to useful metals.

With the translucent aluminas, the trend was away from fine grain in order to maximize light transmission by minimizing grain boundary area. But the fundamental knowledge generated about solid-state sintering has been highly useful in furthering achievement of fine grain. The urge towards fine grain came not only from the users of cutting tools and substrates, but also from the promise of utility for many other potential applications demanding strength for aerospace, deep submergence, armor, and a range of other significant defense needs [23]. In civilian applications, tire studs are mentioned as promising a large usage.

A consequence of the number of potential uses has been a burst of research on the physical ceramics of fine-grain oxide ceramics. The papers included in this book should make us *au courant* in the area of the relationships between fine grain and high levels of strength. Technologically, I am informed that the 100,000 psi alumina ceramic is already here, and not only for use in bits for cutting tools, but also for spark plugs and larger products. The chief limitation to large-scale manufacture and use is now said to be the economic one of reduced cost for already available ultrafine-particle starting materials.

I have sketched this picture of alumina ceramics to demonstrate again the motivation of user needs, stimulating both fundamental research directed to understanding character–property relationships and technological advances in the entire sequence of processing from starting materials to finishing operations. The demands from each new use on properties, character, and processing have drawn forth the understanding and the capabilities needed for the requirements of the succeeding uses.

Conclusion

We see the timeliness of the topic and theme of this volume. Fine-grain ceramics have been realized both in the laboratory and in the mass-production factory. They are successively opening new uses for ceramics. But many of the potential applications are waiting for lowered costs and/or more certain correlation between grain size and wanted properties.

Knowledge about and control over unwanted secondary phases made benefits from full density feasible. Relative freedom from pores, coupled with control over surface finish, permitted benefiting from grain refinement. We may expect that the understanding and control over grain size which is being appraised at this conference will prepare the way for meaningful research on the next critical feature of character.

Great as have been the advances, continuing questioning is needed on whether we adequately understand each significant feature of microstructure (and more broadly of character) and their inter-relationships; on whether we are performing the right tests; on whether the measured properties are being properly related to character.

It is our hope that readers of this book will be inspired to work on these problems and to help bring the potentialities to realization.

Acknowledgments

I am indebted to many colleagues at IITRI, on the Materials Advisory Board Committee on Ceramic Processing, and in the ceramics profession for discussions and suggestions.

References

1. *Ceramic Processing*, Materials Advisory Board of National Research Council, National Academy of Sciences Publication 1576, Washington, D.C. (1968).

2. *Characterization of Materials,* Materials Advisory Board Publication MAB-229-M, National Academy of Sciences–National Academy of Engineering, Washington, D.C. (1967).

3. Reference 1, pp. 184-85.

4. Pincus, A. G., "Utilization of Microstructures in Processing," Chap. 41 in Fulrath, R. M. and Pask, J. A., *Ceramic Microstructures,* John Wiley and Sons Inc., New York (1968).

5. de Wys, E. C. and Leipold, M. H., *The Structure of the Grain Boundary in Ceramics,* NASA-CR-93366 (1968).

6. Reference 1, p. 183.

7. Pincus, A. G., *Critical Compilation of Ceramic Forming Methods,* Air Force Materials Laboratory, Technical Document Report RTD-TDR-63-4069 (Jan. 1964), AD431002 [Condensed version: *Bull. Am. Ceram. Soc.,* 43 (1964) 827, 880; 44 (1965) 1, 145, 209]. Chaps.: Allen, A. W., "Cold Forming Processes"; Fulrath, R. M., "Hot Forming Processes"; Cooper, A. R., "Melt Forming Processes"; Bowers, D. J., "Miscellaneous Forming Methods"; Fuerstenau, D. W. and Mular, A. L., "Characterization of Particulate Systems"; Pincus, A. G. and Pask, J. A., Fundamentals of Forming." Initial bibliographies on "Characterization of Particles and Assemblies of Particles," by Mular, A. L. and Mee, V. A.; "Forming of Ceramics," by Pincus, A. G., Rose, R. E. and Mee, V. A.

8. *Report of the Ad Hoc Committee on Processing of Ceramic Materials,* Materials Advisory Board, National Academy of Sciences/National Research Council, MAB-197-M (1963).

9. Emrich, B. R., *Technology of New Devitrified Ceramics–A Literature Review,* Air Force Materials Laboratory, ML-TDR-64-203, AD608 217 (1964).

10. Brown, S. D., *et al., Critical Evaluation of Ceramic Processing at Subconventional Temperatures,* Air Force Materials Laboratory, AFML-TR-67-194, AD822 145 (1967).

11. Reference 1, Chap. 3, pp. 67-70, 72, 112-21.

12. Reference 2, pp. I-29, V-33.

13. Kuhn, W. E., ed., *Ultrafine Particles,* Proceedings of a Symposium Sponsored by the Electrothermics and Metallurgy Division, Electrochemical Society. John Wiley and Sons, Inc., New York (1963).

14. Stover, E. R., *Critical Survey of Characterization of Particulate Ceramic Raw Materials,* Air Force Materials Laboratory, AFML TR-67-56, AD816 014 (1967).

15. Palmour, H. III, *et al., Raw Materials for Refractory Oxide Ceramics,* Air Force Materials Laboratory, M-TDR-64-100 (1964).

16. Ryshkewitch, E., *Oxide Ceramics,* Academic Press, New York (1960).

17. Bartlett, H. B. and Schwartzwalder, K., "Trends in the Chemical and Mineralogical Constitution of Spark Plug Insulators," *Bull, Am. Ceram. Soc.,* 28 (1949), 462-70. See also: Riddle, F. H., "Ceramic Spark Plug Insulators," *J. Am. Ceram. Soc.,* 32 (1949), 333-46.

18. Navias, L., "Advances in Ceramics Related to Electronic Tube Developments." *J. Am. Ceram. Soc.,* 37 (1954), 329-50.

19. Lintner, R. E., *et al.,* "New Horizons for Industrial Ceramics," *Metal Progr.,* 84, (1963), 109-15, and 101-106.

20. King, A. G. and Wheildon, W. M., *Ceramics in Machining Processes,* Academic Press, New York (1966).

21. Private communications.

22. Coble, R. L., Transparent Alumina and Method of Preparation, U.S. Patent 3,026,210 (1962).

23. Reference 1, pp. 2-3.

SESSION II

PREPARATION OF ULTRAFINE
PARTICLE SIZE MATERIALS

MODERATOR: CAPT. STUART L. BLANK
Army Materials and Mechanics Research Center
Watertown, Massachusetts

2. Fine Grinding—Size Distribution, Particle Characterization, and Mechanical Methods

THOMAS P. MELOY

Melpar, Inc.
Falls Church, Virginia

. *ABSTRACT*

Fine grinding is a complex engineering art about which little is known. The general comminution theories are reviewed. These include comminution of homogeneous substances considering particle shape, and heterogeneous materials broken under uniform stress as well as non-uniform stress. A general theorem is proven, demonstrating that the size distribution in the finer sizes is independent of many parameters that affect the larger size distribution. Recent three-dimensional comminution model work is reviewed, and its implication on the effect of particle characteristics such as shape are discussed. Fine particle grinding by conventional means such as jet and hammer milling are reviewed against the background of modern comminution theory. A discussion of recognitive and non-recognitive properties of fine particles is presented in the light of the need of better particle characterization. Recent work in experimental particle characterization is also reviewed.

Introduction

The fine grinding of powders by mechanical methods is an industrial process important to a wide variety of industries and products. It is an exciting field because it is an operation in which much improvement can be made; interesting because of the diverse problems faced in the different industries; stimulating because of the recent advances in the general theory of comminution and grinding; but exacerbating because so little is known about it. Each year millions of tons of material are ground into the micron size range, but very little is understood about the process, the particles formed, or even the powders that result.

Particles are the product of unit processes which have until recently defied analysis. Unless created by chemical reactions in which the feed and products are monodispersed, each particle is unique in itself and carries a tremendous amount of information. A $1\ \mu$ particle transmits, via the genetic code, information equivalent to the contents of 15 sets of the *Encyclopaedia Britannica*. Par-

17

ticle size, mass, shape, angularity, cornisity, surface area, and surface roughness, are all terms used to describe properties of a particle. With rare exceptions these properties are not defined, nor can they be measured.

This chapter, therefore, is in a way a review of what is not known, of what is not understood; and, in turn, is a plea that the complexity of the problem be recognized and that good fundamental work be started so that many of the necessary experiments and programs can be run, in order that we may truly have a concept of what a powder and particle are.

The following sections contain a discussion of the existing theories of comminution which describe how particles are created; a discussion of characterization of the particles; a discussion of some of the recent modeling work, describing comminution processes in general; and finally, a discussion of some of the problems relating to fine grinding. In a way, this chapter might best be considered a status report.

Theoretical Size Distributions of Single Particles

In the last decade, several attempts have been made to predict, on the basis of mathematical models, the size distributions that would result from the comminution of a single, brittle solid. While these approaches have been moderately successful and have received many of the processes that are occuring during comminution, much remains to be done to clarify several unresolved issues and to broaden the theoretical basis for prediction of results. Much of the trouble in testing these models is that little usable experimental data are available. This is why the need for better particle and powder characterization will be stressed later in this report.

Flaw Model

The first of the derivations was done by J. Gilvarry [1] while working at the Allis-Chalmers Laboratories. He chose a model, considering the activation of volume, surface, and edge flow to be individually and collectively distributed at random. A comminution wave passes through the material activating the flaws. Gilvarry asks the question: "What is the probability of having no volume flaws in volume V, but one flaw in volume dV?" Similarly: "What is the probability of having no flaws in area A, but one flaw in dA; no flaws in length ℓ, but one flaw in $d\ell$?" (See Figure 1.) These, and several key assumptions, lead to the following breakage function (the size distribution that results when a single particle is broken):

$$B(x) = 1 - \exp\left\{-\left(\frac{x}{i}\right) - \left(\frac{x}{i}\right)^2 - \left(\frac{x}{k}\right)^3\right\} \tag{1}$$

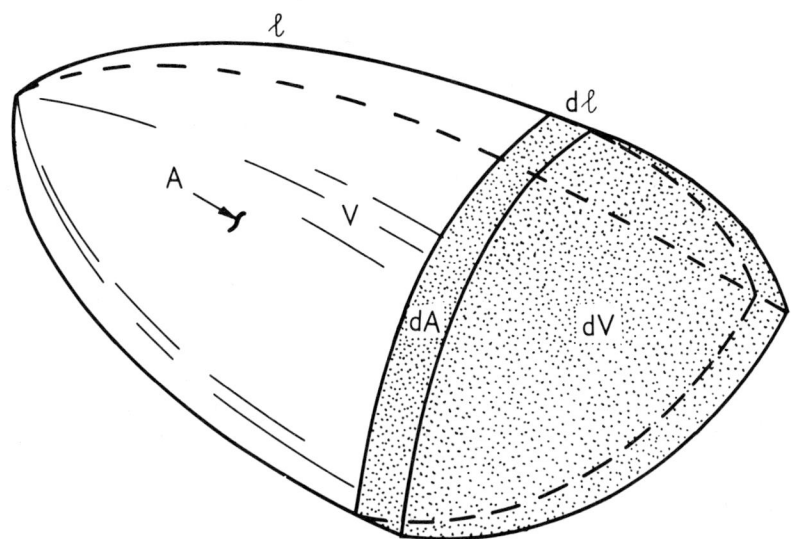

Figure 1. The particle is being formed by three propagating surfaces. A fourth crack at right angles to the other three must be initiated in the Gilvarry model for the particle to be formed. A crack initiated at a flaw in the volume, dV, the surface, dA, or the edge, dl, will complete the particle. See text for equation derived from this model.

The cumulative mass, $B(x)$, finer than size x is a function of the edge flaw density, i, the surface flaw density, j, and the volume flaw density, k, where there is an average of one edge flaw per length, i, one surface flaw per area j^2, and one volume flaw per volume k^3.

Crack Density Model

Working coterminously, Meloy [2] took another approach, utilizing the concept of crack density and avoiding the flaw problem. The basic assumption made is that the crack density at any location is proportional to the peak energy in the solid at the time of comminution. To make the derivation, Meloy asked, "what is the probability of a point, P, being in a particle of size x: $Kx^3 f(x)$, where $f(x)$ is the numerical density function of the number of particles of size x?" He then passed a point through the point and asked, "if $f(x)$ is the density distribution, what is the crack cut distribution along the line: $r(\ell - x)^{r-1}$?" Equating the two probabilities and integrating

$$B(x) = 1 - (1 - x)^r \tag{2}$$

The parameter, r, is the average number of cuts on an imaginary line passed through the original specimen.

Blocky Particles—Flaw Model

One of the difficulties with the derivation of this equation is that r can have a value of less than one. This has been interpreted by some as meaning that r has an average value less than one; that is to say, if 100 lines were passed through the reassembled crystal, then in the case of a lightly comminuted particle some lines would have no cuts, some would have one cut, and a few would have two or more cuts. (See section on Reconciliation of Flaw-Crack Density Model for variations in r values.) To average the cuts per line would give an average value to r, and a meaning to an r less than one. The writer [3] never believed this to be true and has rederived the equation, assuming that the particle fragments were blocky rather than cubical and could be represented by spheres as opposed to cubes. The question is asked, "what is the probability of cutting a sphere of radius x and chord length y?" Going through the now-familiar mathematics results in a breakage function with an added term. The function is:

$$B(x) = 1 - (1 - x)^r - \frac{rx}{2} (1 - x)^{r-1} \tag{3}$$

The result is that r cannot be less than one, a fact aesthetically pleasing; but more important, the assumption that the particles can be approximated by spheres circumvents the former objection that the particles in the reassembled sphere had to have line segments passing through them equal or proportional to the particle size.

Extrusion Flaw Model

Recently, a synthesis of the Gilvarry and Meloy models has been formulated by Austin and Klimpel [4] which is known as the "extrusion" model. The mass of the specimen is extruded into a noodle, and the question is asked, "what is the probability of a volume, surface, or edge flaw not occurring in the length of the noodle, ℓ, but in the length $d\ell$?" (The volume, surface, and length of the noodle are proportional to noodle length.) Using the mathematical approach of Meloy, they arrive at the equation:

$$B(x) = 1 - \left[1 - \frac{(x)}{(x_o)} \right]^{r_1} \cdot \left[1 - \frac{(x)^2}{(x_o)} \right]^{r_s} \cdot \left[1 - \frac{(x)^3}{(x_o)} \right]^{r_v} \tag{4}$$

where r_1, r_s, and r_v are, respectively, the average number of activated edge, surface, and volume flaws, and x_o is the characteristic size of the original specimen. The authors have recently had second thoughts on their approach.

Reconciliation of Flaw-Crack Density Model

As mentioned above in the section on Blocky Particles, r represents the number of cuts on one individual line passed through the crystal. Obviously, if more than one line is passed through a real comminuted crystal, then the number of cuts would vary from line to line. Equation (2) is derived on the assumption that r does not vary.

Assume [5] that r, the number of cuts on a line, is randomly distributed from line to line passed through a crystal. Assume the distribution of r to be Poisson. Summing over all values of r, equation (3) transforms to:

$$B(x) = \frac{1 - \exp(-\bar{r}x)}{1 - \exp(-r)} \tag{5}$$

Where $\bar{r}$ is the average value of r in the crystal.

When equation (5) is compared to equation (1) it becomes apparent that the two are the same. Whether the models are essentially the same or the results merely a coincidence is not known at this time.

Theorem

All the theoretical derivations show that the slope of the cumulative size distribution must be one for single breakage if it is assumed that the comminution density is uniform throughout the specimen, an assumption that is not true. However, a simple derivation [6] can remove that constraint. Let the crack density throughout the specimen be represented by $r(u,v,w)$, where u, v, w are the spatial coordinates in the specimen; thus, the size distribution in any differential volume (dV), (du,dv,dw) is given by $1 - (1 - x)^r (u,v,w)$. By integrating over the volume, one obtains the breakage function for the specimen broken in a non-uniform manner:

$$B(x) = \frac{1}{V} \iiint_V 1 - (1 - x)^{r(u,v,w)} \, du\,dv\,dw \tag{6}$$

For small x, using the binomial expansion of $1 - (1 - x)^r$ and throwing away higher terms of x

$$B(x) = \frac{1}{V} \iiint r(u,v,w) \cdot x \cdot du\,dv\,dw \tag{7}$$

Since the integral is on u, v, w, the variable x may be brought outside the integral

$$B(x) = x \cdot \frac{1}{V} \iiint r(u,v,w) \, du\,dv\,dw \tag{8}$$

The right-hand side of equation (8) is x times the average value of r, or $\bar{r}$. Thus:

$$B(x) = x\bar{r} \tag{9}$$

Equation (9) means that no matter how unevenly the crystal is broken, *the slope in the finer size ranges will still be one*. This is a powerful theorem, for it means that experimental data (which are never perfect) can be used to prove or disprove the validity of the modern comminution models, because the experimental data in the smaller size range are valid, even if the uniform stress conditions are not met. The same theorem can be applied to Gilvarry's work, i.e., use Equation (1) in equation (6). By using techniques of numerical integration and equation (6), size distributions can be made for non-uniform comminution of solids. In a cratering study, Meloy and Faust [7] showed that the size distribution in the upper sizes is much flatter than experienced in normal single-particle work.

Generalized One-Dimensional Inhomogeneous Model

All of the theoretical comminution work published thus far has been for homogeneous materials. Comminution of materials with a definite grain size or any periodic or definable variable susceptibility to comminution have been analyzed by Meloy and Gumtz [8]. Let $\rho(y)$ be the crack susceptibility variation through the sample, and r the average number of cracks along a line passed through the sample. Then, the breakage function for an inhomogeneous substance is:

$$B(x) = 1 - \frac{r}{r+1} \int_0^{(x_o - x)} \left[1 - \int_w^{w+x} \rho(y)\,dy\right]^{r-1} \rho(w)\,dw \frac{-1}{r+1} \left[\int_x^{x_o} (y)\,dy\right]^r \tag{10}$$

As in equation (2), $B(x)$ is the breakage function and x the particle size, but in equation (10), x_o is the size of the original particle. If, as in the derivation of equation (2), the density function $\rho(y)$ is equal to a constant and x_o is set equal to one, then equation (10) reduces to equation (2). This general, one-dimensional solution to homogeneous as well as inhomogeneous comminution, as represented by equation (10), is extremely powerful in its application but difficult to solve in practice in all but the simplest cases.

Discussion of Theoretical Models

All of the comminution models fit the data to a reasonable degree, but all have serious problems, of which the authors are aware. The current flaw theories consider three classes of flaws—edge, surface, and volume—while there are actually eight, four of which are associated with the original surface and four with the interior. For example, the internal flaws like those associated with the surface, are the volume flaws, surface flaws on the surface of the propagating crack, edge flaws at the intersection of two cracks, and corner flaws, which occur at the point of intersection of three cracks. It is well known that the original surface of a solid contains cracks after a short aging period; one of the best examples of this is coal; glass is another. Yet coal is one of the substances used to substantiate the flaw theory by stating that some of the size-distribution anomalies observed in the larger size are due to internal or new surface flaws, while in all probability they are due to large pre-existing cracks in the coal surface. Another problem with the existing theories is an implicit or explicit assumption that particle shape remains constant throughout all size ranges. In the flaw theories, a constant ratio of the cube root of particle volume to length of edge is critical, because the only significant flaws are edge flaws, and if the amount of edge per particle changes (due to a change in particle shape with size), then the predicted cumulative-mass/finer-size distribution will be invalid. Put in another way, volume is easy to conserve in a comminuted particle but edge-length and surface are not.

A more serious problem associated with flaw models is the question of the probability density of particles of volume V. In the initial derivation of the models, it was argued that the number of particles of size x was the probability that a particle of size x occurring, $P(x)$, times the number of possible particles of size $x \frac{(V_o)}{x^3}$, where V_o is the total volume of the specimen and x^3 is the fragment volume. $\frac{P(x)}{x^3} V_o$ is the number of particles in size x that would form if no other size fragments were formed. This type of argument is central to the derivation of the flaw model; yet there is an apparent disparity, for the volume is used over and over in forming fragments of each size. In the crack-density derivation, Meloy avoided this problem—but then his derivation was one of crack density.

Although the flaw models of Gilvarry and Austin have three parameters, only the edge flaws are of importance; *theoretically* because they are dominant mode, and *experimentally* because the data (at least in the smaller size ranges) confirm the edge flaw dominance. The one parameter—crack density, equation (2)—is similar to the extrusion equation (4) and to the flaw model equation (1), when j and k are large compared to i. All three equations—(1), (2), and (4)—are similar in the large sizes and asymptotic in the smaller sizes. This may mean that the two models are different expressions of the same basic process. The crack-density model envisions comminution as a ramulous network of cracks that fill the specimen in response to energy stored in the specimen at the time of comminution. Nothing is said in the crack-density model about how a crack bifurcates or how comminution is initiated.

From the flaw model comes a simple model of how comminution may occur.

A crack is initiated and, as it moves through the specimen, flaws are activated along edges; thus, the cracks bifurcate. This crack branching process occurs over and over and the dominant mode of bifurcation is from edge flaws and, to a much lesser extent, from surface flaws. The key to the flaw model is that cracks are initiated from existing cracks, and thus the whole fragmentation process takes place from one or two initial failure sources. (This is true also of the crack-density model). However, if the mathematical problems of the flaw models are not resolved, the question of how comminution takes place will also be left unresolved.

One of the real problems in the current models is that of crack interference. The speed of crack propagation is below sonic velocities, which means that a crack tip radiates acoustical information on its position and direction. In all probability, the stress fields of two nearby cracks interfere and either suppress crack initiation, control crack propagation direction, or both. This means that flaw or crack site activation does not occur individually and collectively at random, but is influenced by what the neighboring flaws do. When a flaw is activated, one might say that stress relief occurs in the neighborhood of the flaw, thereby prohibiting further flaw initiation. In the crack-density model, where the key to the derivation is the random cuts on a line, the randomness of two nearby cracks is destroyed. This poses a serious but not insurmountable problem in the crack-density model. *This is particularly important in the finer size ranges.*

While the existing comminution models have their problems, as a first attempt at predicting size distribution results from theory, they are as successful as needed, for there is virtually no experimental data that these models do not pre-

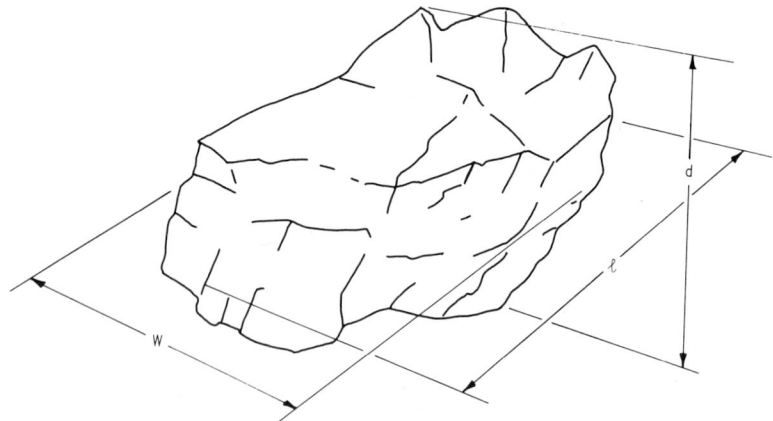

Figure 2. One dimension or number does not characterize a particle. Most current measurements yield an equivalent diameter which says nothing about the particle roughness, elongation, and other parameters which effects its behavior.

dict. What is needed is either more useful experimental data that can be used to evaluate existing models or predictions from an advanced model that can be used as a basis for running definite experiments. The trouble with existing models is that they predict nothing about the variation of particle shape within a given size range. See Figure 2. While not wanting to discuss the potential of the flaw model, it appears to the author that it is entirely possible to predict ℓ, d, and w density distributions for particles from the crack-density model, and, with these data, to test the predicted distribution by screening a sieve fraction with closely graded slot screens. This, of course, is particle and powder characterization—a fertile field for current investigation.

Three-Dimension Models

While some workers may claim their size distribution derivations are for three-dimensional particles, all work of which the writer is aware is one-dimensional in that the size of the particle is characterized by one parameter instead of the three independent parameters necessary if the particles are to be truly characterized as three-dimensional. Consider a three-dimensional model whereby particles are considered rectilinear parallelepipeds, with each of the three characteristic dimensions being a random variable governed by the crack-density model delineated in the sections above.

The first question is the nature of the size distribution, but unfortunately there are no clear criteria as to what size is—size is discussed in the following section on characterization. Also, the work is in progress and theory mathematics are proving obdurate. However, the initial work indicates that when a single particle is fractured, the smaller particles—those which are small compared to the original particle size—are blocky, while the larger particles tend to be more plate-like or needle-like. This tends to confirm the experimental work of Charles [9] and Harris [10], and to open up new avenues of work in particle characterization.

Relatively small changes in particle shape from blocky to needle-like, going from small to large particles, would greatly influence the shape of size/distribution curves. A theoretical cumulative plot to a screen analysis, with a modulus of one, would end up in the larger sizes, having a slope of one-third, assuming that the ℓ/d and ℓ/w ratios vary as the size to the one-third power in the upper size ranges. This may explain the shape of the unusual size/distribution plots found in cratering experiments.

These major changes in the shape of size/distribution plots with minor variations in particle shape mean that almost all of the often very carefully obtained experimental data are worthless because even simple characterization of particle shape has not been considered.

Characterization—Particles and Powders

To speak of a size distribution implies that one knows what a particle is or, at least, what its shape and mass are. A discrete particle is defined intuitively, and that definition is satisfactory for this paper, even when the particle appears to be formed of two formerly separate particles. If constant mass density is assumed, then particle size in relation to one definition of shape can be established by weighing an individual particle—for example, size in the diameter of a sphere yielding an equivalent mass. Particle shape and size are not as clearly defined, measured, or understood as is particle mass.

What is a particle size? According to one definition based on particle mass, particle size is the diameter of a sphere that will yield an equivalent mass. This is not the particle size measured when sieving. When a particle passes through the opening in a sieve, a complex profile of two minor dimensions of the particle is involved. For example, in Figure 3, particles a, b, and c have the same mass

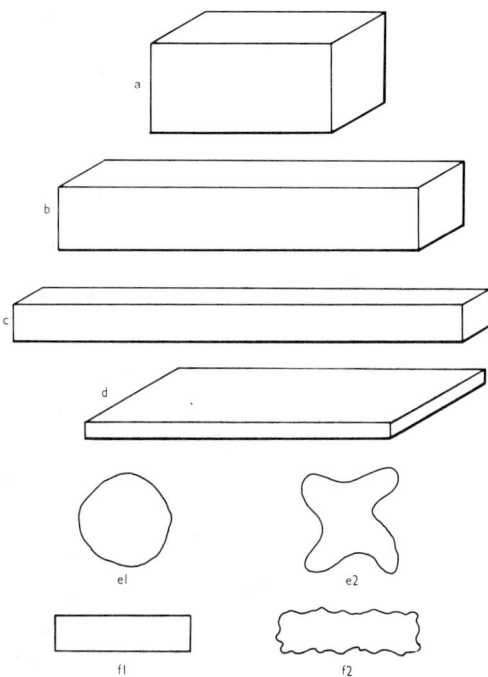

Figure 3. Variations in particle shape are important in particle behavior. Particles numbers b, d, c have the same mass but would pass different sieve sizes, while d, which is lighter, would by far have a larger sieve size. Particles pairs e and f have the same size but would have very different settling characteristics and physical characteristics (e2 and f2 powders would have far higher angle of repose than e1 and f1 powders).

but would have different screen sizes, while d is the lightest but largest. Particles
e1 and e2 are the same size but of different shape, as are f1 and f2. If we know
what the particle shape is, we might be able to define particle size. Not only do
we measure particle "size" by sieving, but also by the particle's settling rate in a
fluid. In the latter case, the size of the particle is defined as the diameter of a
sphere that would settle at the same rate. In this measurement of particle size,
the particle drag coefficient is measured, which determines, in part, the particle
shape, mass, and surface roughness. It is easy to see that a given particle can be
many different "sizes," as would be f1 and f2, measured by sieving and settling.

Certainly our intuitive concept of size derived from everyday concepts of dimensionality is not what is measured by sieving, mass evaluation, or settling. The
very real concept of shape plays a role in every form of measurement, yet a definition of particle shape does not exist that is usable in correlating the measured
sizes obtained by alternative methods.

Characterization

What is really being discussed is particle characterization—a method by which
we may correlate the particle's behavior under experimental conditions to measurable particle characteristics. This is not an easy subject to discuss, because
the existing ideas are vague and ill-defined; yet particle characterization presently
offers a unique opportunity to the investigator to make a contribution to the
field. Now is the time for meaningful experimental work to begin.

When discussing particle characterization, the key question is: How many dimensions are required to define the important characteristics of a particle such
that its behavior is predictable? In the ensuing discussion the particle will be
considered to contain no internal voids and no invagination—only solid particles
with an unbroken external surface, although these surfaces may be smooth or
rough.

Recognitive Characteristics

Two approaches can be used to discuss particle characterization, one of which
might be termed the real, or recognitive approach, and the other, the logical, but
non-recognitive approach. The real approach comes from the real language of
everyday life, where objects and events are defined in words which relate to
prior experience. For example, the size of a particle is thought of (in terms of
real language) as the height, width, and length of the particle, which tries to relate our experience in measuring the size of boxes and cars—where the shape of
the object is known—to the case of measuring the size of a particle where shape

is not known or even understood, and thus height, width, and length have no meaning. The recognitive aspect of real language as it relates to particle characterization means that the dimension being specified is easily recognized in everyday terms, as is the height of a person or the length of a particle, for example.

Non-recognitive Characteristics

The logical, or non-recognitive approach acknowledges the interaction between the particle and the measuring instrument, resulting in a measured dimension which has little relation to our experience (non-recognitive) and may be likened to the very real, but non-recognitive dimension—the spin of an electron. Particle characterization size, as pointed out earlier, has an intuitive, but unmeasurable meaning if real language is used. But if, in logical language, particle size is defined as the size of a square opening through which a particle will pass (as in sieving), then particle size has a logical meaning in that size is operationally defined by the particle–opening interaction, but there is no longer a real (recognitive) understanding of what has been measured. The real language (recognitive) is non-linear in that the characteristic dimension or dimensionality striven for is not linearly related to what is being measured. In other words, in the real language, neither particle size nor shape is measured or defined by the measuring instrument, whereas in the logical (non-recognitive) language, size can be directly related (linearly) to the measuring device (i.e., the rectangular opening through which the particles pass is the particle size for sieving). It now becomes clear that, in the logical language, *three different characteristics* are being measured when *particle size* is determined by: (1) mass equivalent, (2) sieving, and (3) sedimentation. On the other hand, in the real language, the same characteristic size is being measured by all three approaches, but "somehow" they do not correlate (non-linear).

Particle Shape

Shape, like size, is a particle characteristic in the real language and perhaps has greater recognitive appeal; yet attempts to make meaningful measurements are blocked by lack of an approach. For example, in respect to a particle (Figure 2) whose length, ℓ, width, w, and depth, d, can be measured, shape can be characterized by ℓ/d and ℓ/w ratios. But do these ratios really characterize particle shapes? Are they not particle characteristics rather than a definition of particle shape? In the writer's opinion, when using the real language, shape is blocky, chunky, flat, elongated; but a specification of a particle's ℓ/d and ℓ/w ratios is merely partial particle characterization, which may or may not be correlated with particle shape.

· **Particle Characterization vs. Size**

The classical work in particle characterization has been in terms of the real language, which has yielded little; however, if meaningful experimental and theoretical work is to be done in the comminution field, more parameters must be measured and more characterization must be done. The current work on *powder* characterization, as opposed to *particle* characteristics, is dependent on the particle's characteristics remaining constant throughout the powder size range. For example, reverting to the recognitive language, if the particle shape changes from blocky to flat when going from small to large particles, then the meaning of the size/distribution plot obtained from screening is quite different from that of the size/distribution plot obtained when the particle *shape remains a constant size.*

Suggested Experiments

What is needed is some basic characterization work that should consider only non-recognitive characterization. For the initial work, several materials broken in a similar manner should be used. For example, samples of a glass, a quartz, and a feldspar should be prepared by crushing. From these samples should be isolated a 10 X 14 and a 28 X 35 mesh fraction. From *each* of the six subsets (three material and two sieve fractions) a thousand particles should be isolated and identified, and a set of measurements like the following should be made *for each particle* followed by the indicated calculation.

1. Particle mass: equivalent sphere diameter.
2. Laminar settling velocity: drag coefficient and diameter of an equivalent sphere.
3. Turbulent settling velocity: drag coefficient and diameter of an equivalent sphere.
4. Square opening through which the particle will pass: equivalent sphere diameter.
5. Circumscribe circle: equivalent diameter.
6. ℓ, d, and w: diameter of equivalent sphere whose volume equals $\ell.d.w$.
7. Surface area: diameter of sphere of equal surface area.
8. Resistance volume: diameter of equivalent sphere.

With the data accumulated in tests 1 through 8 on the thousand particles in each subset, enough information would be available to characterize the particles non-recognitively in each subset so that, for each test in each subset, a distribution function and an average value of the parameters characteristic of the subset would be obtained. (Two parameters are common in particle distribution functions; the normal has the average and standard deviations, while the Rosin–Rammler and Gaudin–Schuhmann have a slope and size modulus.) With the parameters characteristic of a given subset, comparison can be made between subsets that differ by size and material. At the risk of being redundant, one could compare particle mass diameter for quartz (test 1) in the 10×14 mesh group with those in the 28×35 mesh group, to determine whether they are the same as would be expected on the basis of mesh size; or one could compare the 10×14 mesh particle mass diameter for quartz to that for glass and feldspar. If the particle-mass-diameter to sieve-size ratio is constant from mesh to mesh and material to material, which seems unlikely, then these two measurements can be considered two methods of measuring the same non-recognitive characteristics. Since the ratios are not likely to be the same (at least from material to material), then a significant *powder* characteristic would have been established for materials.

Particle and Powder Characteristics

It is important to differentiate between powder, bulk powder, and particle characterization; for, while the first two are ultimately dependent upon the particles themselves, their characteristics may either not be inferred from the particle characteristics or they may depend on the make-up of the powder (i.e., the powder size distribution). For example, the angle of repose of a bulk powder characteristic is dependent on particle size and shape and powder size distribution. Particle characteristics describe the characteristics of *individual* particles; therefore, a set of the eight measurements on *one* particle would characterize that particle from a powder whose powder characteristics are the particle-to-particle variation within the powder that are often described by distribution functions (i.e., size distribution). Bulk powder characteristics, on the other hand, are measurements of the behavior of the powder as a whole, such as angle of repose and density.

Bulk Powder Properties

To change the information within a subset into size-independent characterization, a correction first must be made for variation in particle size; to do this, the most accurate particle measurement—particle mass diameter—should be used to "normalize" the data. Each of the eight tests results in the calculation of a characteristic diameter, or linear dimension, so that if each of these diameters is divided by the particle *mass diameter*, the result is a dimensionless group expressing the characteristic *diameter ratio* for the latter seven properties. These characteristic diameter ratios for particles are independent of the particle size as well as of how the sample fraction was taken—in this case by screening. With the seven dimensionless groups for each particle (particle characterization) and a thousand particles in each of the six subsets (powder characterization), the stage is set for some bulk powder characterization studies. With the wide availability of computers and generalized statistical programs, the simple approach is to run a regression analysis and determine which powder characteristics are related to which bulk powder characteristics. Having done this analysis, meaningful interpretation can be performed. If one of the powder characteristics varies widely and independently from material to material, but not with size, the following question arises: "Is that non-recognitive property a measure of recognitive property and, if so, which one?" If it is a purely non-recognitive characteristic and is not correlated with a recognitive one, then a most interesting series of questions arises: "What effects does this non-recognitive characteristic have on bulk powder properties? What predictability does this characteristic have on the particle system?" (To correlate recognitive properties with non-recognitive properties, the approach of S. S. Stevens [11] can be used.)

For example, the writer has long believed that particles from a ball mill are both more blocky (equidimensional) and more rounded than particles from a crushed-single-fracture; yet, without a quantitative measure of meaningful powder characteristics, this belief remains an opinion. Without the establishment of not only the significant characteristic curves but also the shape of the characteristic's density function as well, this belief will remain an unsubstantiated opinion. Many ceramic products are dependent, in an unknown way, on the powder characteristics from which their slips are made; yet chemically identical powders that meet "specifications" will yield good or bad results, depending on the source of material. Hence many ceramic manufacturers are wedded to their suppliers because they do not know what powder characteristics to cite in writing material specifications. Yes, he who characterizes particles and powders will greatly advance the technology of particulate systems.

Experimental Characterization Work

Recent work by Kaye [12] and Church [13] on simplified microscopic size characterization, has indicated that not only can more information be obtained for less work by taking multiple measurements on individual particles, but the work also contains more information on particle characteristics and more information on powder characteristics. By measuring on an individual particle such parameters as: Feret's diameter, Martin's diameter, maximum linear diameter, minimum linear diameter, circle of equal diameter, circle of equal area, and random and regular reticle lengths, new insight into particle characteristics is being obtained. It is interesting to note that this work is done in parallel with modeling of the powder such that the particles are represented by ellipsoids whose size and shape are characterized by a linear dimension and an eccentricity. By relating theoretical experiments for the model with actual experiments, efforts are made to characterize the powder. This work is the harbinger of true powder characterization work, which will in turn predict process and bulk powder behavior from laboratory measurements.

Instrument Measurements and Their Meaning—A Warning

There are a variety of expensive instruments currently on the market which claim to measure the "size" of fine particles. By now it should be clear to the reader that there is no such thing as particle size. These instruments therefore non-recognitively characterize the powder in some unknown manner. Furthermore, the powder characterization varies from one type of instrument to the next, and the magnitude of the measurement is sensitive to variations in the powder. To top all the problems, these instruments must be calibrated often, as they drift quite readily.

In short, we do not know the meaning of what most instruments measure: we know that they will give identical measurements for different powders; we do not know how these measurements affect different powders; and we do know that the instruments easily get out of adjustment.

Fine Grinding

To prepare a fine powder from coarser materials by mechanical means, one grinds the material in one of several different instruments—ball mills, sand mills, bead mills, hammer mills, or jet mills. As the fineness of the grinding proceeds, the rate of grinding decreases for the following reasons.

1. The surface on which the grinding takes place doubles every time the particle size is halved.

2. Smaller particles are harder to break because of their size.

3. Smaller particles become entrained in the fluid and are not ground.

4. Smaller particles tend to agglomerate and the agglomerates, not the particles, are broken.

Because of the above-cited reasons, a series of fine-particle mills have been developed to overcome these inherent problems in fine grinding.

Ball Mills

Because of the wide availability of ball mills and the large amount of experience with them, they are widely used in grinding particles into the sub-sieve size ranges. As particles become smaller, the absolute energy required to break a given particle becomes less, and ball size for optimum efficiency decreases due to the need for increasing surface area in which to break the more numerous smaller particles. On the other hand, particle toughness per unit volume becomes greater, and thus a compromise between ball size and ball mass must be reached in order to maintain grinding efficiency. In the smaller size ranges in ball milling, two additional phenomena further exacerbate the efficiency picture. First, particles tend to form flocks which are comminuted back into individual particles and then reformed without breaking up the individual particles in the flock. This dissipates a great deal of energy without creating much size reduction. Secondly, because ball milling is always done in a fluid, the finer particles become suspended due to their large surface area, behave as part of the fluid, and move out of the way of the ball/ball or ball/liner collision. Many people have interpreted this lower limit of grinding in a ball mill as meaning there is a grind limit, but, as will be shown in the section on Shape and Surface, there is no such thing as a grind limit in the micron size range.

To overcome the problems of the ball mill, additives which disperse the particles are added and balls are made smaller but, fundamentally, much longer grind times are used. In some cases dry grinding is more effective in preventing entrainment of the particles, but not usually. Some work on vacuum grinding has been done, but it is inconclusive.

Bead Mills and Sand Mills

Bead and sand mills are ball mills in which the grinding energy is put in by shaking and stirring, respectively. Introducing the energy in this different manner enables higher energy densities per ball to be achieved. In bead mills, smaller balls are used, the mill is filled fuller, and the acceleration is much greater than one G. The smaller balls and increased mill filling yields more grinding area, while the increased G loading allows for greater energy inputs per ball, as well as decreasing the size at which particles are entrained in the grinding fluid. Increased grinding times are also the rule, though this modified ball mill is, when properly tuned, more efficient than the conventional mill. The sand mill is stirred, and the frictional dissipation of energy is actually grinding. The balls or sand have a much larger surface area per unit volume of mill.

Hammer Mills

Apart from the down-time on hammer mills due to replacements, these machines are inherently more efficient than ball mills in grinding almost all sizes of particles. Rapidly rotating vanes, impacting and shattering feed particles, cause a product size distribution of angular particles from single impact events. The value of this grinding method has been recognized in the abrasives industry, because the corners and edges of the particles are not worn by repeated particle contact under pressure as is the case in a ball mill. For fine grinding, the problems in a hammer mill are that the paddle surface area is small relative to the number of particles that must be ground, the windage energy losses are high, finer particles are suspended in the fluid, and the heat build-up on the paddle causes rapid wear.

To grind some materials, hammer mills have been modified for use with special gases or a vacuum. Hydrogen, because of its very low viscosity and high heat conductivity, has been successfully used to grind some materials below 2 μ. By using H_2 gas, windage losses are reduced as is particle entrainment, and the heat build-up on the paddle is also reduced due to the high conductivity of the gas. The small fixed surface area of the hammer will remain a problem for producing quantities of fine powders.

To grind finer, hammer mills have been evacuated to very low pressures. It should be remembered that the viscosity of a gas is independent of its pressure until the particle size is equal to the mean free path of the gas. Problems with heat transfer from the paddle surfaces become severe at low gas pressures, and the problem of maintaining a high-speed rotary vacuum seal in the presence of a fine abrasive powder is very difficult.

Jet and Fluid-Energy Mills

These mills are all of recent origin. The feed material is suspended in a gas and high relative velocities are imparted to the material by expanding the gas. Particles are directed toward another jet or into the flow of a gas also containing particles such that there is high-velocity collision between particles. Particle/particle grinding occurs which differs from the previous grinding methods mentioned. In most of these machines there is a classification action whereby the smaller particles are entrained in the exit gas and the larger particles are thrown to the wall in a vortex and returned to the feed.

Modeling work on this equipment has been done by Hendry [14] which gives a fairly good approximation to the dynamics of the equipment. Some remarkable experimental work has been done by Kürten and Rumpf [15], showing where the areas of comminution are, and indicating not only that the action is one of particle-particle collision but also where in the mill (and even in the jet flow) the actual collisions take place.

Shear Mills

When liquids are violently agitated, high shear forces are set up in the liquid. Suspended solids are subject to these shear forces and can be broken if they are high enough. Shear mills have been used for many years to grind soft substances such as emulsions and, more recently, soft polymer particles.

Grind Limit

There has crept into the literature over the years a concept of a grind limit. Bond attributed this concept to Gates in 1910. The basic concept is that particles, at some cut-off point, become almost impossible to grind due to increasing strength. The genesis of the concept came from the difficulty in observing particles below a certain size.

To put the matter into perspective, the original grind limit for most rocks was set at $10\,\mu$ an observation made with a microscope that had a resolution of somewhere in the neighborhood of $3\,\mu$. As the quality of microscopes and size-sensing devices has increased, the grind limit has progressively—and almost linearly with time—gone from $10\,\mu$ down to $0.01\,\mu$, or roughly $100\,\text{Å}$. At this size, the unit crystal size is approached and, at $10\,\text{Å}$, even molecular sizes are approached. The actual ability to grind particles should go down to atomic sizes somewhere on the order of $5\,\mu$. The second thing to note is that these finer particles have very much higher solubility than the larger particles, due to their very high surface energies. Furthermore, in the presence of a gas, fine particles behave as a gas and thus tend to evaporate.

The impression I hope I have conveyed in the above paragraph is that there is no grind limit; no theory supports the concept of a grind limit, in fact, these theories exclude the possibility. If one plots grind limit versus year, the experimentally observable size of grind limit will be down to $1\,\text{Å}$ in 1997. Let us quietly bury the concept of a grind limit and tiptoe away hoping that we shall never hear of the subject again.

Shape and Surface

Apart from the fact that the section on characterization in this chapter was devoted exclusively to the concept that the non-recognitive properties—shape and size—are unmeasurable and hence unknowable, in this section these properties will be spoken of in their recognitive sense. Hammer mill and jet mill products tend, because of their single impact method of breaking particles, to create daughter fragments which are angular and—in the larger size fractions—less blocky. Ball milling on the other hand, because of its abrasive attritive nature, tends to produce blockier, more rounded particles that would be characterized differently, and in some industrial processes behave differently. Because of the scratching and sliding actions of the ball mill, the surface area will be less for a particle of a given size and, furthermore, it will tend to be more smeared and contaminated. This will play an important role where surface properties contribute significantly to the process. This, of course, again brings up the problem of characterization of particles discussed earlier.

Micron Size Inhomogeneities

Unlike liquids, solids are notoriously inhomogeneous. In many processes large blocks of solids are used, and the average properties of the surface contribute to their behavior. During fine grinding, however, the inhomogeneities can play increasingly important roles as fractions are liberated. For example, glass at approximately the 1μ size range begins to have a continuous phase and harder crystallites. When this material is ground in the sub-micron size range, a power of two different constituents will arise. If one constituent is much harder than the other, one of the materials may be ground up and disappear, either through solubility or entrainment in the fluid media, leaving the harder material.

Summary

It is a little difficult to summarize what is known about fine grinding in a chapter such as this, because what is being said is "we know very little." Work in ball, hammer, jet, and fluid-energy mills is going on, but this work does not have great engineering application at the present moment. Theoretical studies on the breakage of single particles are quite advanced, but once again we see that as yet we have no way of measuring important properties of materials which contribute to their fine-grinding characteristics. For example, such things as size, shape, and angularity are mathematically undefinable and experimentally unmeasurable. While much progress has and is being made, the coming decade should see major break-throughs in the area of particle characterization and the relating of characteristics to process behavior.

References

1. Gilvarry, J. J., "Fracture of Brittle Solids. I. Distribution Function for Fragment Size in Single Fracture (Theoretical)," *J. Appl. Phys.*, 32 (1961), 391-97.
2. Meloy, T. P., "A Three Dimensional Derivation of the Gaudin-Meloy Size Distribution Equation," *Trans. Am. Inst. Mining Engrs.*, 226 (1963), 447-48.
3. Meloy, T. P., *AIChE*, Discussions of Comminution Size Distributions and Particle Characterization.
4. Austin, L. G. and Klimpel, R. R., "Statistical Theory of Primary Breakage Distributions for Brittle Materials," *Trans. Am. Inst. Mining Engrs.*, 232 (1965), 88-94.

5. Marcus, A., Private Communication.
6. Meloy, T. P., *AIChE*, Presentation (1963).
7. Meloy, T. P. and Faust, L. H., "Comminution Model of Lunar Surface," *Proc. of Symp. and Short Course on Computers and Op. Res. in Min. Indust.*, 3 (1966), 44.
8. Meloy, T. P. and Gumtz, Garth D., "The Fracture of Single, Brittle, Heterogeneous Particles—Statistical Derivation of the Mass Distribution Equation," in press.
9. Charles, R. J., "Energy-Size Reduction Relationships in Comminution," *Trans. AIME*, 208 (1957), 80-88.
10. Harris, C. C., On the Limit of Comminution, Private communication.
11. Stevens, S. S., "On the Operation Known as Judgment," *Am. Scientist*, 54 (1966), 385-401.
12. Kaye, B. H., "Some Aspects of the Efficiency of Statistical Methods of Particle Size Analysis," unpublished manuscript, IITRI.
13. Church, T., "Efficiencies Problems Associated with the Use of the Ratio of Martin's Diameter to Feret's Diameter as a Partial Profile Shape Factor," unpublished manuscript, IITRI.
14. Hendry, R., "A Mathematical Model for Fluid Energy Mills," Zerkleinern 34 Vortrage Symposium in Amsterdam 1966 Dechema-Monographien, Band 57, Weinheim/ Bergstrasse (1967), 695-727.
15. Kürten, H. and Rumpf, H., *Chemie-Ingenieur-Technik,* 38 (1966), 331-42.

3. Preparation of Ultrafine Particles of Metal Oxides

KENNETH R. HANCOCK

Chas. Pfizer and Co., Inc.

Easton, Pennsylvania

ABSTRACT

Ultrafine particles of $0.01-10\,\mu$ and their preparation form the basis for a wide range of inorganic compounds which are utilized extensively in such fields as protective coatings, electronics, ceramics, paper, plastics, building materials, etc. Historically, the pigment industry has been a pioneer in this field extending their fine particle technology and products into many areas, other than coatings.

The metal oxides, and specifically those of iron, provide a versatile compound to demonstrate the growth of ultrafine particles. Various mechanisms and preparation techniques are described for producing ultrafine particles by solid state and liquid-solid state reactions. Processing variables such as temperature, time, impurities, concentration, nucleation, fluid dynamics, etc., are shown to have a profound effect on crystal growth. This can be demonstrated by studies of electron micrographs, surface areas, and particle shape and size distributions.

In the 1960 Annual Review on Crystallization [1], contained in *Industrial and Engineering Chemistry,* the statement was made that "the newer aspects of the subject, such as zone-melting and single crystal growth from the melt and from solution, continue to dominate the literature; classical multicrystal growth from solution receives little attention. There is, for example, no comprehensive treatment of the batch process in spite of its wide commercial application. The effects of controlled seeding and cooling upon product size distribution of a batch process have only been touched upon in the literature."

A study of the annual reviews since that date indicates that although crystallization literature is voluminous and has continued its accelerated growth, little that is published relates to commercial processes for growing ultrafine particles which are the basis for many large industrial processes. In the 1968 Annual Review which covers a one-year span (July, 1966, through June, 1967), there are 980 [2] references covering various phases of crystallization such as nucleation, crystallization equilibria, kinetics, variables, vapor phase growth, zone melting, etc. This shows that there is an active interest in the field of crystallization and particle growth.

39

Historically, the pigment industry is one of the pioneers in producing fine particles. The first precipitated white pigment was white lead, which is known to have been used in 400 B.C. Similarly, the iron oxide pigments were used by prehistoric man in cave paintings. The early technology, with a few exceptions, was one of grinding available mineral ores. As the need developed for chemical purity and particle-size consistency, a technology evolved in the chemical industry and the pigment industry to produce ultrafine *synthetic* particles. A few examples of these are chromium oxide, zinc oxide, titanium oxide, iron oxide, lithopone, carbon black, chrome yellow, and a whole host of organic toners, pigments, and dyes. In some cases the chemistry of the process may appear to be simple, the rusting of iron, for example. However, to convert a spontaneous chemical reaction into a controlled process to produce reproducible desirable products requires a high degree of sophisticated chemical technology.

Scope

The preparation of ultrafine particles will be demonstrated by reference to processes which rely on solid-state reactions, liquid-solid reactions, and gas-solid reactions. The major emphasis will be on the liquid-solid reactions. The chemical systems that will be presented are ferric oxide, ferrosoferric oxide, hydrated ferric oxide, chromium oxide, calcium oxide, and calcium carbonate.

Definition of Terms

"Ultrafine Particles" are defined as those particles having an effective diameter of 0.01μ to 10.0μ. Since this range is too broad to be conveniently measured by a single instrument, the method of measurement employed in our laboratories is prescribed.

Range	Instrument
0.01 to 1.0μ	*Electron Microscope**
	Zeiss Particle Counter
	Numinco-N_2 Surface Area
	X-ray Line Broadening
1.0 to 10.0μ	*Andreasen Pipette**
	Electron Microscope
	Optical Microscope
	Zeiss Particle Counter
	Numinco-N_2 Surface Area
	Whitby Centrifuge
	Fisher Sub Sieve Sizer

Certain particles, falling within the above ranges, may present special measuring problems. Magnetic oxides, for example, require special techniques due to their

---

*Preferred.

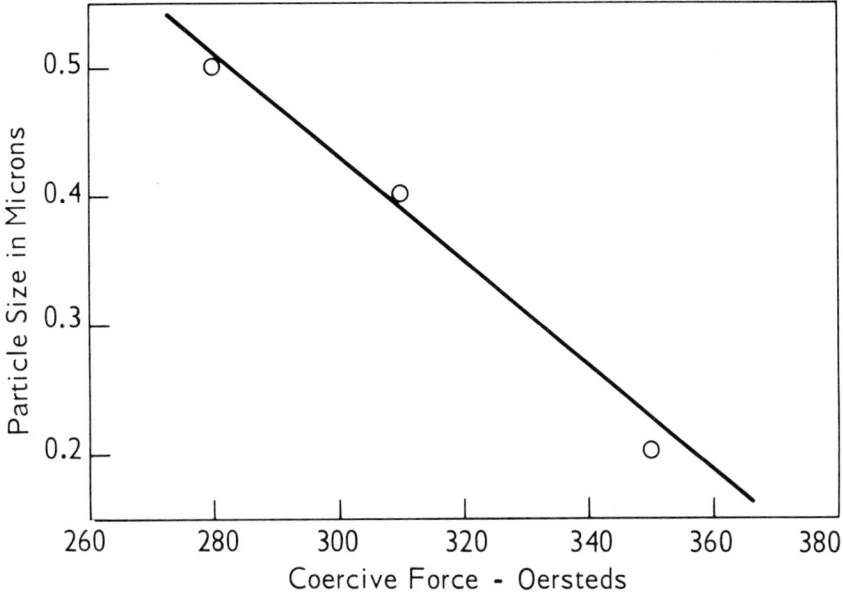

Figure 1. Coercive force and particle size relationship for γFe_2O_3.

flocculating tendencies under the earth's normal magnetic field. Figure 1 shows a relationship between coercive force and particle size. This relationship, as determined with a 60-cycle hysteresis loop tracer at a known flux field, is valid only within the same family of materials [3].

Theory

Solid-State Reactions

One method of manufacturing metal oxides involves converting a metal salt to the oxide by a calcination process. This is essentially a solid-state reaction. The controlling mechanism of indirect heat transfer is conduction [4]. A model may be constructed for the calcination operation

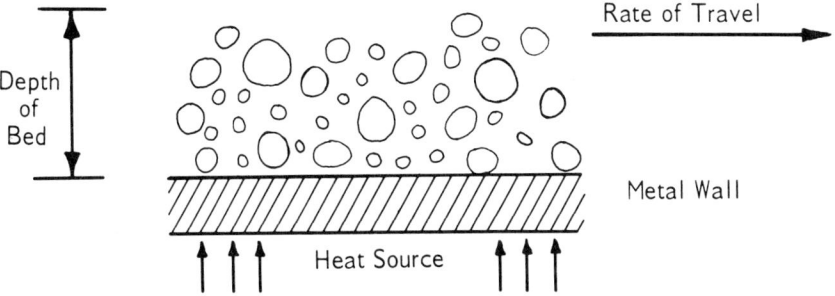

There are many variables which control this reaction and the more important are: (1) Bed depth of the material; (2) Particle size of the material; (3) Particle size distribution of the material; (4) Rate of forward travel of the material; (5) Rate of bed turnover; (6) Thermal conductivity of the solids and gases; (7) Scaling tendencies of the material; (8) Exothermic or endothermic chemical reaction; (9) Type of atmosphere, reactive or inert; (10) Temperature profile of the material.

The rate of heat conduction through a solid may be expressed by the following equation [5].

$$q = K_{avg.} A \frac{t}{x} = BTU/hr$$

where

$$K_{avg} = \text{average thermal conductivity}$$
$$t = \text{temperature span } (^{\circ}F)$$
$$A = \text{cross-sectional area (sq. ft.)}$$
$$X = \text{distance (ft.)}$$

For two solid spherical particles of the same composition, differing only in size, the rate of heat transfer is inversely proportional to the particle diameter. The importance of this factor will be discussed in the section dealing with preparation of particles by solid-state reactions.

Liquid-Solid Reactions

Nucleation from solution is important, since to a large extent it controls the number, size, structure, and morphology of precipitated crystals. A simple set of equations, where A represents a molecule, depicts the nucleation process [6]:

$$A + A \rightleftharpoons A_2$$
$$A_2 + A \rightleftharpoons A_3$$
$$A_{x-1} + A \rightleftharpoons A_x \text{ [critical cluster or nucleus]}$$
$$A_x + A \xrightarrow{\text{nucleation}} A_{x+1}$$
$$A_{x+1} + A \longrightarrow \text{crystal growth}$$

Of the two types of nucleation, homogeneous and heterogeneous, the latter is by far the most common due to the fact that it is almost impossible to eliminate trace impurities. This, of course, is especially true for commercial processes.

The following factors are known to affect precipitation kinetics: (1) Number concentration of initial impurities present; (2) Catalysis by active sites in reaction vessels; (3) Method of mixing reactants; (4) Concentration of reactants; (5) Agitation; (6) Temperature and temperature history.

Ultrafine Particle Growing Techniques

Solid-State Reactions

The thermal decompositon of ferrous sulfate heptahydrate, $FeSO_4 \cdot 7H_2O$, to iron oxide, Fe_2O_3, is a technique employed commercially to produce a range of fine particles from an average size of about 0.25μ to at least 2.0μ. The initial reaction consists of driving off most of the water of hydration to yield $FeSO_4 \cdot H_2O$. Due to heat transfer and material transport considerations further along in the process, it is very important that the above decomposition reaction produce a uniformly sized material. The ferrous sulfate monohydrate is then converted to iron oxide by thermal decomposition in an indirectly heated rotary muffle kiln [7]. A pilot plant rotary kiln used for this purpose is shown in Figure 2.

Figure 2. Pilot plant rotary kiln for solid-state decomposition.

The external heat is adjusted to secure a material temperature between $1300°F$ and $1550°F$ during the calcining operations. By adjusting the time-temperature relationships, a range of particles can be produced. A simple test can demonstrate the importance of the feed size to a rotary muffle kiln. If $1/4''$, $1/2''$, and $3/4''$ ferrous sulfate monohydrate feed particles are subjected to the same time-temperature cycle, the $1/4''$ particle will be almost completely converted to

Fe_2O_3, whereas the 3/4″ particle is only converted on the surface, as shown by the analysis of the calcined product.

Feed Pellet Size	% Fe_2O_3
1/4″	95.2
1/2″	73.9
3/4″	65.1

This process has the versatility of producing at least 15 distinct products, each differing in size distribution by a significant measurable margin. The extremes in particle size are illustrated in Figure 3, and the range of particle

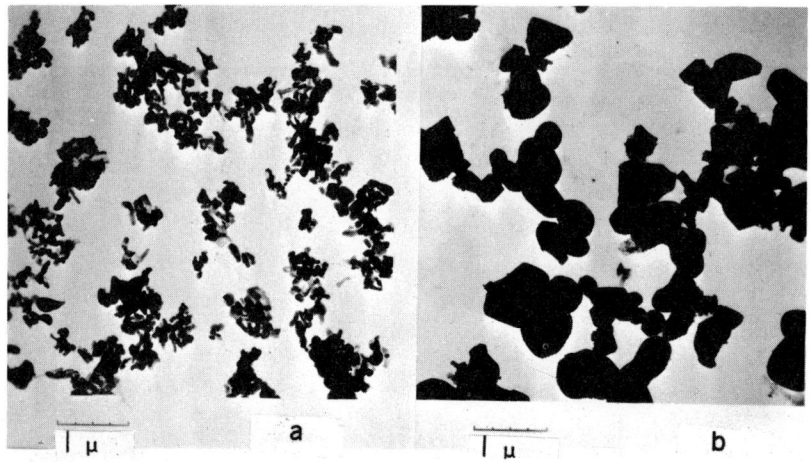

Figure 3. The extremes in particle size as obtained by solid-state reactions. (a) smallest, and (b) largest.

size distributions in Figure 4. The iron oxides produced by this technique find extensive use as the major raw material for hard and soft ferrites. Two reasons for this are high chemical purity and controlled particle size. Figures 5 and 6 show a comparison between a synthetic iron oxide produced in this process and an iron oxide produced by grinding a natural iron ore. This difference between a natural and a synthetic iron oxide can be dramatically shown in its reaction characteristics in ferrite formation. In Figure 7, the Mn–Zn ferrite test pieces were prepared by identical processing. The natural oxide exhibits excessive grain growth. The electrical properties of the test pieces are quite different. For example, the toroid made with the synthetic oxide has a Mu–Q product of 115,000 as compared to only 13,000 for the one containing natural oxide.

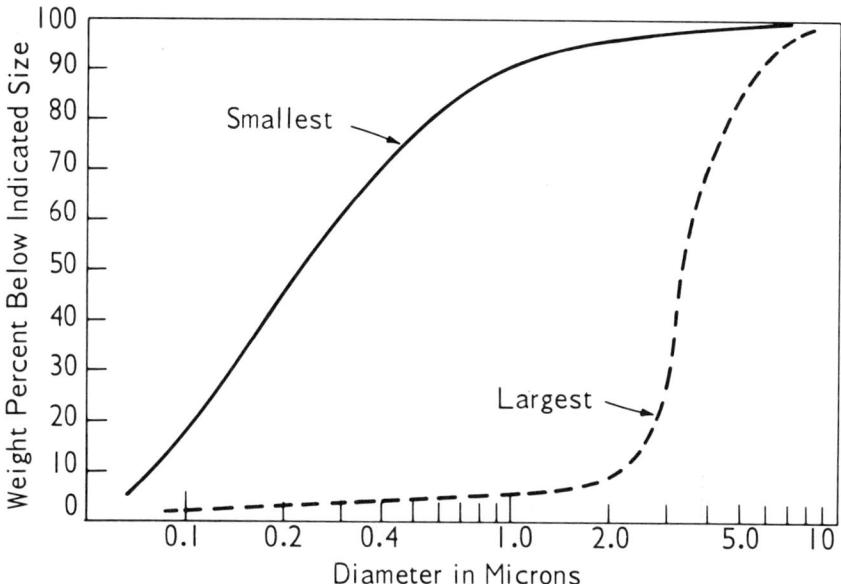

Figure 4. Particle size distribution for synthetic red iron oxides.

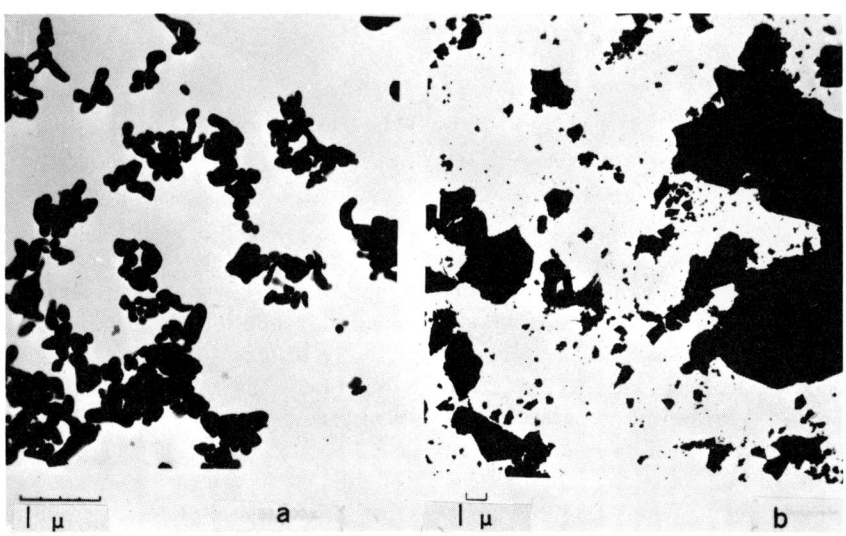

Figure 5. Comparison of particle size between (a) synthetic iron oxide, and (b) ground-natural iron ores.

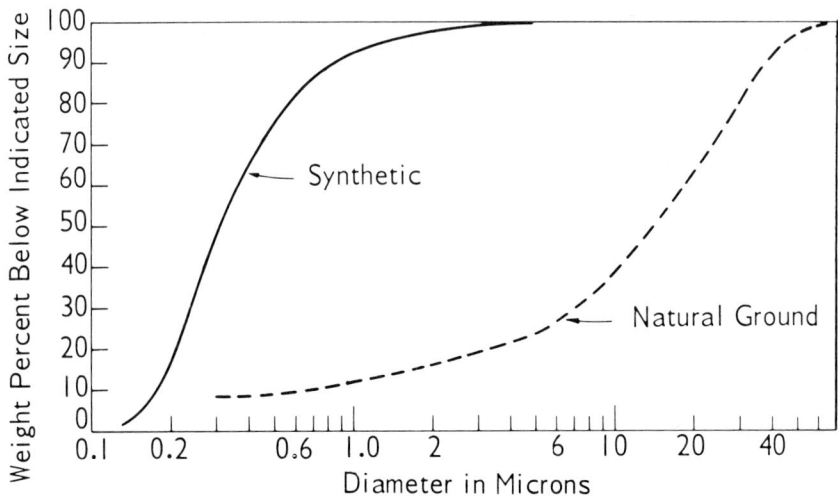

Figure 6. Particle size distribution for synthetic and natural iron oxides.

Figure 7. Fabrication of Mn–Zn ferrites for (a) synthetic, and (b) natural Fe_2O_3.

A somewhat similar solid-state reaction is employed commercially to produce chromium oxide (Cr_2O_3). In this process [8], chromium hydrate is produced from the reaction of sodium chromate and sodium thiosulfate in the presence of H_2O. The reaction is exothermic and the reaction mass goes through a liquid to semi-dry solid state. If the reaction is conducted in a dough mixer, the reaction product, as it dries out and solidifies, can be regulated to produce a granular solid. Once again the size considerations are critical, and a uniform particle of about 20 mesh provides a favorable kiln feed product for ensuing heat transfer considerations. This reaction product is fed to a kiln and calcined at about 1600°F to 1900°F. The kiln discharge product contains a high percentage of soluble salts and must be subjected to a series of purification steps. It should be noted, however, that the basic discreet particle is formed during the calcining operation. Again, a range of particle sizes can be produced by this technique, as shown in Figures 8 and 9.

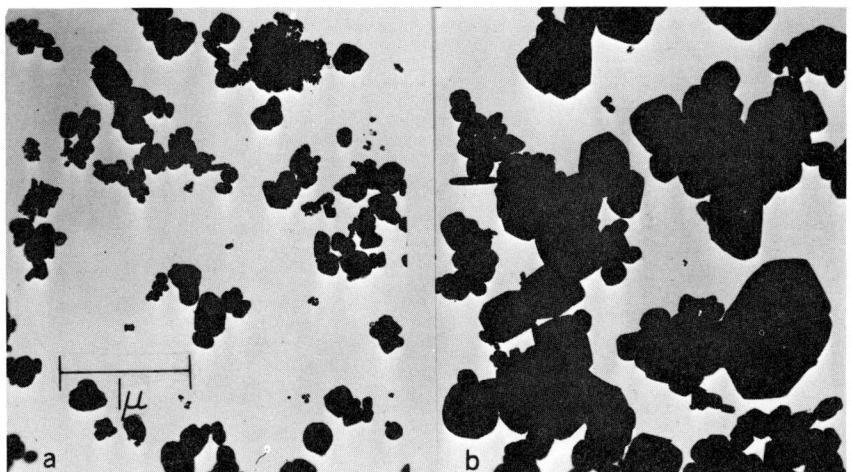

Figure 8. The extremes in particle size for Cr_2O_3 by solid-state reactions. (a) smallest, and (b) largest.

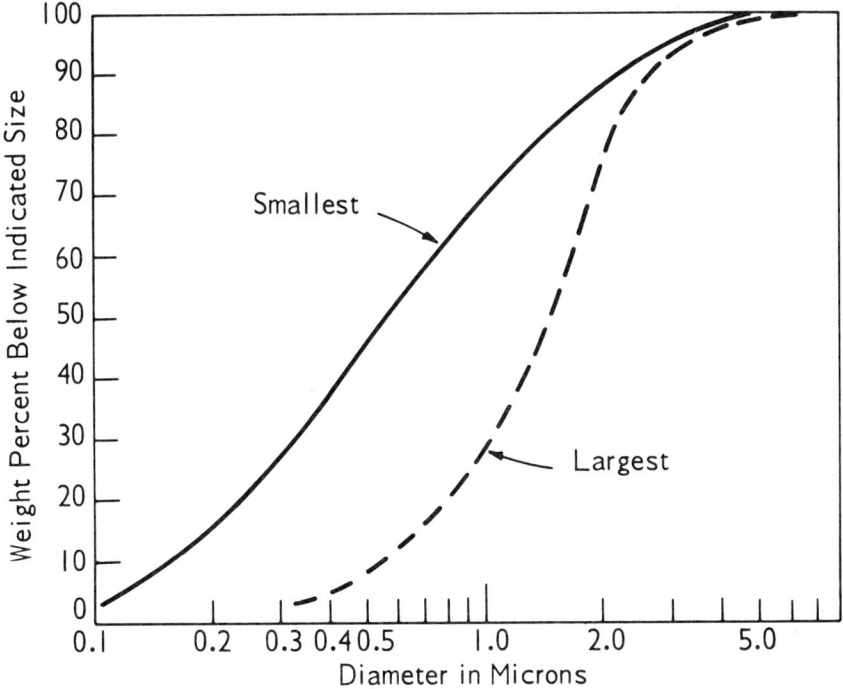

Figure 9. Particle size distribution for Cr_2O_3.

There are numerous other examples of commercial processes which have as their basis the calcination of single or mixed chemicals to produce ultrafine particles. One example worth noting is the fluid bed calcining technique [9]. Here again, the size and size distribution of the feed limestone is a very critical factor in securing successful conversion to CaO and controlling dust losses, as opposed to the much wider latitude employed in rotary or shaft kilns.

Liquid-Solid State Reactions

The growth of ultrafine particles by precipitation reactions provides one of the most versatile techniques for obtaining a wide range of particle shapes and sizes. The thermal decomposition processes quite often produce rounded particles as the particles form under molecularly fluid conditions. This can be seen in Figure 10 by the change in morphology as an acicular iron oxide, when

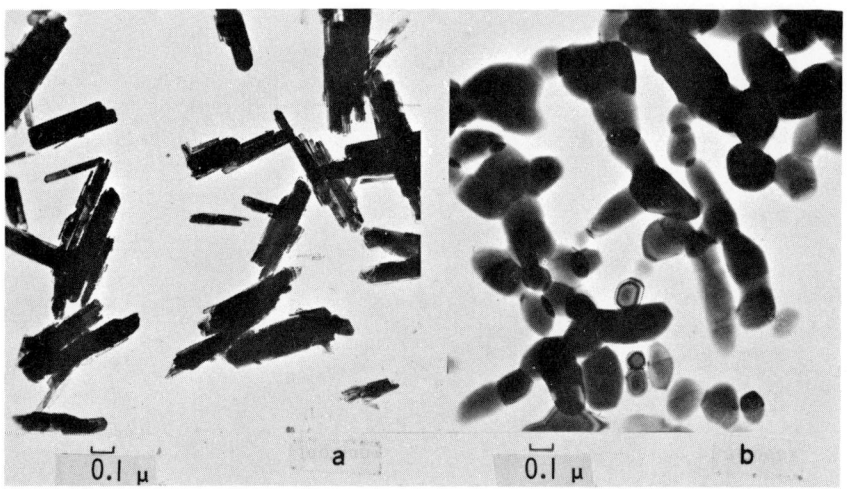

Figure 10. Change in morphology of acicular iron oxide by heat treatment.

subjected to high temperatures, assumes a new crystal shape. The production of hydrated iron oxide $\alpha(Fe_2O_3 \cdot H_2O)$ is a typical example of a commercial process [10, 11, 12] which can produce a wide range of ultrafine particles. The rusting of iron is a very common reaction, but when it occurs spontaneously or in nature it does not produce uniform, fine crystals. The commercial processes are based on the accelerated corrosion or the oxidation of iron salts under closely controlled conditions.

The initial reaction is one of nucleation, wherein a seed nucleus is formed by the reaction of an iron salt and an alkali. The nucleation reaction is normally conducted in an excess of $FeSO_4$, and can be described by the following equations:

$$4NaOH + 2FeSO_4 + H_2O \longrightarrow 2Fe[OH]_2 + Na_2SO_4 + H_2O$$

$$2Fe[OH]_2 + 1/2\ O_2 \longrightarrow \frac{Fe_2O_3 \cdot H_2O}{nucleus} + H_2O$$

The nucleus is transferred to a precipitator containing $FeSO_4$, H_2O, and iron, usually in the form of thin-gauge stampings. The seed particles are continuously circulated over the bed of iron in an oxidizing environment and, as the iron corrodes, the reaction products precipitate on the nuclei causing the particles to grow in size. The chemical reactions involved are:

$$2Fe^{++} + 1/2\ O_2 + 3H_2O \rightleftharpoons Fe_2O_3 \cdot H_2O + 4H^+$$

also

$$2FeSO_4 + 3H_2O + 1/2\ O_2 \longrightarrow Fe_2O_3 \cdot H_2O + 2H_2SO_4$$

$$H_2SO_4 + Fe \longrightarrow FeSO_4 + H_2$$

A study of the process, which appears to be fairly simple chemically, reveals that a host of variables affect the products of the reaction. Some of the more

Figure 11. Pilot seed preparation unit and precipitation tank.

important variables are: (1) Rate of circulation; (2) Quantity of oxygen; (3) Temperature; (4) Type and quantity of iron; (5) pH; (6) Quantity of initial seed nuclei; (7) Size and shape of seed nuclei; (8) Presence of other ions—impurities.

In Figure 11, a typical small pilot seed preparation unit and a precipitation tank are shown. Adjacent to the precipitator is the scrap chamber which holds the scrap iron during the particle growth process. Figure 12 shows a picture of the seed and its growth to a very fine and large yellow iron oxide.

Figure 12. Particle growth from (a) seed, to (b) small, and (c) large $Fe_2O_3 \cdot H_2O$.

The basic technology, as previously described, can be used to prepare a rather large variety of particles having the same chemical composition. In Figure 13, a very large particle is shown which results when very low concentrations of nuclei are employed. The other extreme is demonstrated in Figure 14, which shows an ultrafine particle that is so fine it has almost zero hiding power when incorporated in a paint vehicle system. A crystal growth phenomenon known as twinning, generally caused by impurities, is shown in Figure 15.

The previous examples have dealt with the mineral species goethite ($\alpha Fe_2O_3 \cdot H_2O$). Another less stable mineralogical species, having the same chemical composition, is lepidocrocite ($\gamma Fe_2O_3 \cdot H_2O$). This species is preferably prepared from ferrous chloride [13] in equipment similar to that previously described, but at lower temperatures and in conjunction with a chemical agent, such as zinc chloride, to retard the formation of goethite. A comparison between Figures 12 and 16 will show the difference in particle morphology between lepidocrocite and goethite. The lepidocrocite particle, due to its length/width ratio and platy acicular characteristic form, is a unique raw material for the

Figure 13. Large particles of $Fe_2O_3 \cdot H_2O$ resulting from low nuclei concentration.

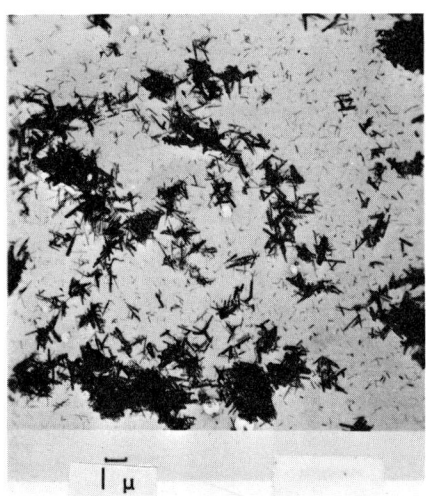

Figure 14. Ultrafine particles precipitated $Fe_2O_3 \cdot H_2O$.

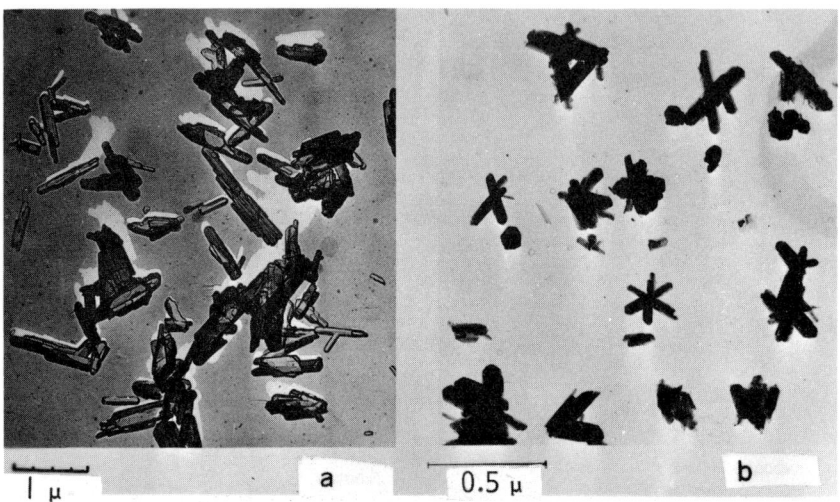

Figure 15. Twinning and dendritic growth of $Fe_2O_3 \cdot H_2O$ generally caused by impurities.

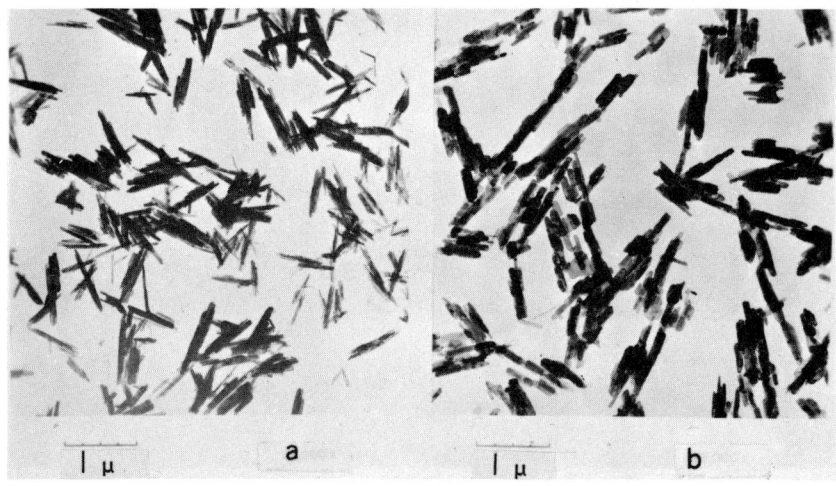

Figure 16. Particle morphology of (a) geothite ($\alpha Fe_2O_3 \cdot H_2O$), and (b) lepidocrocite ($\gamma Fe_2O_3 \cdot H_2O$).

production of magnetic iron oxides (γFe_2O_3) for magnetic tape recording. It has previously been stated that solid-state calcination techniques have certain disadvantages. It is difficult in commercial practice to produce discrete, sharply defined, closely sized ultrafine particles by this technique. The liquid-solid reactions are better suited to meet this goal. Hence a number of methods were developed to produce red iron oxides (Fe_2O_3) by precipitation techniques [14, 15, 16, 17, 18].

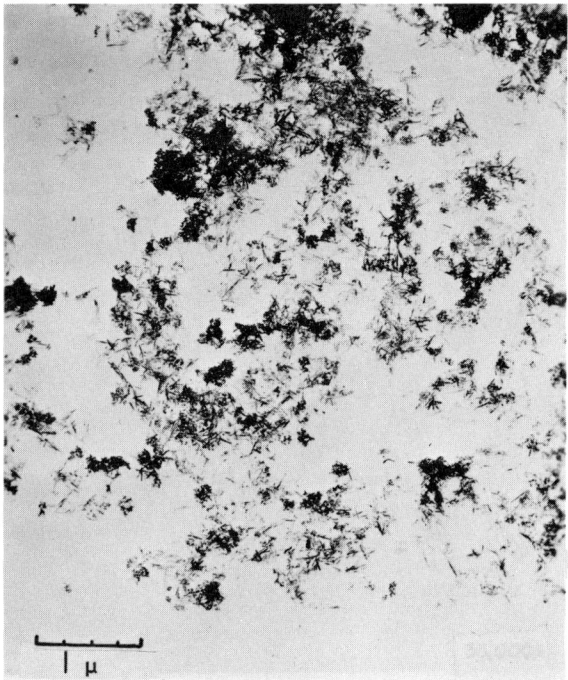

Figure 17. Seed nuclei of precipitated red Fe_2O_3.

Most of the precipitated red iron oxide manufactured in the United States is produced by a process very similar to the yellow iron oxide process [19]. The major difference lies in the seed preparation procedure which produces hematite (Fe_2O_3) nuclei, in addition to goethite ($\alpha Fe_2O_3 \cdot H_2O$) nuclei. The processing of the seed is much more critical with respect to control of pH and temperature. Temperatures below 170°F will produce undesirable acicular particles. In Figure 17 the seed nucleus is shown, and in Figure 18 a small and large particle product are shown.

The reaction is much more affected by impurities in addition to requiring more critical processing conditions. In Figure 19, a well-developed crystal is compared with a product having inhibited crystal growth due to impurities. One ceramic application of iron oxides is in the production of yttrium-iron garnets. Dr. Spencer [20] of Bell Telephone Laboratories at Murray Hill, New Jersey. required an iron oxide which was purer than analytical reagent grade. By purifying all the basic raw materials and exercising every possible precaution, a precipitated iron oxide was produced which had total impurities of less than 100 ppm and contained less than 5 ppm SiO_2. Spencer's work indicated that ferromagnetic resonance losses were greatly reduced, and harder, more brittle

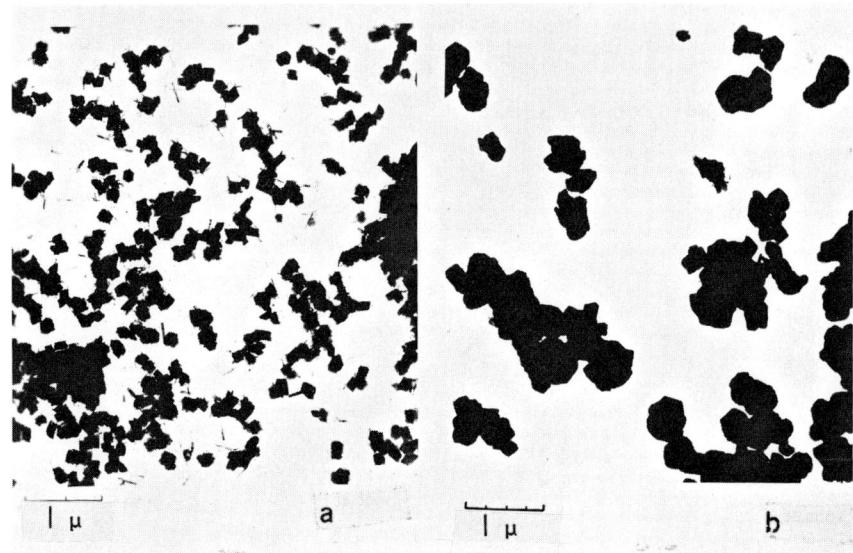

Figure 18. Small (a), and large (b) particles of precipitated Fe_2O_3.

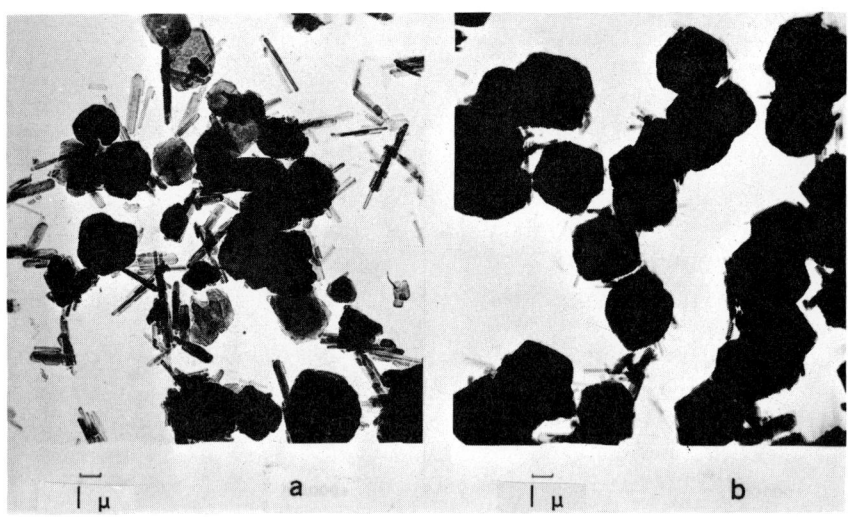

Figure 19. Crystal growth behavior of precipitated Fe_2O_3 from (a) impure, and (b) pure raw materials.

crystals were produced as purity was increased. This oxide was converted into a garnet of the type ultimately used in the Telstar communications satellite.

Still another class of precipitated iron oxides is the cubic ferrosoferric or magnetic black iron oxide. Commercially it is prepared [21] by precipitating Fe_3O_4 [$FeO \cdot Fe_2O_3$] from a ferrous sulfate solution at somewhat below the boiling point at an alkaline pH. Time-temperature relationships control the size of the particles.

Figure 20 compares very fine and fine Fe_3O_4 crystals. A technique for growing crystals to even larger sizes involves the use of an incremental precipi-

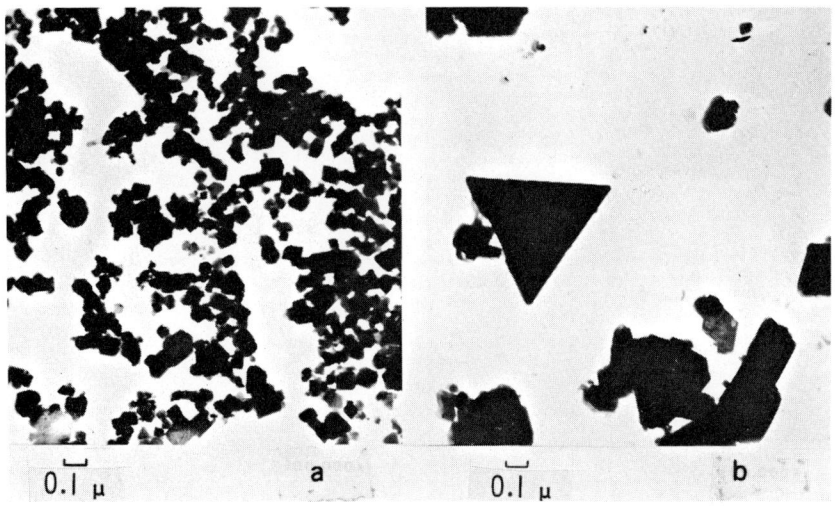

Figure 20. Comparison of (a) very fine, and (b) fine Fe_3O_4 crystals.

tation method wherein the product of the preceding precipitation forms the nucleus for the succeeding precipitation. A final example of the versatility of the technology can be demonstrated by reference to various forms of calcium carbonate. Figure 21 shows four distinctly different crystal shapes all having the chemical composition of calcium carbonate [$CaCO_3$]. The stubby, prismatic form was developed to provide a superior pigment because of its reflective surfaces and ease of dispersion. To achieve this, a process [22] was developed which employs nuclei of calcium hydroxide having a discrete particle size of about 0.01 μ but arranged in clusters of about 10 μ. These nuclei, when carbonated under extremely close temperature programming and specified carbonation rates, produce the novel prismatic form in the figure.

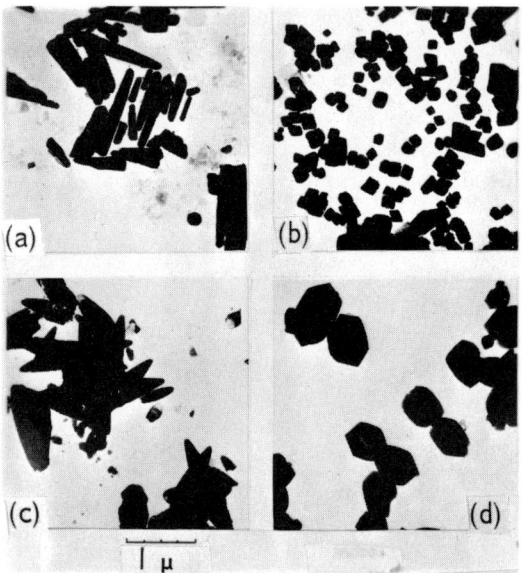

Figure 21. Crystal shapes of various forms of calcium carbonate ($CaCO_3$); (a) acicular, (b) cubic, (c) scalenohedral, and (d) prismatic.

Gas-Solid Reactions

Final areas of metal oxide preparations are the gas-solid techniques. The cloud of dust over a basic open-hearth furnace steel mill is evidence of iron oxide particles being formed in an *uncontrolled* reaction. Figure 22 shows the wide distribution and spherical nature of the particles. The electron-micrograph preparation procedure rejects very large particles and these do not show. Particles as large as 78 μ are actually present in the basic open-hearth furnace oxide. An example of a *controlled* reaction is the production of TiO_2 [23] by the flame-roasting of titanium tetrachloride. Iron oxides can be prepared in a similar technique by processing in an electric arc, producing the spherical particles shown in Figure 23.

A more glamorous example of a gas-solid state metal oxide reaction is the production of magnetic iron oxide from non-magnetic Fe_2O_3 [24, 25]. In this process the particle-size control is exercised in the selection of the raw material. To secure the proper shape anisotropy, an acicular particle is selected. This particle is subjected to a reducing gas such as H_2 in a rotary kiln at temperatures between $600°$ and $1000°F$. The reactions are:

$$\alpha\, Fe_2O_3 \cdot H_2O \xrightarrow{\triangle} Fe_2O_3$$

$$\alpha\, 3Fe_2O_3 + H_2 \xrightarrow{\triangle} 2Fe_3O_4 + H_2O$$

$$2Fe_3O_4 + 1/2O_2 \xrightarrow{\triangle} 3\gamma Fe_2O_3$$

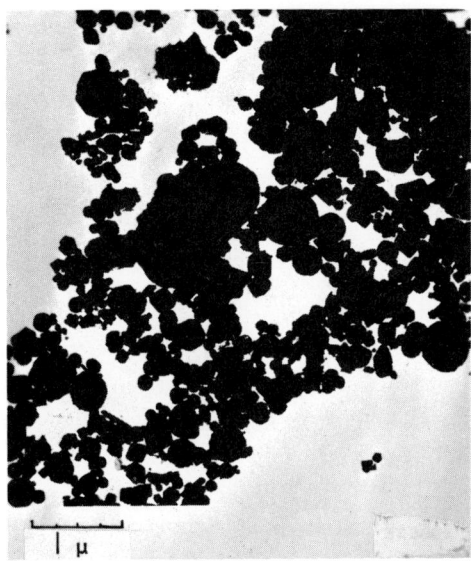

Figure 22. Nature and distribution of basic open hearth particles of iron oxide.

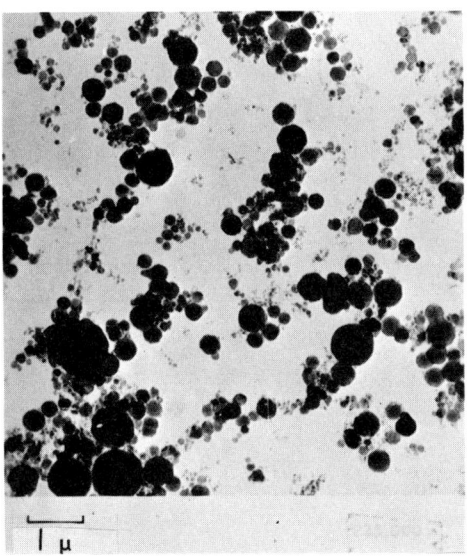

Figure 23. Arc-particulated iron oxide.

The change in particle morphology can be noted in Figure 24 in the electron micrographs. The porous structure is due to the removal of the molecule of water of hydration. The product of this reaction is the magnetic oxide used in almost all recording materials today.

Figure 24. Electron-micrograph of magnetic iron oxide ($\gamma\, Fe_2O_3$).

TABLE I

Physical-Chemical Properties of Synthetic Ultrafine Particles as Affected by Particle Size

Property	0.01 μ	0.1 μ	1.0 μ	10.0 μ
Preparation Difficulty	← Increases →			
Chemical Reactivity	Decreases →			
Chemical Purity	Increases →			
Catalytic Activity	Decreases →			
Surface Area	Decreases →			
Vehicle Demand	Decreases →			
Packing Density	Increases →			
Hiding Power	← Decreases →			
Settling Rate	Increases →			
Agglomerating Tend.	Decreases →			
Visc. of Suspensions	Decreases →			
Color		Darker →		
Ease of Dispersion	Increases →			
X-Ray & Electron Diffraction	Sharper →			

Discussion

One might question the need for such a variety of particle shapes and sizes, since the few typical metal oxide processing examples presented in this paper have the capability of producing over 200 separate and distinct particle shapes or distributions. The multiplicity of specialized end-uses depend upon their physical attributes as much as their chemical composition. Hence, as the demands of industry change, new forms of these physical-chemicals will continue to evolve.

Particle size, shape, and distribution have a profound affect on various physical-chemical properties. The following table summarizes some of these properties of the compounds discussed in this paper as they are affected by particle size.

Summary

Solid-state, liquid-solid state, and gas-solid state reactions have been presented as examples of methods of producing ultrafine particles.

Solid-state reactions, if they are dependent upon heat transfer mechanisms, must be controlled so that the heat transfer will proceed at the same rate if uniform ultrafine particles are to be produced.

In liquid-solid reactions, the major controlling influence stems from the nuclei. Subsequent operations or variables can alter the shape, size, and distribution of the ultrafine particles. Without the desired nuclei present in the correct quantity, the growth process tends to proceed in an uncontrolled manner. Commercial precipitation processes are similar to single-crystal growing techniques with respect to the influence of impurities on promoting or retarding growth on certain crystal faces.

Gas-solid reactions cover a wide range and are difficult to generalize. Those reactions which involve minor crystal structure changes, such as the transition of αFe_2O_3 to γFe_2O_3, depend to a large extent upon the starting particle for their ultimate size and shape characteristics.

Within the size range of particles described in this paper, and considering the commercial aspects, the precipitation process possesses the ability to control the growth of ultrafine particles with the greatest precision.

Acknowledgments

The writer wishes to thank the Chas. Pfizer & Co., Inc. for permission to publish this paper. Thanks are also due to Mr. Mark Hoffman for his valuable suggestions, and to Mr. Harold Stanley and Mr. Walter Bonstein for their assistance in preparing the electron-micrographs and particle-size data.

References

1. Schoen, Herbert M., "Annual Review of Crystallization," *Ind. and Eng. Chem.,* 52 (1960), 173.
2. Palermo, J., "Annual Review of Crystallization," *Ind. and Eng. Chem.,* 60 (1968), 65–93.
3. Crittenden, E. C. Jr., Hudimac, A. A. and Strough, R. I., "Magnetization Hysteresis Loop Tracer for Long Specimens of Extremely Small Cross Section," *The Review of Scientific Instruments,* 22 (1951), 872–77.
4. Perry, J. H., *Chemical Engineers Handbook,* 4th ed., Sect. 11, p. 42.
5. Perry, J. H., *Chemical Engineers Handbook,* 4th ed., Sect. 10, p. 5.
6. Walton, A. G., *The Formation and Properties of Precipitates,* Interscience Publishers, New York (1967).
7. Ayers, J. W., U.S. Patent 2,394,579 (assigned to Chas. Pfizer & Co., Inc.).
8. Frayne, C. G., U.S. Patent 2,560,338 (assigned to Chas. Pfizer & Co., Inc.).
9. White, F. S., U.S. Patent 3,293,330 (assigned to Chas. Pfizer & Co., Inc.).
10. Penniman, R. and Zoph, N., U.S. Patent 1,368,748 (assigned to National Ferrite Co.).
11. Plews, G., U.S. Patent 2,111,727 (assigned to Chas. Pfizer & Co., Inc.).
12. Ayers, J. W., U.S. Patent 2,255,607 (assigned to Chas. Pfizer & Co., Inc.).
13. Ayers, J. W. and Stephens, R. A., U.S. Patent 3,015,628 (assigned to Chas. Pfizer & Co., Inc.).
14. Toxby, T., U.S. Patent 2,620,261 (assigned to Chas. Pfizer & Co., Inc.).
15. Bennetch, L., U.S. Patent 2,785,991 (assigned to Chas. Pfizer & Co., Inc.).
16. Bennetch, L., U.S. Patent 3,009,821 (assigned to Chas. Pfizer & Co., Inc.).
17. McCallum, R., U.S. Patent 3,267,041 (assigned to W. R. Grace Co.).
18. Marsh, B., U.S. Patent 2,716,595 (assigned to Chas. Pfizer & Co., Inc.).
19. Ayers, J. W., U.S. Patent 2,937,927 (assigned to Chas. Pfizer & Co., Inc.).
20. Spencer, E. G. and Remeika, J. P., "Effects of Chemical Purification on Physical Properties of Magnetic and Other Compounds," 1964 Proceedings of the Intermag. Conference.
21. Ayers, J. W., U.S. Patent 2,133,267 (assigned to Chas. Pfizer & Co., Inc.).
22. Waldeck, W. F., U.S. Patent 3,320,026 (assigned to Chas. Pfizer & Co., Inc.).
23. Wendell, U.S. Patent 3,372,001 (assigned to Cabot Corp.).
24. Farbenfabriken, Bayer Aktiengesellschaft, English Patent 923,038.
25. Ayers, J. W. and Stephens, R. A., U.S. Patent 3,015,628 (assigned to Chas. Pfizer & Co., Inc.).

SESSION III

CHARACTERIZATION OF ULTRAFINE PARTICLES AND ULTRAFINE-GRAIN CERAMICS

MODERATOR: JOSEPH A. PASK
University of California
Berkeley, California

4. On The Problems of Characterizing Fine Powders

B. H. KAYE* and M. R. JACKSON

IIT Research Institute
Chicago, Illinois

ABSTRACT

Problems associated with the size analysis of powders in the range 0.1 μ to 5 μ are discussed. Special attention is focussed on felvation and centrifugal analysis with respect to shape-dependent information which could be obtained from these techniques.

· Introduction

With the advent of advanced ceramic materials manufactured from powders containing sub-micron particles, emphasis has been focussed on the need for reliable size analysis of fine and ultrafine materials. Techniques developed and used successfully for analysis of larger particles have proved inadequate when applied to these small particles, due to the drastic changes in the ratio of surface to volume which occur as the particle size is reduced. The ratio of surface to volume is an important parameter controlling the behavior of a powder system, since the surface and mass forces control the particle interactions and attractions which ultimately govern the handling properties of the powder.

The problems specific to the size analysis of fine and ultrafine particles are discussed in this chapter, and a description and discussion is presented of a centrifugal and a sieving technique suitable for analysis of small particles.

· Handling Problems

As the particle size of a powder is reduced, the powder becomes more cohesive and lower in bulk density, causing the powder to become sticky and fluffy. Under these conditions, handling becomes a problem even in analysis, since flow is restricted. Dispersion is difficult and the smallest particles may become ineffective, either because of actual loss by dusting, or by association with larger par-

---

*Presently: Laurentian University, Sudbury, Ontario, Canada.

ticles. Thus, preparation of an analytical sample from a larger quantity presents difficult problems which are unique to each powder system.

It is known that viscous shear is one of the most effective mechanisms in dispersion preparation; however, the extent to which this may be used is dependent upon the mechanical properties of the particles. If the particles are friable, shear forces must be controlled to ensure that fracture is inhibited, whereas if particles are abrasive, care must be exercised to prevent abrading material from the shear cell walls. To this end, shear may be imposed by many methods ranging in severity from gentle brushing to high shear between plattens.

Since each particulate system is individual, it is not the intention in this chapter to itemize methods found suitable by experiment for specific powders. Rather, a method of approach will be briefly indicated which is suggested as a criterion for use in analytical size measurement of powders. Such a method depends upon obtaining measured size distributions which remain invariant as changes are made in the dispersion technique, thus indicating a reproducible stable state of dispersion.

General Problems of Analysis

The analysis of powders composed of particles smaller then about $1\ \mu$ presents an enhancement of physical properties displayed by larger powders. This may invalidate direct extrapolation of test methods suitable for large particles to analysis of fine and ultrafine powders.

Direct observation of large particles is possible by optical microscopic methods. As the particle size is reduced to less than $1\ \mu$, the apparent size is largely controlled by the diffractive interaction of the particles with light, and accurate observations become difficult. Below about $0.50\ \mu$, direct observation of the particle becomes impossible, although under optimum lighting conditions the presence of particles may be detected. For particles smaller than $0.50\ \mu$, it is necessary to resort to electron-microscopy for individual particle observation.

Electron-microscopic analysis will permit the observation of particles whose size is measured in Ångstrom units. However, at the magnifications needed for this resolution, the area of the field of view is small and may be only several hundred or thousand Ångstrom units across. Under these conditions, only a small number of particles can be examined at one time and statistically, there is a bias against the observation of the larger particles which may approach or exceed the field of view in diameter. Thus, analysis of wide size range powders becomes difficult when a given statistical confidence is required over the entire size range.

The physical properties of cohesion are greatly enhanced as the particle size of a powder is reduced. Thus, a coarse sand may be free-flowing, but when milled to a fine size may exhibit angles of repose approaching $90°$. Under these conditions, it is extremely difficult to disperse powders so that individual particles are presented singly to a sensing zone for analysis. This may involve the clustering of particles on a microscope slide so that it is impossible to state with certainty whether or not particles within a group are separate, adjacent, or tightly sintered together. Similarly, particles presented to a sieving surface may cluster together by virtue of the cohesive forces so that they are unable to pass through the sieve apertures, and during sedimentation, the particles—if undispersed—settle as agglomerates, or may cluster during sedimentation to form agglomerates which settle at velocities faster than those of the separate particles.

Cohesion may also cause the association of small particles on large particles so that the number of free small particles is reduced and the true concentration is masked. This is particularly important in sieving procedures where a reduction in free small particles may result in a significant reduction in the fraction capable of passing through the sieve aperture.

In a similar manner, the adhesive properties of particles are enhanced as they become finer, due to an increase in the surface-to-volume ratio. As particles encounter surfaces, such as sieve plates, the surface forces cause cohesion resulting in a build-up of powders around the sieve apertures. This may completely blind the sieving surface. In a similar manner, adhesive powders may accumulate on the walls of containing vessels, causing preferential loss of the finest and most adhesive particles.

In the following discussion, the application of sieving and sedimentation techniques to the analysis of fine powders is presented, with particular emphasis on the importance of the fine-particle properties on the efficiency and accuracy of the analytical method.

· Sieve Analysis

Traditionally, the lower limit of particle size analysis by sieving techniques is between 40 μ and 50 μ. With the advent of electro-formed and punched plate micromesh sieves, this range has been extended down to about 1 μ. At this size, the adhesive and cohesive forces preclude the use of traditional wet or dry sieving techniques, and a method has been presented [1] for the selective presentation of particles by size to the sieve.

In this technique, a heterogeneous powder is selectively fractionated by fluidization and elutriation in fluid upward toward a sieve surface. Careful control of the elutriation conditions results in a selective presentation of particles to the sieve apertures in such a manner that the smallest particles are presented first, and are carried by the fluid flow streams to the apertures in the sieve. The smallest particles pass through the holes. As the sieving process is continued, the elutriation velocity is increased so that larger particles are presented selectively to the sieve aperture. Ultimately, a condition is reached when the particles which are elutriated are of sufficient size to be retained by the sieve. At this time, the particles capable of passing through the sieve have done so and the sieving process is complete.

This process has been called felvation, to draw attention to the fact that it is a new process combining the advantageous facets of fluidization, elutriation, and sieving.   These facets have been combined to minimize the effects of adhesion and cohesion which impart the physical characteristics to a powder in the 1–10 μ size range.

The felvation process is shown schematically in Figure 1. Four major areas of powder handling are shown: a shear zone, a fluidization bed, an elutriation column, and the sieving plane. In use, the powder to be analyzed is placed before the shear zone as shown. Fluid containing a dispersant is introduced into the system at a low flow rate to carry the particles through the shear zone. In this zone the particles undergo viscous shear, which aids dispersion of agglomerates.

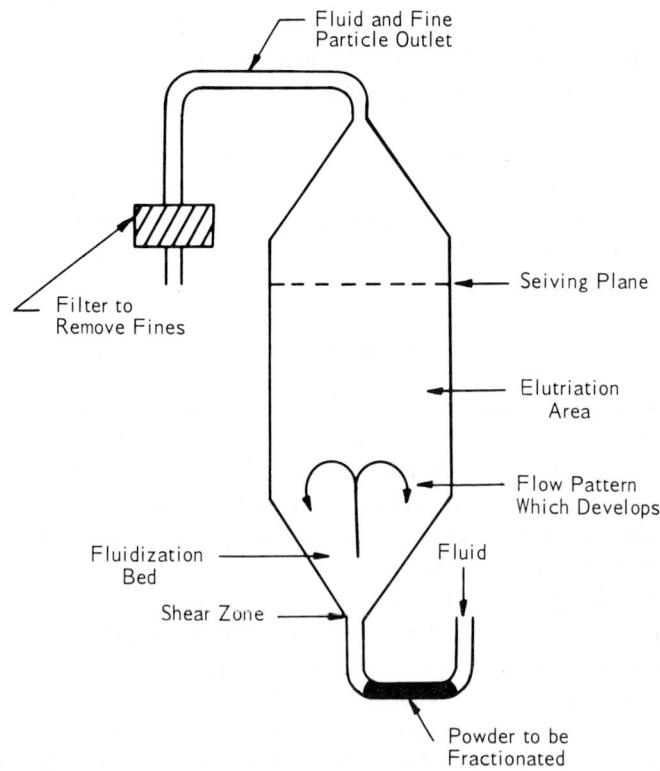

Figure 1. Felvation process.

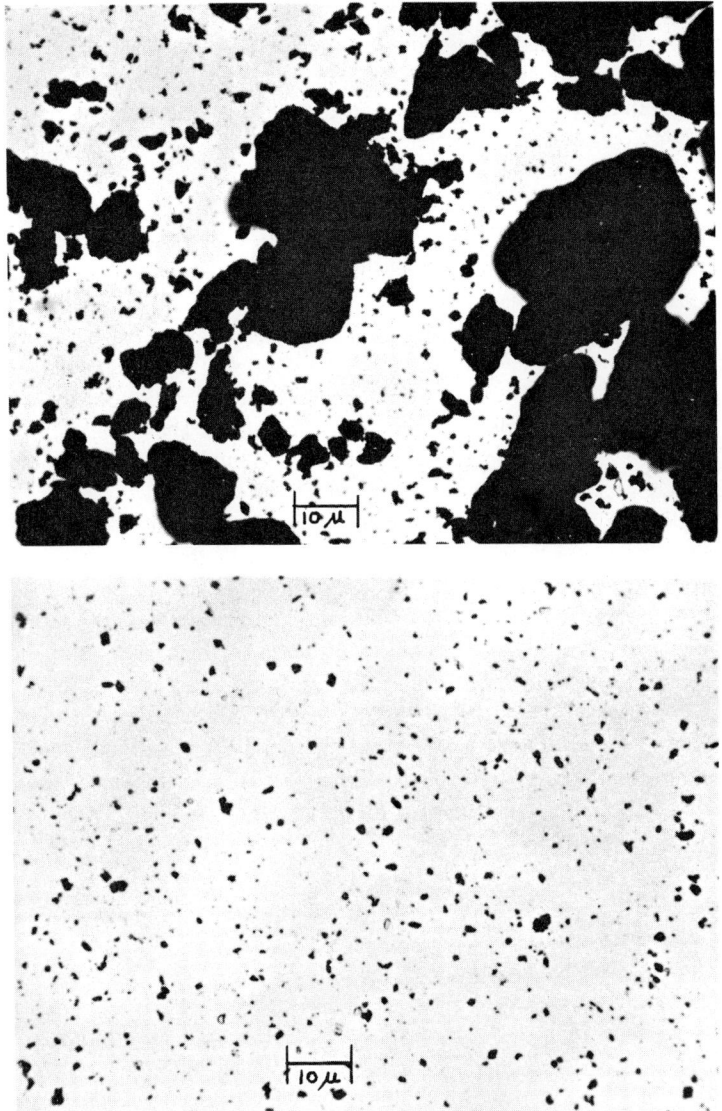

Figure 2. Fractionated chromium powder. Above: Chromium powder through 325 mesh fraction. Below: Chromium powder narrow size range fraction.

If a dispersion is added rather than a dry powder, this shear helps maintain the states of dispersion. Within the fluidized bed area, the upward flow rate is less than that in the shear tube, allowing a turbulent fluidized bed of particles to develop. In the bed, shear forces act to separate the small particles associated with larger particles and agglomerates. The small particles are flushed from the bed as they are released, while the larger ones settle toward the shear zone where they are subjected to further disrupting shear forces. A reverse vortex flow pattern is established within the fluidized bed, which allows all the particles to be presented to the shear zone and to the upper surface of the bed for release into the elutriation column.

Particles are either retained in or passed up the elutriation column, depending upon their mass and the drag forces of the upward fluid flow. Thus, the elutriation column acts as a size separator which presents particles to the sieve apertures selectively. The smallest particles are presented first, and are capable of passing through immediately. The flow stream cushions the particles from the edges of the apertures and thus minimizes contact and subsequent retention of particles on the sieve itself. In addition, the flow rates may be adjusted so that a dilute condition exists within the elutriation column. Thus, once dispersed, a low probability exists that particles will undergo random encounter and subsequent re-agglomeration.

By utilizing the advantageous aspects of the shear, fluidization, elutriation, and sieving concepts, a process has been developed to largely overcome the difficulties of sieving peculiar to powders within the size range of 1-10 μ in diameter. The effectiveness of this approach is best evaluated for each powder system analyzed. A suitable method for evaluation, which may also be used to indicate completeness of sieving under the test conditions, is to pulse the fluid column after completion of the felvation process. This imposes additional shear forces on the powder retained by the sieve, and observation will indicate whether significant further agglomerate dispersion is occurring. Figure 2 shows the cleanliness of fraction which may be achieved by felvation.

Sedimentation Analysis

Sedimentation analysis by gravity has been used for many years for analysis of powders within the size range down to a few microns [2, 3]. The lower useful size limit of these methods is imposed because the sedimentation velocity of small particles is so small that the assumptions made of undisturbed separate streamlined fall under gravity are invalidated. The assumptions fail because the particle terminal velocity under gravity is comparable with that of convection currents caused either by fluid displacement by sedimenting particles, or by unavoidable thermal gradients. In addition, Brownian motion becomes important in that it may cause random collisions between particles and hence increase the probability of agglomerate formation.

Centrifugal methods have been developed [4, 5, 6] in which the sedimentation velocity is increased under the influence of the centrifugal force vector. In these methods the increased sedimentation force diminishes the adverse effects of convection currents and Brownian motion, and allows analysis of sub-micron particles. One of the centrifugal methods has design features which tend to offset the agglomerative character of the powders throughout the entire analytical technique [5]. A brief description of this method is presented to indicate the useful application of the de-agglomerative processes.

The centrifugal disc photosedimentometer, shown schematically in Figure 3, comprises a hollow cylindrical disk-shaped vessel having viewing windows concentric with the horizontal axis. In use, the disk is partially filled with a clear liquid through which the particles are to be sedimented and rotated at a speed sufficiently high to centrifugally drive the liquid of the circumferential walls of the disk, thus leaving a hollow concentric cylinder at the center of the disk bounded by a liquid surface. A suspension of the particles to be analyzed is injected into the hollow cylinder and ultimately onto the liquid surface. Upon contact with the liquid, the particles sediment radially through the liquid at velocities related to their Stokesian size. The presence of particles passing the viewing windows is detected optically as a reduction in the light intensity transmitted in a forward direction by the liquid. This reduction in energy is related to the quantity of material in the viewing zone. The size of particles reaching this viewing zone is calculated using Stokes' law of viscous sedimentation, thus permitting the construction of a size/distribution curve.

It has been shown [5] that the volume concentration of particles in the injected suspension must be small so that the interparticle distance is large, thus minimizing the probability of random particle encounters and agglomeration. Thus, a newly prepared dispersion of the powder to be analyzed may be diluted to the point where significant re-agglomeration does not occur within the time required for sample injection. Upon injection, the sample is impacted upon the

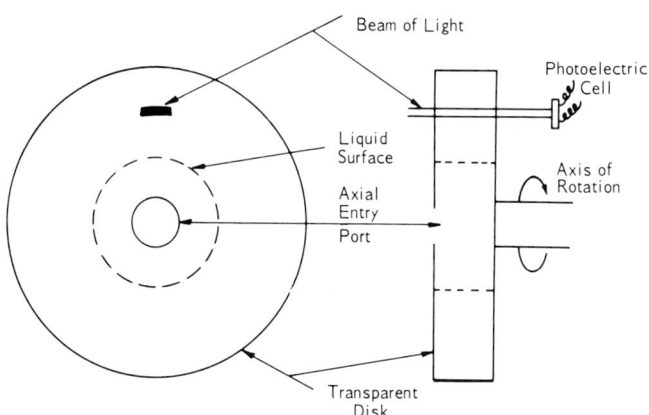

Figure 3. Essential system of Kaye disk centrifuge.

rapidly rotating liquid surface causing shear, which aids de-agglomeration of chance agglomerates formed after sample preparation. After the particles enter the sedimentation liquid they travel radially outward, becoming even less concentrated and thus further inhibiting re-agglomeration. Throughout sedimentation, the particles travel in paths parallel to the side walls of the disk and contact of the major portion of the sample with any walls before passing the viewing zone is minimized. This virtually eliminates any probability of adhesion of particles upon the containing walls until after the analysis.

The points detailed show how, at each stage of the analysis, the conditions tend to reduce the probability of particle loss by adhesion or agglomerates. A detailed analysis of other methods of sedimentation analysis is beyond the scope of this chapter, but comparative analysis would show that other methods involve a high concentration of particles at some stage of the analysis. This tends to promote rather than inhibit agglomeration and hence, inaccurate analysis. The technique does not, however, in its present form reduce the importance of the interaction of light with the particles and, where particles approach the wavelength of light in size, it is desirable to take account of their extinction coefficient. Where only comparative data of size are required, the measured times of arrival of particles at the viewing zone are satisfactory, as shown in the analysis of latex emulsions [6].

This method is particularly applicable to a study of dispersion state and technique, since repetitive determinations of size distribution may be made of samples having different dispersion treatments. In this way, dispersion techniques may be determined which are known by experiment to provide a reproducible size distribution, thus indicating the non-agglomerative nature of the analyzed material.

Conclusion

A discussion has been presented of some of the difficulties encountered in the analysis of fine and ultrafine powders. These problems are not typical in analysis of coarse powders, since their origin is based upon the physical properties of the particles related to the ratio of surface to mass forces. The degree to which this ratio effects the handling, and hence the analytical procedures of fine and ultrafine powders, is dependent upon the individual properties to each powder system, making it necessary to treat each analysis on its own merits. Guides have been presented, with examples in sieving and sedimentation, of some of the more common factors attributed to analytical difficulty. These stress the fact that analysis of difficult powders—and use for fine powders—of technique acceptable for analysis of coarse powders, should be undertaken with extreme caution.

References

1. Kaye, B. H. and Jackson, M. R., "A New Technique for Fractionating Powders," *Powder Technology,* 1 (1967), 43–50.
2. Robinson, C. S., "Some Factors Influencing Sedimentation," *2nd Eng. Chem.,* 18 (1926), 869–71.
3. Andreasen, A. H. M., "Uber die Feinheitsbestimmung und ihre Bedutung fur die Keramische Industrie," *Berichte Deutsche Keramische Gesellschaft,* II (1930), 675.
4. Whitby, K. T., "A Rapid General Purpose Centrifugal Sedimentation Method for Measurement of Size Distribution of Small Particles," *J. Air Pollution Control Assoc.,* 5 (1955), 120–26, 132.
5. Kaye, B. H. and Jackson, M. R., "Size Analysis of Latex Emulsions Using a Centrifugal Disc Photosedimentometer," *Powder Technology*, (1967), 81–88.
6. Beresford, J., "Size Analysis of Organic Pigments Using the ICI–Joyce Loebl Disc Centrifuge," *J. Oil Col. Chem. Assoc.,* 50 (1967), 594–614.

5. Optimum Procedures for Determining Ultrafine-Grain Sizes

J. E. HILLARD, J. B. COHEN, and W. M. PAULSON
Northwestern University
Evanston, Illinois

ABSTRACT

A review is given of microscopic and X-ray methods for estimating grain size in compacted or sintered ceramics. Particular emphasis is given to two aspects of grain-size analysis that are often neglected. The first is the definition of an "average" size and the type of average that each procedure yields. The second is the sampling error and how this can be minimized for a particular procedure.

It is shown that from microscopical measurements it is possible to determine three different average grain sizes. A theoretical and experimental study has also been made of an X-ray procedure that involves the measurement of the variance in diffracted intensity from different parts of the specimen.

Introduction

For the purposes of the present review, we will consider the term "ultrafine" to encompass the range 0.1 to 10 μ in grain size. The discussion will be limited to the sizing of grains in compacted or sintered specimens (although some of the procedures to be described are also applicable to loose powders). The procedures fall into two classes. The first require observations under the microscope, and are only applicable to the upper end of the size range unless electron-microscopy is used. The second utilize X-ray diffraction. We have made a detailed theoretical and experimental evaluation of one diffraction method that is particularly suitable for the range of sizes under consideration.

We will not treat the determination of grain-size *distributions*. To have done so adequately would have meant devoting a great deal of space to an analysis that is infrequently performed. Instead, we have considered it more useful to examine in detail two aspects of grain-size analyses that are often neglected. The first is the exact definition of the "average" size that the different procedures yield. In too many cases average sizes are reported without any real indication of the type of average or, equivalently, the method used to obtain the average. The second aspect is the sampling errors associated with the measurements. A

73

knowledge of these is important, because they determine not only the reliability of the results but also the number of observations that have to be made. Wherever possible, therefore, we will give expressions that permit an a priori estimate of the standard deviation of each analysis.

The principal symbols that will be used in the derivations are listed in Table I.

TABLE I

A_o	Cross-sectional area of X-ray beam.
A_t	Total area examined.
D	Lineal dimension specifying grain size.
$D_{m,n}$	Average grain size [defined by equation (2)].
K	Mean-caliper diameter.
$\overline{L}$	Average intercept length.
L_L	Lineal fraction; i.e., fractional length of test line (or figure) intercepted by grains.
L_t	Total length of test line.
N'	Total number of intersections of test line with grain boundaries counted.
N''	Total number of grain profiles counted.
N'''	Total number of intercept lengths measured.
N_A	Number of grain profiles per unit area ($= N''/A_t$).
M	Magnification.
P	Number of points applied in a point count.
P_L	Number of intersections per unit length of test line ($= N'/L_t$).
S	Total boundary area of grains.
S_V	Grain boundary area per unit volume.
V	Total volume of grains.
V_V	Volume fraction of grains.
a	Average cross-sectional area of grain.
g	ASTM grain-size number [equation (11)].
j	Multiplicity of reflection in X-ray diffraction.
k', k''	$\sigma(N')/\sqrt{N'}, \sigma(N'')/\sqrt{N''}$.
k'''	$\sigma(L)/\overline{L}$.
$\overline{\ell}$	Average intercept length in an individual grain.
q	Set of parameters defining grain shape.
s	Surface area of grain.
v	Volume of grain.
Ω	Divergence of X-ray beam.
α, β, γ	Shape constants defined by equations (6), (7), and (14).
ϵ	Correction factor for preferred orientation.
μ	Linear absorption coefficient.
$\mu_n'(x)$	nth moment of x about the origin.
θ	Diffraction angle.
ω	$\sigma^2(Y)/\overline{Y}^2$

Specification of Grain Size

Most measurements of grain size yield a single parameter—namely an average size. We shall see that this average takes the form of a ratio of two of the moments of the size distribution; i.e., integrals of the form:

$$\mu_n'(D) = \int_O^{D_M} D^n f(D)dD, \quad (0 \leqslant D \leqslant D_M) \tag{1}$$

in which n is an integer and $f(D)dD$ is the probability that a grain will be of size $D \pm (dD/2)$. In terms of the moments, we can define an infinite set of averages, $D_{m,n}$ by:

$$D_{m,n} = \left[\mu_m'(D)/\mu_n'(D) \right]^{[1/(m-n)]}, (m \neq n) \tag{2}$$

It will be noted that $D_{m,n} = D_{n,m}$. If n is held constant, then, for a given distribution of sizes, $D_{m,n}$ increases with increasing values of m.

As a numerical illustration, several averages (mostly ones that we will be encountering later) are listed in Table II that would be observed in a hypothetical specimen containing grains all having the same shape but of two different sizes;

TABLE II

Numerical Values and Descriptions of
Various Grain-Size Averages for a
Hypothetical Specimen*

m	n	Description	$\frac{D_{m,n}}{(\mu)}$
1	0	Weighted by number.	1.09
2	1	Weighted by linear size.	1.82
3	0	Constant grain size giving same number of grains per unit volume.	2.22
3	1	Constant grain size giving same number of grain profiles per unit area of the plane of polish.	3.16
3	2	Weighted by surface area or constant grain size having same total boundary area per unit volume.	5.50
4	2	–	7.11
4	3	Weighted by volume.	9.18
6	3	–	9.69

*The numerical values of $D_{m,n}$ are those computed from equations (1) and (2) for a specimen composed of grains of the same shape but of two different sizes; one set being 1 μ and the other 10 μ, present in the ratio of 100 small grains to one large grain.

one set being 1μ in size and the other 10μ, present in the ratio of 100 small grains to one large grain. It will be noted that the average sizes almost cover the complete range from the smallest to the largest grain present. Consequently, it is not meaningful to talk about an average size unless the type of average is specified. It also follows that two different methods of measurement will not, in general, yield the same size.

Ideally, the investigator should select the particular average that will best correlate with the property of the material in which he is interested. For example, if the objective were to relate the grain size to the initial rate of a transformation that was nucleated heterogeneously at the grain boundaries, then the appropriate average would be $D_{3,2}$. All specimens having the same value for this average would have the same rate of transformation even if they did not have the same distribution, $f(D)$, of sizes. As another example, this time in the sizing of loose powder, consider the description of metallographic polishing powders. In this case it would be desirable to use an average, such as $D_{6,3}$, that very heavily weighted the largest particles present. Even if these were few in number, they would determine the maximum scratch size.

One other problem is the choice of the linear dimension, D, to be used in specifying the grain size. As in the selection of the type of average, the choice for D will depend on the application of the measurement. Some of the possibilities are: average chord length, maximum chord length, and diameter of sphere having the same area or volume as the particle.

Procedures Utilizing Microscopy

In this section it will be assumed in the derivations that observations are made on a two-dimensional section through the material, either optically with reflected light or by electron-microscopy using replicas. For some of the analyses it is necessary to distinguish between what will be termed "contiguous" and "non-contiguous" grains. The former correspond to grains in a single-phase polycrystalline material and the latter to unconnected grains (or particles) dispersed in a matrix.

Basic Relationships

Determination of $D_{3,1}$

The first procedure we shall discuss is one involving the measurement of the number of grain profiles, N_A, per unit area. (We will use the term "profile" to denote the two-dimensional trace of the grain with the plane of polish.) Let q stand for a set of one or more parameters defining the shape of a grain, and let $f(D,q)dDdq$ be the probability that a grain (in three-dimensional space) is of size $D \pm (dD/2)$ and of shape $q \pm (dq/2)$. The density of profiles on a plane is then given by:

$$N_A = N_V \int_q \int_0^{D_M} K(D,q)f(D,q)dDdq \qquad (3)$$

in which N_V is the total number of grains per unit volume and $K(D,q)$ is the mean-caliper diameter of a grain of size D and shape q. The latter quantity is defined as the average distance between two parallel tangent planes for random orientations of the grain. A general expression is available [1] for calculating the mean-caliper diameter, and values have been tabulated for various common shapes.

If $v(D,q)$ is the volume of a grain, then the volume fraction of grains is given by:

$$V_V = N_V \int_q \int_0^{D_M} v(D,q)f(D,q)dDdq \qquad (4)$$

Eliminating N_V from equations (3) and (4) we obtain:

$$(N_A/V_V) = \left[\int_q \int_0^{D_M} K(D,q)f(D,q)dDdq \right]$$

$$X \left[\int_q \int_0^{D_M} v(D,q)f(D,q)dDdq \right]^{-1} \qquad (5)$$

This equation can be simplified if we assume that the grains all have the same shape. In this case we can set:

$$K(D,q) = \alpha D \qquad (6)$$

and

$$V(D,q) = \beta D^3 \qquad (7)$$

from which it follows that α and β are numerically equal to the mean-caliper diameter and volume, respectively, of a grain of size $D = 1$. Substituting equations (6) and (7) into equation (5) and applying equation (1) we obtain:

$$(N_A/V_V) = \alpha\mu_1'(D)/\beta\mu_3'(D) \qquad (8)$$

and hence, by equation (2):

$$D_{3,1} = [(\alpha/\beta)(V_V/N_A)]^{1/2} \tag{9}$$

Thus, for grains of constant shape, the average $D_{3,1}$ can be determined from a measurement of N_A and V_V (the estimation of V_V will be discussed later). If the grains are not of the same shape, the observed average will, as indicated by equation (5), depend upon the shape distribution as well as the size distribution.

For contiguous grains, $V_V = 1$, and equation (9) reduces to:

$$D_{3,1} = [(\alpha/\beta)(1/N_A)]^{1/2} \text{ (contiguous grains)} \tag{10}$$

Also, if the grains are contiguous, the result can be reported, if desired, in terms of an ASTM grain-size number, g. By definition:

$$n = 2^{(g-1)} \tag{11}$$

in which n is the number of grain profiles per square inch at a magnification of 100 X. This expression can be written in the following more general form:

$$g = -9.60 + 3.22 \log_{10}(M^2 N''/A_t) \tag{12}$$

where N'' is the number of profiles in an area of $A_t \text{cm}^2$ of the image at a magnification M.

Determination of $D_{3,2}$

If a lineal test figure is applied to the plane of polish, then the number of intersections, P_L, that a unit length of the figure makes with the traces of the grain boundaries is given by

$$P_L = (1/2)S_V \tag{13}$$

in which S_V is the total boundary area per unit volume. It should be emphasized that this relationship is perfectly general and that there are no assumptions involved about the form of either the structure or the test figure.

Although we could derive an equation analagous to equation (5) we shall, for simplicity, go straight to the assumption of constant grain shape. Thus, if $s(D)$ is the boundary area of a grain size D, we can set:

$$s(D) = \gamma D^2 \tag{14}$$

where γ is the surface area of a grain of unit size. For non-contiguous grains:

$$S_V = N_V \int_0^{D_M} s(D)f(D)dD = \gamma N_V \mu_2'(D) \tag{15}$$

Combining this result with equations (1), (4), (7), and (13) we obtain

$$2(P_L/V_V) = \gamma\mu_2'(D)/\beta\mu_3'(D) \qquad (16)$$

and hence, by equation (2):

$$D_{3,2} = (\gamma/2\beta)\,(V_V/P_L)\,\text{(non-contiguous grains)} \qquad (17)$$

If the grains are contiguous, $V_V = 1$ and, in addition, each boundary is shared by two grains. In this case:

$$D_{3,2} = (\gamma/4\beta)\,(1/P_L)\,\text{(contiguous grains)} \qquad (18)$$

the extra factor of 2 appearing because of the shared boundaries.

An expression for $D_{3,2}$ can also be obtained in terms of the average intercept length. From the well-known relationship:

$$V_V = L_L \qquad (19)$$

where L_L is the fractional length of the test figure intersecting the grains, it follows that:

$$V_V/P_L = \begin{cases} \overline{L}/2 \text{ (non-contiguous grains)} \\ \overline{L} \quad \text{(contiguous grains)} \end{cases}$$

in which $\overline{L}$ is the average intercept length in the grains. Thus, both equations (17) and (18) can be written:

$$D_{3,2} = (\gamma/4\beta)\overline{L} \qquad (20)$$

For an intercept analysis, it is advantageous to use the average intercept (or chord) length, $\overline{\ell}$, as a measure of the size, D, of a grain. For an individual grain of any shape:

$$\overline{\ell} = 4v/s. \qquad (21)$$

where v and s are the volume and boundary area of the grain. With this choice of D, it follows from equations (7), (14), (17), and (21):

$$D_{3,2} = 2V_V/P_L \quad (D = \overline{\ell}, \text{ non-contiguous grains}) \qquad (22)$$

from equation (18),

$$D_{3,2} = 1/P_L \quad (D = \overline{\ell}, \text{ contiguous grains}) \qquad (23)$$

and from equation (20),

$$D_{3,2} = \overline{L} \quad (D = \overline{\ell}) \qquad (24)$$

It will be noted that the geometrical factors are absent from equations (22)–(24). They are therefore valid for grains of any shape or for any mixture of shapes. This, of course, is the advantage of setting $D = \overline{\ell}$. Furthermore, under this condition $D_{3,2} = 4V/S$, where V and S are the total volume and boundary area of the grains.

If desired, the results of an intercept analysis on contiguous grains can be reported in terms of an equivalent ASTM grain-size number, g_e, by use of the following relationship [2]:

$$g_e = -10.00 + 6.64 \log_{10}(N'M/L_t). \tag{25}$$

in which N' is the number of intersections of the boundary traces with a test figure of total length $L_t$cm applied to a field viewed at a magnification M. The value of g_e given by equation (25) is that which would be observed in a profile count on an imaginary specimen with the same boundary area as the specimen under test, and in which the grains were tetrakaidecahedra of constant size. A nomograph based on equation (25) is available [3] to facilitate the calculation of g_e.

Determination of $D_{2,1}$

Eliminating $\mu'_3(D)$ from equations (8) and (16) we obtain for non-contiguous grains:

$$\mu'_2(D)/\mu'_1(D) = 2\alpha P_L/\gamma N_A \tag{26}$$

Hence, from equation (2):

$$D_{2,1} = (2\alpha/\gamma)(P_L/N_A) \text{ (non-contiguous grains)} \tag{27}$$

and, for contiguous grains,

$$D_{2,1} = (4\alpha/\gamma)(P_L/N_A) \text{ (contiguous grains)} \tag{28}$$

Determination of $D_{\infty,0}$

From the discussion of the moments of the distribution it is apparent that as $m \to \infty$, $D_{m,0} \to D_M$; i.e., $D_{\infty,0}$ is the size of the largest grain present. This can be determined from observations on the largest grain profile seen on the plane of polish. The rationale for this, of course, being that the largest profile will originate from an intersection of the plane across the diameter of the largest grain present.

Determination of V_V

For the sizing of non-contiguous grains, it is necessary to determine the volume fraction, V_V, that they occupy. The most efficient means [4, 5] of doing this is by point counting; i.e., a grid of points is superimposed on the image of the structure, and a count made of the fraction of points, P_P, falling on the grains whose volume fraction is being estimated. Then:

$$P_P = V_V \tag{29}$$

Sampling Errors in Metallographic Procedures

We will first consider the standard deviations in the measurement of P_L, N_A, and V_V and then examine how these propagate into the determination of the various averages.

Estimates of $\sigma(P_L)$, $\sigma(N_A)$, and $\sigma(V_V)$

It has been shown theoretically [6] that the standard deviation $\sigma(N')$ in the number, N', of interceptions that a lineal test figure makes with grain boundaries on the plane of polish is given by:

$$\sigma(N') = k'\sqrt{N'} \tag{30}$$

in which k' is a constant depending on the structure and the form of the test figure. For contiguous, reasonably equiaxed grains, $k' \approx 0.6$ for a straight line or for a test figure in the form of a circle on which the number of intersections is greater than 6. For non-contiguous grains, k' will probably have a higher value, although it is unlikely to exceed unity.

Since, by definition,

$$P_L = N'/L_t \tag{31}$$

in which L_t is the total length of the test figure, we have from equation (30)

$$\sigma(P_L) = \sigma(N)/L_t = k'\sqrt{N'}/L_t \tag{32}$$

or, in terms of the relative error,

$$\sigma(P_L)/P_L = k'/\sqrt{N'} \tag{33}$$

The standard deviation in the number of grain profiles, N'', falling within a given area of the plane of polish will follow a relationship similar to equation (30) with a proportionality factor of k'', for example. Thus, by analogy with equation (33) we have:

$$\sigma(N_A)/N_A = k''/\sqrt{N''} \tag{34}$$

in which N'' is the total number of grain profiles counted during the analysis. For contiguous, equiaxed grains it was found experimentally [2] that $k'' \approx 1.0$. The same value probably holds also for non-contiguous grains.

Finally, we have to consider the volume fraction measurement. As previously stated, the best method of estimating V_V of non-contiguous grains is by a point count. Provided the point density is such that, on the average, there is one point or less per grain, then:

$$\sigma(V_V) = \left\{ V_V(1 - V_V)/P \right\}^{1/2} \qquad (35)$$

in which P is the total number of points applied. This expression has been established theoretically [4] for the limits $V_V \to 0$ and 1, and empirically [5,7] for intermediate values.

Propagation of Errors in the D's.

In determining how the standard deviations in the measurements propagate into the estimates of the averages, we will make use of the relationship that if

$$X = g(x_i)$$

then

$$\sigma^2(X) \approx \sum_i \sigma^2(x_i) \, (\partial X/\partial x_i)^2 \,_{x_i = \bar{x}_i} \qquad (36)$$

which holds exactly if X is a linear function of the x_i's, and is a good approximation for other functions if $\sigma(x_i) \ll \bar{x}_i$.

Considering first $D_{3,1}$; applying equation (36) to equation (9) we obtain:

$$\sigma^2(D_{3,1}) = (D_{3,1}^2/4) \left\{ [\sigma^2(N_A)/N_A^2] + [\sigma^2(V_V)/V_V^2] \right\}$$

Substituting from equations (34) and (35) we obtain, after rearrangement:

$$\sigma(D_{3,1})/D_{3,1} = (1/2) \left\{ [(k'')^2/N''] + [(1 - V_V)/PV_V] \right\}^{1/2} \qquad (37)$$

If the grains are contiguous, then $V_V = 1$, and equation (37) reduces to:

$$\sigma(D_{3,1})/D_{3,1} = k''/2\sqrt{N''} \qquad (38)$$

If the grain size is reported as an ASTM grain-size number g, we obtain from equations (12), (34), and (36):

$$\sigma(g_e) = 1.44 \, k''/\sqrt{N''} \qquad (39)$$

Considering next $D_{3,2}$; applying equation (36) to equation (17) and substituting from equations (30) and (35), we obtain for non-contiguous grains:

$$\sigma(D_{3,2})/D_{3,2} = \left\{ [(k')^2/N'] + [(1 - V_V)/V_V]PV_V \right\}^{1/2} \qquad (40)$$

in which N' is the total number of intersections counted. For contiguous grains, equation (18) yields:

$$\sigma(D_{3,2})/D_{3,2} = k'/\sqrt{N'} \qquad (41)$$

In the case of non-contiguous grains, it may be more efficient to measure the average intercept length, L, and to determine $D_{3,2}$ from equations (20) or (24), since this avoids the separate measurement of V_V and P_L. It follows from equations (20) or (24) that:

$$\sigma(D_{3,2})/D_{3,2} = \sigma(\overline{L})/\overline{L} \qquad (42)$$

From the well-known rule for the standard deviation of an average:

$$\sigma(\overline{L}) = \sigma(L)/\sqrt{N} \qquad (43)$$

where N is the total number of intercept lengths measured. Substituting equation (43) into (42) we obtain:

$$\sigma(D_{3,2})/D_{3,2} = [\sigma(L)/\overline{L}]/\sqrt{N} \qquad (44)$$

The ratio $\sigma(L)/\overline{L}$ will depend on the shape and size distribution of the grains. It has a minimum value of 0.35 (for spherical grains of constant size).

With respect to the equivalent ASTM grain-size number g_e, we obtain from equations (25), (33), and (36):

$$\sigma(g_e) = 2.88k'/\sqrt{N'} \qquad (45)$$

The remaining average to be considered is $D_{2,1}$. Following the same procedure as in the derivation of equation (37), we obtain from equations (27) or (28):

$$\sigma(D_{2,1})/D_{2,1} = \{[(k')^2/N'] + [(k'')^2/N'']\}^{1/2} \qquad (46)$$

in which, as before, N' is the total number of intersections counted, and N'' the number of grain profiles.

Experimental Procedures

The various microscopical relationships derived in the previous sections are summarized in Table III. In order to use some of these relationships, it is necessary to know certain shape factors. Values of these factors are listed in Table IV in the form of the ratios: $(\alpha/\beta)^{1/2}$, $(\gamma/4\beta)$, and $(4\alpha/\gamma)$ for eight different shapes and for two definitions of D; one being $D = K$, the mean-caliper diameter, and the other $D = \ell$, the mean intercept length.

It is apparent from Table III that the determination of an average grain size requires the measurement of one or more of the following parameters: $P_L, N_A, \overline{L}$, and V_V. For maximum efficiency, the measurements should be made directly under the microscope rather than on micrographs. In the discussion that follows we shall assume that observations are made optically; modifications will obviously be required if electron-microscopy is used.

The first step in any of the analyses is to decide on the accuracy required. The expressions for $\sigma(D)/D$ listed in Table III can then be used to determine the total

TABLE III

Summary of Microscopical Relationships

	$D_{3,1}$	$D_{3,2}$	$D_{2,1}$
	Contiguous Grains		
D	$[(\alpha/\beta)(1/N_A)]^{1/2}$	$(\gamma/4\beta)(1/P_L)$ or $(\gamma/4\beta)\bar{L}$	$(4\alpha/\gamma)(P_L/N_A)$
$\sigma(D)/D$	$k''/2\sqrt{N''}$ $(k''\approx1.0)$	$k'/\sqrt{N'}$ $(k'\approx0.6)$ or $k'''/\sqrt{N'''}$ $(k'''\approx0.6)$	$\{[(k')^2/N'] + [(k'')^2/N'']\}^{1/2}$ $(k'\approx0.6, k''\approx1.0)$
	Non-Contiguous Grains		
D	$[(\alpha/\beta)(V_V/N_A)]^{1/2}$	$(\gamma/2\beta)(V_V/P_L)$ or $(\gamma/4\beta)\bar{L}$	$(2\alpha/\gamma)(P_L/N_A)$
$\sigma(D)/D$	$(1/2)\{[(k'')^2/N''] + [(1-V_V)/PV_V]\}^{1/2}$ $(k''\approx1.0)$	$\{[(k')^2/N'] + [(1-V_V)/PV_V]\}^{1/2}$ $(k''\approx1.0)$ or $k'''/\sqrt{N'''}$ $(k'''\approx0.6)$	$\{[(k')^2/N'] + [(k'')^2/N'']\}^{1/2}$ $(k'\approx k''\approx1.0)$

number of measurements required. The values listed for the k's are applicable to specimens containing grains that are reasonably equiaxed and uniform in shape. If measurements are to be made on a large number of similar specimens, it will be worthwhile to determine the k's experimentally. It should also be emphasized that the expressions for the standard deviations include only the sampling errors.

TABLE IV

Values of the Geometrical Factors Appearing
in Equations (10), (18), and (28) for Various Grain Shapes*

	$D = K$			$D = \bar{\ell}$		
	$(\alpha/\beta)^{1/2}$	$\gamma/4\beta$	$4\alpha/\gamma$	$(\alpha/\beta)^{1/2}$	$\gamma/4\beta$	$4\alpha/\gamma$
Sphere	1.382	1.500	1.273	0.9213	1	0.8488
Hemisphere	1.648	2.008	1.353	0.8207	1	0.6736
Tetrahedron	2.538	3.352	1.922	0.7572	1	0.5734
Cube	1.837	2.250	1.500	0.8165	1	0.6667
Octahedron	1.856	2.160	1.596	0.8596	1	0.7388
Dodecahedron	1.552	1.780	1.353	0.8719	1	0.7603
Icosahedron	1.557	1.729	1.402	0.9005	1	0.8108
Tetrakai-decahedron	1.545	1.776	1.344	0.8700	1	0.7570

*The factors have been computed from values of K, s, and v, tabulated in [1].

In addition, there will be experimental errors due, for example, to lack of resolution.

The intercept density, P_L, can be measured by using a reticule in a focusing eyepiece to superimpose a test figure on the image of the structure. The best shape of the test figure is a circle, since this automatically averages over all orientations in the plane of polish. A stage micrometer is used to determine the circumference of the test circle at the specimen plane. The magnification should be such that there is a minimum of six intersections per application. The circle is applied to successive non-overlapping regions until enough counts have been accumulated to yield the required accuracy.

An alternative procedure is to traverse the specimen a known distance under the cross hairs of a microscope and count the number of boundaries intersected. This will probably be quicker than the use of a test circle if a large number of counts is required.

With either procedure, an unresolved contact with a boundary should be recorded as one intersection, and if the circle (or traverse) apparently passes through a triple junction, 1.5 intersections should be recorded. The proportion of such unresolved contacts will then afford some measure of the experimental error.

The reticule can also be used to determine N_A. However, the average number of grains falling within the circle will have to be kept small, otherwise the eye will lose track of the profiles counted. Alternatively, the count can be made on the screen of a projection microscope. In either case, profiles that overlap the boundaries of the test field should be counted as one-half.

The determination of $\overline{L}$ requires the use of a device, such as the Hurlbut counter, that will record the total length of traverse in the non-contiguous grains whose size is being estimated. The traverse length divided by the number of profiles intercepted yields $\overline{L}$.

The optimum procedure for determining V_V by point counting has been described in detail elsewhere [5]. It is recommended that the points be applied by use of a reticule on which a square grid is inscribed. It is useful to have a set of reticules ranging from 3 X 3 to 9 X 9 grids. The magnification and point density should be such that any grain is not occupied by more than one or two points. (The use of higher point densities will increase the counting time with very little improvement in accuracy.) The grid is applied to non-overlapping regions of the plane, the number of applications being determined by the error allowable. Points falling on a grain boundary are counted as one-half.

Comparison of Microscopical Methods

Procedures have been described for determining three averages of the grain size: $D_{3,1}$, $D_{3,2}$, and $D_{2,1}$. Unless all the grains are of the same size, these averages will differ from one another. As previously pointed out, the choice of which to use depends on the application of the measurement. However, there may be circumstances under which there is no basis for selection or, if there is only a small variation of size within a specimen, the choice may be unimportant. In such cases, the investigator will naturally select the procedure that is the most efficient in the sense of requiring the least effort for a given accuracy. Inspection of Table III shows that, for a given number of observations on contiguous grains, $D_{3,1}$ has the smallest standard deviation. However, the measurement involved, N_A, is more time-consuming than the determination of the intercept density, P_L, needed for $D_{3,2}$ because one has to keep track of the profiles already counted. Experimentally [2], it has been found that the longer counting time for N_A more than offsets the lower standard deviation. In fact, the estimation of $D_{3,2}$ is at least 50 percent more efficient than the estimation of $D_{3,1}$. In addition, if the grain size, D, is defined by the average intercept length, $D_{3,2}$ has the important advantage of being independent of grain shape.

In the case of non-contiguous grains it is believed that the determination of $D_{3,2}$ from a measurement of $\bar{L}$ will be the most efficient. However, this has not been checked experimentally.

X-Ray Methods

There are several procedures utilizing X-ray diffraction or absorption that are available for grain-size determination. One [8,9] depends on the property that the integrated peak intensity is a function of grain size when the grains are non-contiguous and have an absorption coefficient differing from that of the matrix. However, this method is insensitive for grain sizes less than 50 μ and is therefore outside our range of interest. At the other extreme, there are small-angle scattering and line-broadening [10] techniques. These have an upper limit of about 0.1 μ.

An ingenious X-ray absorption method has been proposed by Suoninen [11] that is applicable to non-contiguous grains in the size range of interest. He has shown that the intensity, I, transmitted by a slice of material containing spherical grains of diameter D is given by:

$$I = \exp\left\{-\left(\frac{3V_V}{2D} + \mu_2\right)d\right\} \sum_{m=0}^{\infty} \left(\frac{3dV_V}{D^3}\right)^m \frac{[\Phi(D\Delta\mu)]^m}{m!} \qquad (47)$$

in which

$$\Phi(D\Delta\mu) = [1 - (1 + D\Delta\mu)\exp(-D\Delta\mu)]/(\Delta\mu)^2$$

where V_V is the volume fraction of the grains, μ_2 the linear absorption coefficient of the matrix, $\Delta\mu$ the difference in absorption coefficient of matrix and grains, and d the thickness of the slice (usually $< \approx 10^{-2}$ cm). The summation in equation (47) is continued until the terms become negligible. In order to eliminate the error due to an uncertainty in d, radiations of two different wavelengths can be used and D determined from the ratios of the transmitted intensities.

If the grains are not of constant size, then the measurement yields the average, $D_{3,2}$. Suoninen compared X-ray and metallographic measurements on a series of lead-brass alloys containing lead particles with diameters ranging from 1-7 μ. The average given by metallographic measurements was $D_{3,1}$ and was always smaller (by 5-25 percent) than the size obtained from X-ray measurements. This difference is to be expected since, as previously discussed, $D_{3,2} \geqslant D_{3,1}$.

Although this procedure is obviously capable of yielding reliable results, it does suffer from the following disadvantages: (1) it is necessary to prepare a thin slice of the material; (2) it is limited to non-contiguous grains; and (3) a separate measurement has to be made of the volume fraction.

The final X-ray procedure to be considered involves the measurement of the variation in the power of the diffracted beam from different parts of the specimen. Since this method has wide applicability, we will treat it in detail.

X-Ray Variance Procedure

Theory

When a volume of the specimen is irradiated with a monochromatic X-ray beam the number of grains in a diffracting position will only be a small fraction (< 0.01) of the total number of grains in the volume (assuming that a very strong texture is not present). As a consequence, there will be a statistical variation in the number of diffracting grains as the specimen is moved under the beam, and this in turn leads to a variation in the diffracted power. This effect has been analyzed by Alexander, Klug, and Kummer [12], De Wolff [13], and De Wolff, Taylor, and Parrish [14], primarily with the objective of estimating the errors in intensity measurements produced by this effect and how the errors could be minimized. However, the latter authors pointed out that the effect could also be used for measuring grain size. This application was later considered in more detail by Warren [15], who showed how the procedure could be simplified by the appropriate choice of experimental conditions. We will first list the expressions given by Warren and then derive a correction for bias introduced by counting statistics. Finally an analysis will be made of the sampling errors and the conditions necessary for optimization.

Because of the small probability that a grain will be in a diffracting position, the number of such grains will follow a Poisson distribution as the specimen is moved from one counting position to another. From the properties of this distribution, it is possible to derive the variance that will be observed in the diffracted power. Warren assumed that the depth of X-ray penetration was large compared with the grain size and that the specimen was of infinite thickness for the radiation employed. Under these conditions he obtained for contiguous grains [15]:

$$\sigma^2 (Y)/\overline{Y}^2 = (4\pi\mu/jA_o\Omega\epsilon) [\mu_2'(v)/\mu_1'(v)] \tag{48}$$

in which

$$\sigma^2(Y) = \text{variance in integrated peak intensity, } Y$$
$$\mu = \text{linear absorption coefficient}$$
$$j = \text{multiplicity of reflection}$$
$$A_o = \text{cross-sectional area of X-ray beam}$$
$$\Omega = \text{spherical angle defining the divergence}$$
$$\epsilon = \text{correction factor for preferred orientation (= 1}$$
$$\text{for randomly oriented grains)}$$
$$\mu_n'(v) = \text{nth moment of distribution of grain volumes } v.$$

If it is assumed that all grains are of the same shape, then:

$$\mu_1'(v) = \beta\mu_3'(D) \tag{49}$$

and

$$\mu_2'(v) = \beta^2 \mu_6'(D) \tag{50}$$

in which β is the volume of a grain of unit size. Substituting equations (49) and (50) in equation (48) and utilizing equation (2), we obtain:

$$D_{6,3} = \{(A_o\Omega\epsilon/4\pi\mu\beta) [\sigma^2(Y)/\overline{Y}^2]\}^{1/3} \tag{51}$$

For non-contiguous grains occupying a volume fraction V_V it is easily shown that

$$D_{6,3} = \{(V_V jA_o\Omega\epsilon/4\pi\mu_a\beta) [\sigma^2(Y)/\overline{Y}^2]\}^{1/3} \tag{52}$$

in which μ_a is an average linear absorption coefficient defined by

$$\mu_a = V_V\mu_g + (1 - V_V)\mu_m \tag{53}$$

where μ_g and μ_m are the absorption coefficients for the grains and matrix respectively. If the non-contiguity is due to porosity, then $\mu_m \approx 0$ and $\mu_a \approx V_V\mu_g$. When this is substituted in equation (52), the V_V's cancel out. This is a very useful property since it means that for porous specimens (or for loose powders) it is not necessary to determine the volume fraction of grains or the density of the sample.

In both equations (51) and (52) it is assumed that the depth of penetration is large compared with the grain size. If the depth of penetration is much smaller than the grain size then, in effect, only the surface of the specimen is being sampled. Under this condition it can be shown that in place of equation (48) we have:

$$\sigma^2(Y)/\overline{Y}^2 = (4\pi V_V \sin\theta/jA_o\Omega\epsilon)\ [\mu_2'(a)/\mu_1'(a)] \tag{54}$$

where θ is the diffraction angle and a the average area of intersection of a grain with a plane. This expression holds for both non-contiguous and (with $V_V = 1$) contiguous grains. In terms of the shape factors defined by equations (6) and (7), the value of a for a grain of unit size is (β/α). Hence:

$$\mu_1'(a) = (\beta/\alpha)\mu_2'(D) \tag{55}$$

and

$$\mu_2'(a) = (\beta/\alpha)^2\ \mu_4'(D) \tag{56}$$

We can thus write equation (54) as:

$$D_{4,2} = \{(jA_o\Omega\epsilon\alpha/4\pi V_V\beta\sin\theta)\ [\sigma^2(Y)/\overline{Y}^2]\}^{1/2} \tag{57}$$

It is apparent from equations (51), (52), and (57) that $D_{6,3}$ or $D_{4,2}$ (depending on the depth of penetration) can be determined for both contiguous and non-contiguous grains from a measurement of the variance in the integrated intensity. It is to be noted that since only the ratio $\sigma^2(Y)/\overline{Y}^2$ is required, it is not necessary to measure intensities on an absolute basis.

Correction for Bias

One of the minor disadvantages of this method is that any random errors in the measurement of Y not only increase the sampling error but, what is worse, they produce a positive bias in the estimation of D. With modern equipment and with the short counting times required by this method, the only significant addition to the variance will be from counting statistics. Since these will follow a Poisson distribution, it is easy to make a correction for the bias.

Let Y be the number of counts observed from the diffraction peak in a time t, and let Y_b be the number of counts in the estimation of the background during a time t_b. The quantity Y will be the true number of counts Y_p from the peak plus background; i.e.,

$$Y = Y_p + (t/t_b)Y_b \tag{58}$$

and

$$\sigma^2(Y) = \sigma^2(Y_p) + (t/t_b)^2\ \sigma^2(Y_b) \tag{59}$$

There will be two contributions to $\sigma^2(Y_p)$; one $\sigma_g^2(Y_p)$ from the grain-size effect (and this is the variance we are trying to estimate), and the other from counting statistics. Since the counts follow a Poisson distribution, the latter contribution is equal to $\overline{Y}$. Thus:

$$\sigma^2(Y_p) = \sigma_g^2(Y_p) + \overline{Y}_p$$

Also, $\sigma^2(Y_b) = Y_b$. Substituting these values in equation (59) and rearranging, using equation (58), we obtain:

$$\sigma_g^2(Y_p)/\overline{Y}_p = [\sigma^2(Y) - \overline{Y} + (t/t_b)Y_b] \ [1 - (t/t_b)] \ [\overline{Y} - (t/t_b)Y_b]^{-2} \quad (60)$$

The right-hand side of equation (60) should be used in place of the observed variance in the equations for $D_{6,3}$ or $D_{4,2}$. To recapitulate the meaning of the terms in equation (60): $\sigma^2(Y)$ and $\overline{Y}$ are the variance and average computed from the observed number of counts *without* correction for background, and Y_b is the background count. It is also to be emphasized that Y and Y_b are the actual number of counts recorded and not the counting rates.

Sampling Error and Optimization of Analysis

The calculation of the sampling error in the determination of $D_{6,3}$ or $D_{4,2}$ involves the application of equation (36) to equations (52) and (57) after applying the correction for bias given by equation (60). The derivation is straightforward but lengthy, so we will only give the results. Assuming $\overline{Y}$ and $n \gg 1$:

$$\sigma \ (D_{6,3})/D_{6,3} = (1/3) \ (2Q/n)^{1/2} \quad (61)$$

and

$$\sigma(D_{4,2})/D_{4,2} = (1/2) \ (2Q/n)^{1/2} \quad (62)$$

in which n is the number of different positions on the specimen at which the intensity was measured, and

$$Q = 1 + (1/\omega\overline{Y})^2 + (5/2)\omega \quad (63)$$

where

$$\omega = \sigma_g^2 \ (Y)/\overline{Y}^2 \quad (64)$$

We now have to determine the conditions that optimize the analysis; i.e., the counting strategy that minimizes the time required to achieve a given sampling accuracy. It will be assumed that the time for the analysis is proportional to the total number of counts, $N \ (= n\overline{Y})$, but is otherwise independent of n. (In other words, we are neglecting the time taken by the computation and in advancing the specimen from one counting position to the next.) A choice obviously has to be made of the division of N between n and $\overline{Y}$. It is apparent from equations (61) and (62) that the optimum division will be that which minimizes Q/n for a fixed N. Rewriting equation (63) in terms of N, dividing through by n, and then equating the derivative with respect to n to zero, we obtain for the optimum value of n:

$$n = N\omega \, [1 + (5\omega/2)]^{-1/2} \tag{65}$$

Substituting this value back in equation (63) we find:

$$[Q/n]_{min} = (1/N\omega) \{ [1 + (5\omega/2)]^{-1/2} + [1 + (5\omega/2)]^{1/2} \} \tag{66}$$

Usually, in practice, $\omega \ll 1$, thus to a good approximation equations (65) and (66) can be written

$$n \approx N\omega = N\sigma^2 \, (Y)/\overline{Y}^2 \tag{67}$$

and

$$[Q/n]_{min} \approx (2/N\omega) = (2/N) \, [\overline{Y}^2/\sigma^2(Y)] \tag{68}$$

Since $\overline{Y} = N/n$, the condition given by equation (67) can alternatively be written as:

$$\overline{Y} \approx 1/\omega = \overline{Y}^2/\sigma^2(Y) \tag{69}$$

Substituting equation (68) into equations (61) and (62) we obtain for the sampling error of an analysis in which the optimum value is used for n:

$$\sigma(D_{6,3})/D_{6,3} \approx (2/3) \, [\overline{Y}/\sigma(Y)]/\sqrt{N} \tag{70}$$

and

$$\sigma(D_{4,2})/D_{4,2} \approx [\overline{Y}/\sigma(Y)]/\sqrt{N} \tag{71}$$

It is apparent from equation (48) that, for a given grain size, the ratio $\sigma(Y)/\overline{Y}$ depends on the diffraction parameters A_o and Ω. It would thus appear from equations (70) and (71) that we could improve the efficiency of the analysis by selecting values for these parameters that maximize the ratio $\sigma(Y)/\overline{Y}$. However, this is not true if we assume that the intensity (or power per unit area) of the beam is independent* of A_o. In this case, an increase in $\sigma(Y)/\overline{Y}$ is accompanied by a proportionate decrease in the diffracted power, and thus by an increase in the time to achieve a given N. A similar argument applies to changes in Ω and the other parameters. In other words, the quantity $\overline{Y}\sqrt{N}/\sigma(Y)$ that determines the sampling error is invariant with respect to changes in A_o, Ω, and j. However, it may be desirable to keep A_o small so as to minimize the total specimen area required for the analysis. We shall also see later that there are other experimental reasons for keeping A_o small.

The two other factors affecting the time required to achieve a given sampling accuracy are the beam intensity and the grain size. The time will be inversely

---

*This is not strictly true, because the incident beam is not uniform in intensity.

proportional to the intensity, so that the Bragg peak having the maximum intensity should be used. It is also advantageous to use a microfocus tube.

With respect to grain size, it is apparent from equations (51) and (68) that N is proportional to $(1/D^3)$. Since the grain size does not affect $\overline{Y}$, the time for the analysis will also vary as $(1/D^3)$. It is this extremely strong dependence on size that sets a lower limit (albeit an economic one) to the grain size that can be determined by this method. The lower limit obviously depends on the material, the diffraction conditions, and the time one is prepared to spend on the analysis. However, we can get rough estimate from measurements to be described later that were made on an Al_2O_3 powder. The X-ray source was a regular Co tube operated at 8 ma and 46 kv. The cross-sectional area (A_o) of the beam was 2.6×10^{-3} cm^2 and $\Omega = 8.8 \times 10^{-5}$. Under these conditions, the diffracted power from a 110 peak was 800 cpm and for a 5 μ powder $[\sigma^2(Y)/\overline{Y}^2] = 5.3 \times 10^{-3}$. Assuming that a relative accuracy of 10 percent is acceptable, we obtain from equation (70), $N = 8,400$ cpm and, according to equation (67) this should be divided among 44 areas, so that $\overline{Y} \approx 170$ cpm. With a counting rate of 800 cpm, the total counting time would be approximately 10 min for a grain size of 5 μ. Assuming that the maximum permissible time for the analysis is 60 min, the smallest value of $D_{6,3}$ that could be determined under these conditions to a 10 percent accuracy is $5 \times (10/60)^{1/3} = 2.8 \mu$. Obviously, a smaller size can be determined if one is prepared to spend more time, but the stability of the equipment may then become significant. It should also be mentioned that many materials give considerably larger diffracted intensities than Al_2O_3 and, for these, measurements of sizes less than 1 μ should be experimentally feasible, especially with microfocus tube.

The results with respect to efficiency and optimization of the analysis can be summarized as follows.

The time required to achieve a given accuracy is: (1) inversely proportional to the square of the relative error that can be tolerated; (2) minimized by the value of n satisfying equation (67); (3) insensitive to the cross-sectional area and the divergence of the beam; (4) inversely proportional to the diffracted intensity; (5) proportional to $1/D^3$, a property that sets an economic limit to the minimum size that can be determined.

Experimental Procedure

In order to use this procedure it is first necessary to determine the various diffraction parameters that appear in equations (51), (52), or (57).

If the receiving slits are very wide relative to the incident beam then [15]:

$$\Omega = w\ell/4R^2 \sin \theta \tag{72}$$

where R is the distance from the sample to a receiving slit of width w and height ℓ, and θ the angle between the incident beam and the surface of the specimen (it is assumed that the usual condition is used with the normal to the surface of the specimen bisecting the incident and diffracted beams). In the foregoing expression for Ω, the divergence of the incident beam is neglected, as is the divergence due to the natural width of the reflection. The dimensions ω and ℓ of the receiving slit can either be measured directly or determined from the size of the image on a photographic film placed behind the slit. A film can also be used to determine the cross-sectional area, A_o, of the beam. If it is not possible or convenient to place the film at the specimen surface, the area of the beam at two different distances can be determined and the results extrapolated to the specimen surface.

Alternatively, a calibration can be made using samples of known grain size. (This has the advantage of eliminating errors arising from assumptions in the basic equations.) Preferably, loose and approximately spherical powder particles should be used for the calibration. These can be sized either microscopically or by one of the other methods described by Kaye and Jackson in this volume. Care must be exercised that the powder particles are single crystals, since the X-ray method estimates the grain or crystallite size and not the particle size if the particles are polycrystalline.

As previously stated, the specimen should be of infinite thickness with respect to the radiation used. The nomograph given in Figure 1 can be used to determine the minimum thickness required to satisfy this condition. The same nomograph will also indicate whether $D_{6,3}$ or $D_{4,2}$ is being estimated.

A photograph of a suitable experimental set-up is shown in Figure 2. In this case, the specimen is mounted on a spinner. Although in this measurement the specimen is not rotated during counting, the electric motor can be used to drive the specimen from one counting position to the next.

The steps in the analysis are as follows:

1. A value for the maximum sampling error, $\sigma(D)/D$ that can be tolerated is selected.

2. The nomograph in Figure 1 is used to determine the minimum specimen thickness and whether $D_{6,3}$ or $D_{4,2}$ will be determined.

3. The value of $\sigma(D)/D$ together with a preliminary estimate of $\sigma(Y)/Y$ are substituted in equations (70) or (71) to determine the total number of counts N required.

4. The optimum values of n and $\overline{Y}$ are calculated from equations (67) and (69).

5. The diffracted power, Y, of a Bragg peak is then determined at n different and non-overlapping areas of the specimen. Provided conditions are such that a broad peak is observed, it will suffice to set Y equal to the peak height instead of the integrated area under the peak.

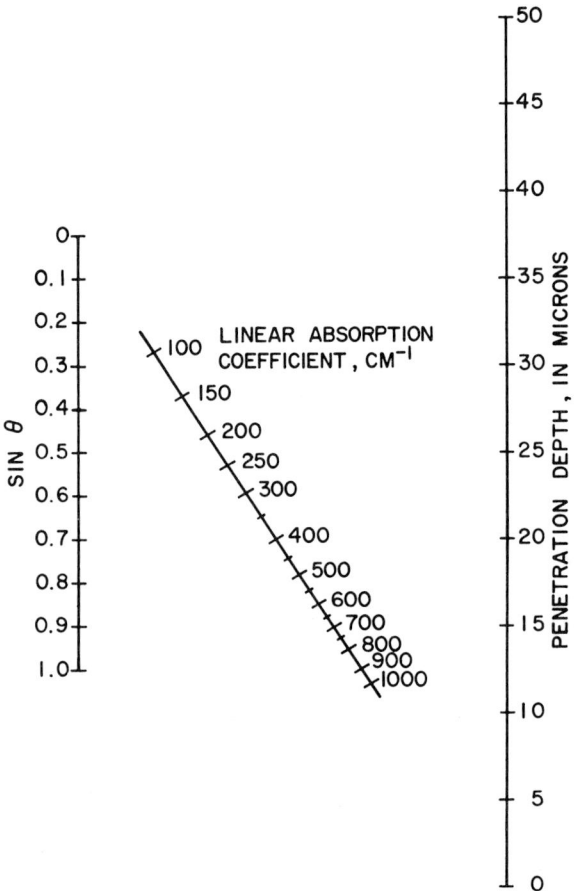

Figure 1. Nomograph for determining the penetration distance at which the intensity of the beam has decreased by 90 percent of its original value. A straight edge is set at the sinθ value of Bragg reflection on the left-hand scale, and at the absorption coefficient on the middle scale. The penetration distance is then read off the right-hand scale. The thickness of the specimen should be at least twice this distance when using the variance procedure. Also, if the penetration distance is appreciably smaller than the grain size, this procedure estimates $D_{4,2}$, equation (51), instead of $D_{6,3}$.

6. The correction, ϵ, for preferred orientation is determined, if necessary, by using the standard techniques for estimating texture.

7. The background Y_b is determined from the average of counts on each side of the peak.

8. The observed average and variance are computed from:

$$\overline{Y} = (1/n) \sum_{i}^{1} Y_i$$

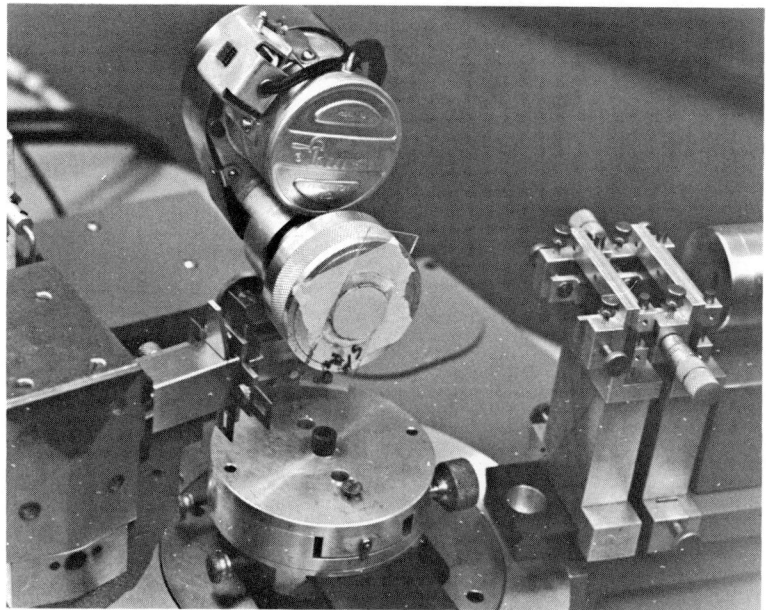

Figure 2. Picker diffractometer showing sample on a home-built spinner and variable divergence slits.

and

$$\sigma^2(Y) = [1/(n-1)] \sum_i [Y_i - \overline{Y}]^2$$

$$= - [n/(n-1)] \overline{Y}^2 + [1/(n-1)] \sum_i Y_i^2$$

9. These values are substituted in equation (60) to obtain an unbiased estimate of $\sigma_g^2(Y_p)/\overline{Y}_p^2$ from which the average grain size is obtained by means of equations (51) or (57), using shape factors determined from the ratios listed in Table IV. If the grains are non-contiguous and the absorption factor of the matrix is not negligible, then it will be necessary to determine the volume fraction V_V. This can be obtained, after calibration, from $\overline{Y}$, and no additional measurement is therefore required.

It is apparent that the analysis can be automated at several different levels of sophistication. The first is to employ a printout counter and automatic advance of the specimen so that the instrument can be unattended. The next step would be to have the printout in a form that can be fed directly into a computer for calculation of the variance. The ultimate would be to run the diffractometer on line with a computer. In this case it would be feasible to utilize sequential sampling techniques that would significantly improve the efficiency.

Experimental Test

Warren [15] applied this method to a 5 μ powder of KCl and obtained reasonable agreement with the size determined under the microscope. We have made measurements on alumina powders having nominal sizes (determined by settling techniques) varying between 3 μ and 25 μ. The results of the X-ray measurements are listed in the second and third columns of Table V. The lineal

TABLE V

Comparison of Measurements on Al_2O_3

| Nom. Size (μ) | X-Ray | | Coulter Counter |
	$D_{6,3}$ (μ)	$\sigma(D)/D$ %	$D_{6,3}$ (μ)
3.0	2.6	31	9.5
5.0	5.2	12	6.7
17.5	11.3	15	20.0
25.0	15.4	15	21.3

dimension used to define D was the diameter of a sphere having the same volume as the particle. The fourth column lists values of $D_{6,3}$ determined from the cumulative size distribution obtained with a Model A Coulter counter. For this analysis the alumina was dispersed in an HCl solution of pH 3.5–4.0 and then added to a 5% NaCl electrolyte. The counter was calibrated with ragweed pollen of 18.8 μ diameter.

It will be seen that the X-ray values agree with the nominal ones for the 3 μ and 5 μ powders and that the Coulter counter gives an extremely high value (9.5 μ) for the 3 μ sample. The particles in the latter were observed microscopically to be in the form of plates. These tend to enter the counter with the flat faces parallel to the plane of the orifice. They therefore cause a much larger increase in resistance than spheres of the same volume, and therefore have a much larger apparent diameter. In the case of the larger sizes, the discrepancy may be due to some of the powder particles being polycrystalline. As previously noted, the X-ray method estimates a grain size rather than a particle size.

Summary

By way of a summary, we list the relative advantages and disadvantages of the microscopic and X-ray variance analyses. The advantages of the former are: (1) a reasonably accurate estimate can be obtained quickly once the specimen has been prepared. For example, a sampling accuracy of 10 percent on contiguous grains requires fewer than 100 observations; (2) three different average sizes, $D_{3,1}$, $D_{3,2}$, and $D_{2,1}$ can be determined; (3) the efficiency of the analysis is independent of grain size; and (4) for optical measurements, nothing more than a bench microscope and reticule are required in the way of equipment. The disadvantages are: (1) a polished surface has to be prepared; (2) the analysis is destructive; (3) the analysis is limited to grain sizes $> \approx 1\ \mu$ unless electron-microscopy is used; and (4) in comparison with the X-ray method, the analysis is difficult to automate.*

The advantages of the X-ray variance method are: (1) the analysis is non-destructive; (2) the surface of the specimen need not be polished nor need it be flat; (3) a standard diffractometer can be used without modification; and (4) the measurement is easily automated. The primary disadvantage is the strong dependence of the analysis time on grain size. Also, considerably more computation is required than with microscopic methods. However, since most laboratories have access to a computer, this is not an important factor. The other characteristic of the variance method—the heavy weighting given to the large grains—may either be an advantage or a disadvantage depending on the application of the results.

Acknowledgments

The experimental data were initiated as part of an advanced diffraction course at Northwestern University. We are indebted to Mr. C. Johnson of Buehler, Ltd., for supplying the alumina powders. This work was supported by the Advanced Research Projects Agency of the Department of Defense through the Materials Research Center at Northwestern University.

References

1. Hilliard, J. E., *Proceedings of the Second International Congress for Stereology, Chicago, 1967*, H. Elias, ed., Springer-Verlag, New York (1967).
2. Hilliard, J. E., *Metal Progress*, 85 (1964), 99.
3. Hilliard, J. E., *General Electric Research Laboratory Report No. 62-RL-3133M* (December 1962).
4. Hilliard, J. E. and Cahn, J. W., *Trans. Met. Soc. AIME*, 221 (1961), 344.
5. Hilliard, J. E., *Quantitative Microscopy*, R. T. De Hoff and F. N. Rhines, eds., McGraw-Hill, New York (1968), 45–76.

---

*The automatic scanning devices presently available are at a primitive level in discriminating between phases.

6. Hilliard, J. E., *Recrystallization, Grain Growth and Textures*, H. Margolin, ed., American Society for Metals, Metals Park, Ohio (1966), 267.

7. Gladman, T. and Woodhead, J. H., *J. Iron and Steel Inst.*, 194 (1960), 189.

8. De Marco, J. J. and Weiss, R. J., Reported in [9].

9. Cooper, M. J., *Phil. Mag.*, 11 (1965), 969.

10. Warren, B. E., *Progress in Metal Physics*, 8 (1959), 147.

11. Suoninen, E., *Acta Polytechnia Scandinavica (Physics Nucleonic Ser.) Ph.*, 54 (1968).

12. Alexander, L., Klug, H. P. and Kummer, E., *J. Appl. Phys.*, 19 (1948), 742.

13. De Wolff, P. M., *Appl. Sci. Research*, 7 (1958), 102.

14. De Wolff, P. M., Taylor, J. M. and Parrish, W., *J. Appl. Phys.*, 30 (1959), 63.

15. Warren, B. E., *J. Appl. Phys.*, 31 (1960), 2237.

6. The Role of Composition in Ultrafine-Grain Ceramics

M. H. LEIPOLD
University of Kentucky
Lexington, Kentucky

and

E. R. BLOSSER
Battelle Memorial Institute
Columbus, Ohio

ABSTRACT

The importance of composition and impurities to the behavior of ultrafine-grain ceramics is discussed with emphasis on: (1) significance of large grain boundary area per unit volume; (2) nature and distribution of impurities; (3) source and elimination of impurities; and (4) analysis of impurities.

The importance of large grain boundary area per unit volume in ultrafine-grain ceramics is considered in light of our present lack of data and models for the extent, structure, and composition of any grain boundaries in ceramics. However, much evidence exists suggesting that these boundaries occupy a large volume and behave markedly differently than the bulk lattice, which would have a substantial effect in ultrafine-grain materials.

The nature and distribution of impurities in such ceramics is discussed, emphasizing the importance of anions and their especial importance in fine-grain material. These anions are expected to have large and varied effects, as indicated by theory and by meager experimental data, especially with reference to densification during fabrication. The sources of impurities in ultrafine-grain ceramics, primarily anions, are discussed in relation to the surfaces of the fine powders from which these ceramics are produced. Possible means of purification of such powders before and during further processing are presented, again with emphasis on the form and location of the impurities in the powders.

Analytical procedures for the determination of the quantity and distribution of impurities in such ceramics are discussed, along with the reasons for the long neglect of anion impurities, emphasizing the previous lack of suitable analytical techniques. Consideration is given to the success of mass spectroscopy, and its associated sensitivity, sample preparation, and handling problems.

99

The development of ceramic materials with extremely fine grain size ($< 1 \mu$) has been a constant goal of research and development scientists for many years. The desirability of such materials has been based on expected improved properties, especially mechanical, extrapolated on the basis of available theory and experiment from larger grain materials [1,2,3]. However, consideration of the fact that ultrafine ceramics must generally be produced from even finer starting powders leads to serious questions concerning the validity of the extrapolation, because of expected differences other than grain size in these ceramics. One major difference would be changes in chemical impurities, both total and distribution. Such changes may not be theoretically necessary, but they are a major practical consideration in the fabrication and properties of such ultrafine material.

The problem of chemical characterization of these fine-grain materials is substantial because satisfactory analytical techniques are under development and have not been widely used. When we introduce the problem of distribution of impurities within these fine-grain ceramics, the small scale over which the distribution (grain size) would vary complicates the picture even further.

Another part of the difficulty in adequately describing the role of composition in the behavior of ultrafine-grain ceramics lies in our almost total lack of understanding of the nature of typical grain boundaries in any ceramics. We do not even know how far the influence of the grain boundary extends in the lattice of the neighboring grains. Any width to the grain boundary implies greater and greater significance to grain boundary behavior as the grain size decreases. (See Figure 1.) We know substantially nothing about the details of this increasing fraction of our material as we enter the ultrafine region.

If we limit our consideration of these grain boundaries to their composition, we must consider the behavior of impurities with respect to the two volumes— the grain boundary volume, V_g, and the total volume, V_t. Although our understanding of this behavior is far from complete, there is ample evidence that some impurities reside at grain boundaries (V_g) to our present limit of detection [4,5]. In addition, the specific distribution depends on the nature of impurity, the previous thermal history, state of oxidation, etc. [6]. For example, in the case of cation impurities in MgO, there appears to be agreement as to the segregation of calcium and silicon to grain boundaries, but with iron, some investigators have found segregation while others have not [5,7]. However, this previous work did not differentiate between Fe^{+2} and Fe^{+3}, and these have been shown to behave differently in studies of mechanical behavior [8] and may likewise behave differently with respect to distribution. Other cations may be uniformly distributed, dependent upon the particular system; there is little doubt, however, that many of the common impurities lie at grain boundaries in ceramics.

When we consider the total amount of these impurities, and how this amount will vary with grain size, we must look to their source. Cation impurities in

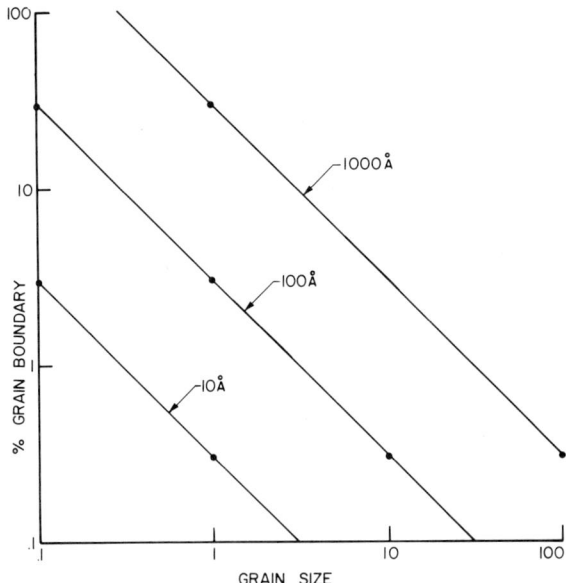

Figure 1. Volume fraction grain boundary as a fraction of grain size and grain boundary thickness.

ceramics largely result from the impurities of the starting material, and to some extent to subsequent processing during fabrication. Because we control the quality of the starting material and processing procedures, the total level of these impurities in the ultrafine powders would not be expected to be vastly different than in conventional types. The elimination of grinding as a fabrication procedure, for example, can improve cation purity. In ultrafine ceramics, the considerably larger proportion of grain boundary ($V_g/V_t > 0.01$) would permit greater distribution of these cations, with a lower average concentration in any boundary zone. Thus, the grain boundaries would be purer with respect to cations than in conventional ceramics. Any behavior of the material (e.g., intermediate temperature diffusion constants) which is influenced by grain boundary composition would thus vary as a function of this grain boundary volume. This effect has been demonstrated by comparing the cation impurity distribution in moderate- and coarse-grained ceramics [5]. In ultrafine materials, however, we have not been able to observe distributions on the finer scale. Thus, summarizing the situation with respect to cations in ultrafine ceramics, we may expect constant bulk concentrations of cations comparable to more conventional ceramics, with considerably lower concentrations per unit grain boundary because of the increased grain boundary volume.

When we transfer our attention from cations to anions, our level of certainty is much poorer, even with conventional grain sizes. Generally used analytical techniques are not sensitive to quantities of anions in polycrystalline ceramics and consequently their presence is often ignored. However, in many cases, especially in research-grade material, the quantity represents major impurities [9,10,11]. Here we have no direct analytical evidence for proposing existence of these anions at grain boundaries to the same extent as cations, but there is indirect evidence for the proposal. First, there is an effect of anions on the sintering of ceramics [12, 13, 14]. Second, coupling the expected structural similarity between ceramic surfaces and ceramic grain boundaries with the affinity of anions for such surfaces suggests that anions should show affinity for grain boundaries also. Thus, we are left with the conclusion that anions do exist as impurities in ceramics, and the expectation of finding them segregated to grain boundaries.

If one considers the source of these anion impurities, it is necessary to invoke the assumption that the more common refractory ceramic oxides reduce surface energy largely through absorption of gases because of the nature of bonding and atomic species in these structures. These gases arrive at the surfaces as a result of (1) reaction with the environment in which the particles were produced (e.g., decomposition of a hydroxide or a chloride), and (2) reaction with the ambient atmosphere after particle production is complete. These gaseous impurities will exist on the surface of the finer particles used for the production of ultrafine-grain ceramics and form the source for the commonly observed anion impurities (Cl, S, F, OH, CO_2) in ceramic oxides. Thus, the increasing surface area of the fine particles will introduce anions in increasing amounts as grain size is reduced. This is in contrast to cations, which should remain relatively constant. Note also that in order to maintain the fine grain size in the finished body, low-temperature sintering is usually employed, e.g., hot pressing. Such lower temperatures are further conducive to retention of anion impurities. It is for this reason that emphasis must be placed on anions for proper characterization with respect to chemistry of ultrafine-grain ceramics.

Since the anion concentration and the quantity of grain boundary both increase with reduced grain size, the anion concentration per unit volume grain boundary could be more nearly constant as one decreases grain size than the cation concentration. Indeed, there may be interaction between these quantities of anions and cations to maintain charge balances, but considerably more data are required before such possibilities become meaningful proposals.

This hypothesis of a higher level of anions is fine-grain ceramics is supported by experiment. The majority of analyses which are sensitive to these anion impurities do indeed indicate' their presence when the starting material is submicron in size [9,10,11]. A typical example may be seen in Table I. It must be emphasized that routine analyses (e.g., emission spectroscopy) do not indicate anions, and in at least one case—hydrogen or hydroxyl ions—there has been no satisfactory general technique. These analyses for anions are bulk concentrations only, as no distributional technique is available even for large-grain ceramics.

TABLE I

Anion Impurities In Dense Magnesium Oxide Compacts
ppm Atomic by Spark Source Mass Spectroscopy

	$0.5\,\mu$ As-Pressed[1]		Single Crystal[2]	$0.05\,\mu$ As-Pressed[3]	$0.05\,\mu$ Reheated[4] $(10\,\mu)$
H, OH	4,000–5,200		–	33,000[5]	1,000
C	3,000–400		–	2,000	1,000
N	300	–30	–	300	50
F	300	–60	<0.5	10	1
P	3		<1.0	5	2
S	500	–70	3.0	100	100
Cl	30	–32	4.0	100	60

[1] Kanto Chem. Work Mg(OH)$_2$ vacuum calcined and hot pressed [20]. Second number by corroborative technique.

[2] Single crystal not desorbed in mass spectrometer; therefore H, C, N unreliable.

[3] Hot isostatic pressed [21].

[4] Hot isostatic pressed and reheated to 1750°C in air.

[5] Brucite detectable.

The technique of spark source mass spectroscopy used here offers a considerable advantage in this type of analysis; e.g., high and relatively constant sensitivity, complete analyses [9]. Other techniques will undoubtedly also prove useful in these problem areas; for example, activation analysis using thin sections and appropriate photographic film for distribution. However, limited sensitivity for some species (e.g., S) and interferences present problems. Consequently, the emphasis here has been on the use of spark source mass spectroscopy with its development directed toward a universal analytical tool comparable to the electron-beam microprobe, but with good sensitivity for the complete spectrum of atomic species. However, even certain phases of this technique are still under development, primarily in the range of specimen preparation. The results given here were obtained by sparking crushed material mixed with a conductor, or by sparking a dense piece against a metal counter electrode. Agreement between the two methods has been generally satisfactory. Surface contamination was reduced by baking at 150°C in the mass spectrometer source before analysis. More extensive discussion on the details of the technique are available elsewhere [9,10,11].

It is obvious from Table I that, in the realm of sub-micron grain size, the anion levels increase tremendously although the cation levels were generally low. Total cations were less than 400 ppm atomic for the single crystal and 100 ppm atomic for the polycrystals. The lower anion levels obtained on single-crystal material offer quantitative validity to the technique and to the supposition that these impurities reside at grain boundaries. Further, the failure to eliminate such anions from polycrystalline ceramics by reheating indicates their stability at some concentration in the material. At higher levels ($> 10,000$ ppm) the quantities can be reduced thermally, as has been shown repeatedly, as with removal of fluorine from LiF-doped MgO [19].

Considering these anion impurities individually, varying emphasis may be placed on each—at least in the case of MgO. Certainly the importance will change with other oxides and environments and so these can only be considered as typical. Hydrogen (most likely in the form of hydroxyl ions) appears to be the most substantial anion impurity. To justify the presence of hydroxyl ions in fine-grain ceramics (approximately 0.5 μ, dense MgO), several semi-quantitative experimental observations may be introduced. First, infra-red transmission indicates strong absorption at 2,700 mμ [15]. For example, in a 0.5 μ, dense MgO sample (0.12 cm thick), approximately 35 percent transmission was obtained when compared with single-crystal MgO. Using data for the adsorption of OH in glass [16], this would suggest a hydroxyl ion concentration of $\approx$ 2,700 ppm atomic. Second, mass spectrographic analysis (Table II) shows significant peaks

TABLE II

Hydroxyl (ppm Atomic) Extracted from Dense
0.5 μ MgO Compact After Crushing to 100 μ

Condition of Extraction	After Crushing	After Readsorption 30°C and 50% RH	Calculated Monolayer [18]
100°C 10^{-5} torr	300	200	40
500°C 10^{-5} torr	4,700	180	80
500°C 20 torr (O_2)	80	10	–
900°C 10^{-5} torr	180	1	20
900°C 20 torr (O_2)	20	1	–
Total	5,200	380	140

(perhaps 3,000 ppm atomic when compared to other impurities) at m/ϵ of 1 and 17 from samples baked and desorbed *in situ*. Third, spin resonance work suggests the existence of hydroxyl-Mg^{+2} vacancy complexes in MgO [17]. This last finding eliminates the charge balance difficulty raised in substitutional hydroxyl ions (identically sized) for oxygen in a primarily ionic lattice like MgO. Of course, cation impurity atoms of different valence would also be effective to balance charges.

However, when one attempted to quantitatively calibrate the response observed by mass spectroscopy and infra-red spectroscopy, a severe problem was encountered; no analytical technique for substitutional hydroxyl ions in oxides had been developed.

Consequently, considerable effort has been made in the development of a procedure for hydroxyl ion determination in oxides (initially MgO). This work is by no means complete; however, sufficient data have been developed for useful discussion and to indicate that this may be a satisfactory procedure of bulk OH^- concentrations. Summarizing these results suggests that extraction of the hydroxyl ions may be made in the temperature range of $100-900°C$. (See Table II.)

The procedure used consisted in crushing a dense piece of 0.5μ-grain MgO to 100μ-particle size, and immediately placing a known amount in an all-Vycor system for which blanks had been determined. The quantity of gas evolved while heating the specimen in a trapped vacuum system and in a partial atmosphere of oxygen was determined by volumetric techniques, and the nature by mass spectroscopy. The oxygen used has been dried by molecular sieves at $-10°C$. Repeat analyses on the same material agree within ± 5 percent.

The maximum amount of hydroxyl ions expected to be desorbed from the surface area of this sample would be near two orders of magnitude less, as determined by previous studies of surfaces of MgO [18]. However, to explain the necessary mobility of the hydrogen within the grains at these temperatures, it is necessary to propose either an interstitial mechanism or an oxygen-to-oxygen transfer. The latter might occur at grain boundaries where a relatively continuous network of charge balancing impurity cations exist. Considerably more effort is required to clarify this picture. However, there appears to be no doubt concerning the existence of significant quantities of hydroxyl ions in fine-grain ceramics ($\approx 10^3$ ppm atomic in 0.5μ MgO).

The other substantial anion impurity observed in MgO is carbon, and here the picture is even less clear. Size and bonding considerations make direct substitution of C^{+4} or $CO_3^=$ into the anion lattice less likely, and again good analytical confirmation of the amount and location of this species is lacking. However, experimental evidence for its existence at a high level in dense fine-grained ceramics by means of mass spectroscopy (Table I) Knudsen cell effusion [18] cannot be ignored. Here again, considerable additional research and development is necessary to clarify the picture.

Considering other anion impurities (S, F, Cl, N, P) in these fine-grain materials, less difficulty is encountered, at least in this case. (See Table I.) Two reasons exist for this: first, large sources of these impurity anions are not available in this particular fabrication procedure and, consequently, the amounts are less although still substantial (note that here, where their available sources are limited, the quantities of these anions still approximately equal the cations). Second, there are adequate analytical procedures available to confirm these data. There is need, however, to confirm the analysis of such impurities in samples where they may be higher as, for example, with LiF-doped MgO or powders produced from oxidation of a halide, to ensure that such processes do not merely substitute the trade of one anion impurity for another. Table III shows mass spectroscopy data determined for a dense (transparent) MgO which had been hot

TABLE III

Impurity Analysis ppm Atomic for Dense 1–5 μ MgO*

OH	600	S	200
Li	40	Cl	200
F	2,000	K	200
Na	80	Ca	60
Al	20	Cr	20
Si	400	Fe	20
P	20	Ni	40

*Sample courtesy S. E. Hatch, Eastman Kodak Company.

pressed, using LiF as the pressing aid. Direct comparisons are difficult because of differing grain sizes, but the presence of residual fluorine as the major impurity and lower hydroxyl concentrations is obvious. Thus one function of fluorine as a sintering aid with MgO may be replacement of hydroxyl.

Finally, it must be repeated that in all of these cases we know nothing about the specific location of these anion impurities either generally (e.g., at grain boundaries), or in detail (e.g., associated with cation defects, etc.). Obviously, general questions concerning total amounts and sources must be answered before we can concern ourselves with precise locations, interactions, and detailed effect on properties.

For this reason, it is not within the scope of this chapter to consider the detailed effect of these particular impurities on the behavior of fine-grain ceramics. Too little is known concerning the specific effect of quantity and nature of impurities on properties of even normal grain size ceramics. When we consider how the relative purity of the grain boundary will change as we continually decease the grain size, it is probably true that we have not previously observed ceramics with comparable distributions of impurities and comparable behavior. However, as we do make this innovation, we must be alert to changes in other parameters—here, composition—likely to accompany the grain size change.

Acknowledgment

This research was in part performed for the Jet Propulsion Laboratory, California Institute of Technology, sponsored by the National Aeronautics and Space Administration under Contract NAS 7-100, and in part for the Aerospace Research Laboratories, OAR, Wright-Patterson Air Force Base, Ohio.

References

1. Carneglia, S. C., "Petch Relations in Single Phase Oxide Ceramics," *J. Amer. Ceram. Soc.,* 48 (1965), 580.

2. Spriggs, R. M. and Vasilos, T., "Effect of Grain Size on Transverse Bend Strength of Alumina and Magnesia," *ibid*, 46 (1963), 224-28.

3. Bentle, G. G. and Kneifel, R. M., "Brittle and Plastic Behavior of Hot Pressed BeO," *ibid*, (1965), 570.

4. Wuench, B. J. and Vasilos, T., "Origin of Grain Boundary Diffusion in MgO," *ibid*, 49 (1966), 433.

5. Leipold, M. H., "Impurity Distribution in MgO," *ibid*, 46 (1966), 398.

6. Leipold, M. H., "Addenda to Impurity Distribution in MgO," *ibid*, 47 (1967), 628.

7. Obst, K. H., Horn, H. C. and Munchberg, W., "Berrteilung von Sintermagnesiten mit Mikroskopie und Elextronenstrahl–Mikroanalyse," *Sonderdruck ans Tonind-Ztg.*, 90 (1966), 415.

8. Fine, M. E., "Precipitation in Ceramics," *Proc. of the OAR Res. Appl. Conference* (April 1966).

9. Socha, A. J. and Leipold, M. H., *J. Am. Ceram. Soc.*, 49 (1965), 463.

10. Blosser, E. R., "Development of Chemical Analysis Techniques for Advanced Materials," *Annual Report Accession #N68-17382, Final Report,* in print, NASA Scientific and Technical Information Facility, College Park, Maryland.

11. Leipziger, F. D., "Development of Chemical Analysis Techniques for Advanced Materials Research Program," Accession N66-26248, NASA Scientific and Technical Information Facility, College Park, Maryland.

12. Rice, R. W., "Production of Transparent MgO at Moderate Temperature and Pressures," Paper 5w62, Annual Meeting Amer. Ceram. Soc., Abstract in *Bull. Am. Ceram. Soc.*, 41 (1962), 27.

13. Rhodes, W. H., *et al.*, Report CR–67–01(F), Army Materials and Mechanics Research Center, Watertown, Mass.

14. Adams, R. B. and Stuart, W. I., "Effect of Adsorbed Sulfate and Fluoride on the Density of Hot-Pressed Beryllium Oxide," *J. Am. Ceram. Soc.*, 50 (1967), 685.

15. Chapman, A. T., Georgia Institute of Technology, Atlanta, Ga., Private communication (June 1968).

16. Scholze, H., "Water in Glass Structure," *Glass Industry*, 40 (1959), 301.

17. Kirklin, P. W., Auzins, Peteris and Wertz, J. E., "A Hydrogen Containing Trapped Hole in Magnesium Oxide," *J. Phys. Chem. Solids*, 26 (1965), 1067.

18. Anderson, P. J., Horlock, R. F. and Oliver, J. E., "Interaction of Water with Magnesium Oxide Surface," *Trans. Faraday Soc.*, 61 (1965), 2754.

19. Rice, R. W. and Hunt, J. G., "Identifying Parameters of Hot Extrusions," NASA Contract NAS 7-276, Office of Scientific and Technical Information, Attention: AFSS–A, Washington, D.C. (May 1967).

20. Leipold, M. H. and Nielsen, T. H., "Hot-Pressed High Purity Polycrystalline MgO," *Bull. Am. Ceram. Soc.*, 45 (1966), 281.

21. Leipold, M. H. and Nielsen, T. H., "Fabrication and Characterization of Isostatically Hot-Pressed MgO," *J. Am. Ceram. Soc.*, 51 (1968), 94.

7. Microstructure of Fine-Grain Ceramics

N. J. TIGHE*
National Bureau of Standards
Washington, D. C.

ABSTRACT

This chapter describes the use of transmission electron-microscopy to characterize the microstructure of fine-grain ceramics. Observations have been made on a number of polycrystalline materials including alumina, magnesia, zirconia, metal-ceramic composites, and rock specimens.

Thin sections were prepared by ion bombardment. In these sections grain boundaries, pores, impurity precipitates, and dislocations could be observed directly. Crystalline second-phase material formed as grains and small precipitates could be identified by means of electron diffraction. The method of specimen preparation and the results obtained from the observation of the specimens will be discussed.

Introduction

Studies of the microstructure of ceramics generally have employed conventional techniques of optical microscopy or electron-microscopy of surface replicas [1, 2]. This chapter presents results obtained using transmission electron-microscopy of thin foils. The electron microscope has the resolution and magnification necessary for studying microstructure of ultrafine-grain ceramics. Grain boundaries, dislocations, faults, inclusions, and voids can be imaged by diffraction and phase contrast, and phases can be identified from their selected area diffraction patterns [3–6].

The use of transmission electron-microscopy in studies on polycrystalline ceramic materials has been hampered by the difficulty of preparing satisfactory specimens. These specimens must have regions thin enough to transmit electrons, i.e., a few thousand Ångstroms or less, and they must be made in a reproducible manner so that specific areas of a bulk material can be sampled. The most promising method is one in which material is removed by bombardment with energetic ions. This method has been used for studies on alumina [7, 8] and on magnesia [9]. Chemical thinning methods have been used with limited success for magnesia [10] and alumina [11, 12]. However, satisfactory samples are difficult to prepare because chemical dissolution rates are dependent on crys-

*Presently at Dept. of Metallurgy, Imperial College of Science and Technology, London, England.

tallographic orientation; also, grain boundaries and precipitates can be etched preferentially [13]. A mechanical polishing technique developed for glass samples [14] has been used with some success to prepare samples of glass-ceramics [15], quartz [16], and UO_2 [17]. The method seems to be more satisfactory for glass-ceramics than for the polycrystalline compacts. Some specimens of a glass-ceramic have been obtained by crushing a sample and retrieving thin fragments [18].

Specimens for the present study were prepared by ion bombardment thinning. The capability and versatility of the method are illustrated by examples of microstructures taken from the familiar ceramics—alumina, magnesia, silica, and zirconia—in compacts of different grain size and porosity.

Experimental

Materials

The ceramic materials which are included in this paper were of various compositions and grain sizes. Brief materials descriptions are given in Table I.

TABLE I

Materials Thinned by Ion Bombardment

Material	Composition	Fabrication Method/ Description	Density or Porosity	Grain Size	Reference
Al_2O_3 I	$Al_2O_3 + \frac{1}{4}\%$ MgO	sintered	3.98g cm^{-3}	$\sim 50 \mu m$	[7]
Al_2O_3 II	99.7% Al_2O_3	sintered	3.78g cm^{-3}	variable	[7]
Al_2O_3 III	$Al_2O_3 + \frac{1}{4}\%$ MgO	hot-pressed	0.3%	$\sim 1 \mu m$	[8]
Al_2O_3 IV	$Al_2O_3 + \frac{1}{4}\%$ MgO	hot-pressed deformed 3% in bending at 1400°C	0.3%	$\sim 1 \mu m$	
Al_2O_3 V	$Al_2O_3 + \frac{1}{4}\%$ MgO	press-forged 1900°C $117 \, MN/m^2$	0.3%	$\sim 50 \mu m$	
MgO I	98% MgO	sintered sea-water periclase brick	19%	variable	[9]
MgO II	MgO + 2% NiO	extruded alloy		$\sim 10 \mu m$	
ZrO_2	$ZrO_2 + 6\%$ Y_2O_3	sintered		$\sim 25 \mu m$	
SiO_2 I	$SiO_2 + $ Mica	natural quartzite rock		$\sim 100 \mu m$	
SiO_2 II	SiO_2	natural flint rock		$\sim 10 \mu m$	

Specimen Preparation

The foils prepared for electron-microscopy were in the form of disks ~3 mm in diameter. The disks were cut from mechanically polished slices of the bulk samples with a tube tool held either in an ultrasonic drill or by hand, depending on the slice thickness and on the material. The thickness of the disk before ion bombardment thinning varied from ~150 μm for porous periclase brick to ~45 μm for alumina and quartzite.

The ion bombardment apparatus has been described elsewhere [7, 19, 20]. Specimens are thinned by bombarding each surface at a glancing angle with beams of argon ions. The periphery of the disk is masked by the specimen holder so that the foil becomes "dished" in the center. Bombardment is continued until large areas less than a few thousand Ångstroms in thickness are obtained. The rate of thinning is 1/2–2 μm/hr, depending on the operating conditions and on the material.

Results

Grain Boundaries

In the electron microscope, grain boundaries with misfit angles greater than a few degrees can be imaged as a set of equal thickness fringes or as lines, depend-

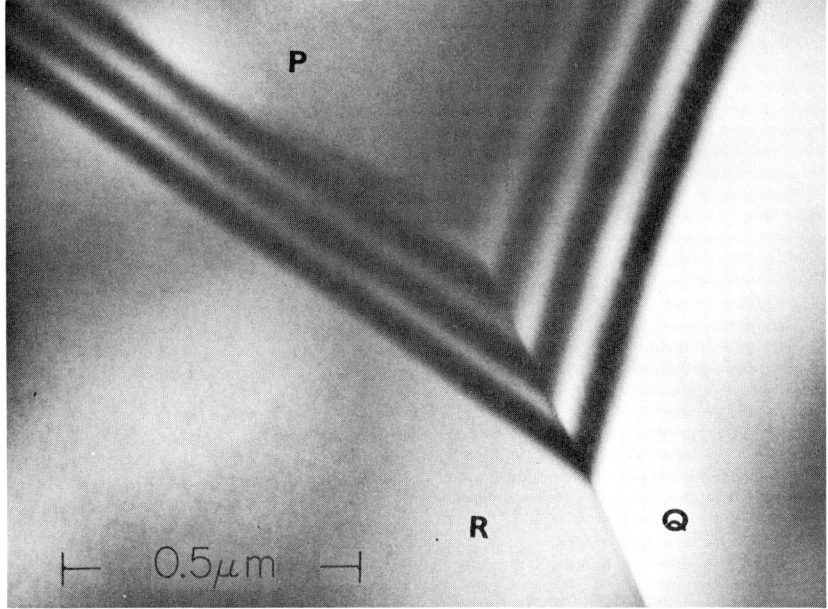

Figure 1. Triple junction of grains in alumina I (after [7]).

ing on the orientation of the boundary with respect to the electron beam and on the operating reflection g. The fringe contrast is the same as that described for a wedge or an interface inclined to the electron beam [3-6]. The triple junction in Figure 1 shows fringe contrast from inclined boundaries, and a line between grains Q and R where the boundary is parallel to the electron beam.

The fringes follow the contour of the boundary at equal distances from the specimen surfaces. Irregular specimen surfaces or non-planar boundary interfaces will be reflected in the appearance of the fringes. The serrated boundary in Figure 2, seen in magnesia I, is a good example of a non-planar interface. The

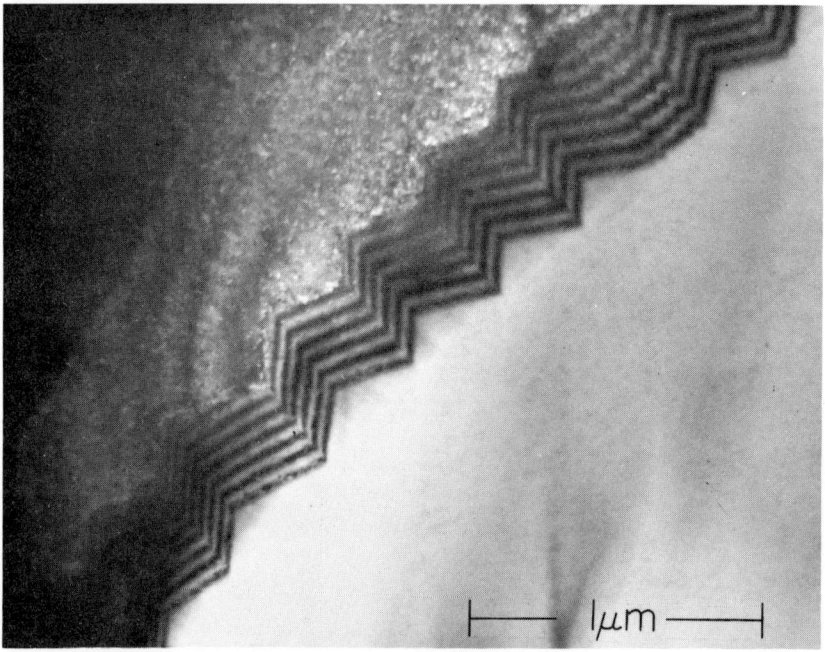

Figure 2. Serrated grain boundary in sea-water periclase brick (magnesia I).

boundary in Figure 3, found in alumina V, shows the effect of an irregular specimen surface which resulted from poor thinning conditions. The irregular grain boundary interface resulted from the press-forging operation. Irregularities in the fringe spacing can result when a strong Bragg reflection in each grain causes superposition of thickness fringes along their common boundary. This is illustrated in Figure 4 (alumina IV), where the occurrence of a pore along the grain boundary allows identification of the fringes resulting from diffraction in each grain.

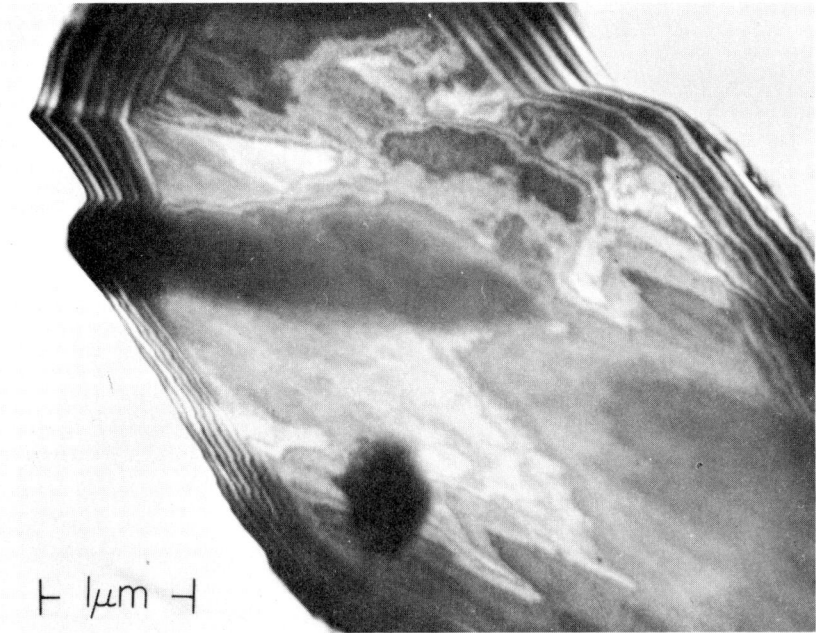

Figure 3. Grain boundary in alumina V, with uneven thickness fringes resulting from both an irregular boundary interface and an irregular specimen surface.

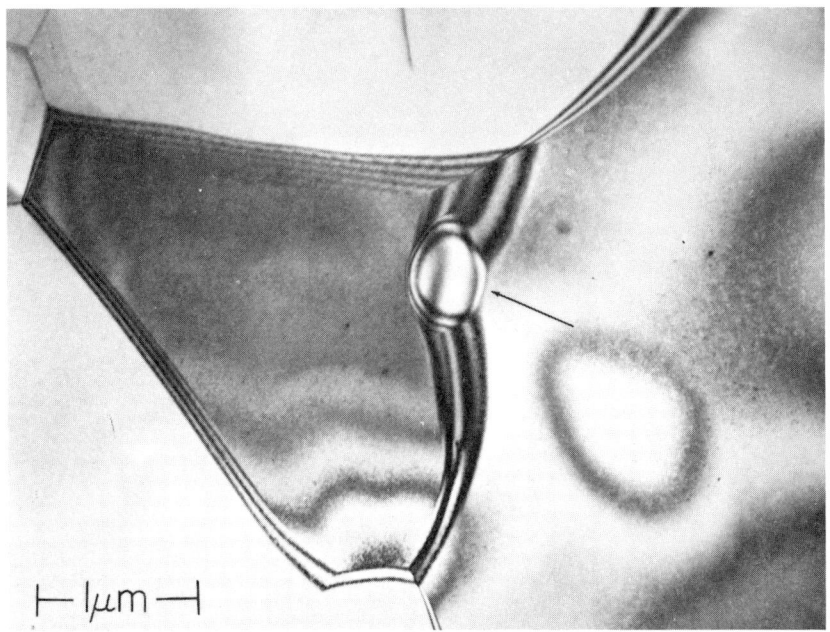

Figure 4. Grains in alumina IV showing superposition of extinction fringes along a grain boundary.

Boundaries between Phases

The fringe contrast at boundaries between grains differing in composition can be nearly the same as that between grains differing in orientation. In Figure 5, the boundary between a monticellite grain ($CaMgSiO_4$) and an MgO grain in magnesia I is imaged as fringes. In Figure 6, the boundary between an Al_2O_3 grain and an $MgAl_2O_4$ grain in alumina I is seen in a fringe contrast, and misfit dislocations are visible. In this figure (as in Figure 5), the second phase usually was detected because of sub-structures in the grain such as the fringe faults (arrowed), which were different from those in the major phase.

Diffraction contrast at the boundary between a glass phase and a crystalline phase arises only from the crystalline phase, although the distinctness of thickness fringes could be affected by the absorption of the glass. Strip A in Figure 7 is an amorphous phase joining two MgO grains in periclase brick (MgO I). One of the MgO boundaries is faceted, the other is straight. When faceted MgO grains were found in this material, there was usually a glass phase along the grain boundary [9].

When the glassy phase has devitrified or contains a fine crystalline precipitate, the fringe contrast is affected. This can be seen in Figure 8, which is of periclase

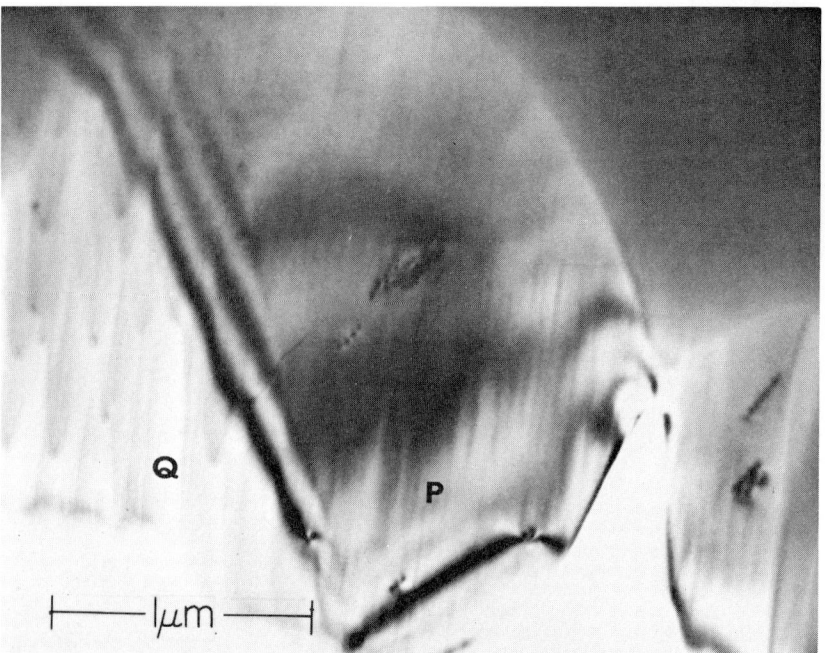

Figure 5. Monticellite grain P in sea-water periclase brick; grain Q is MgO.

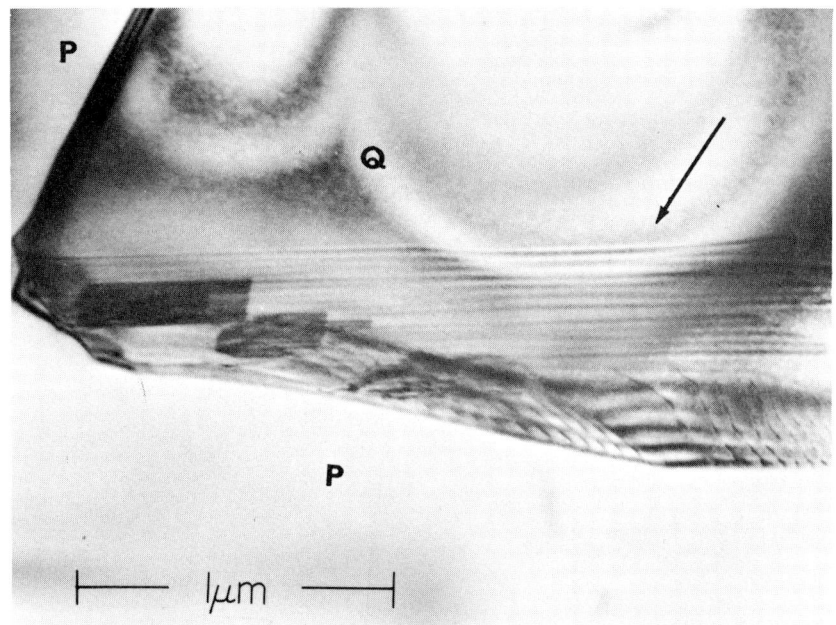

Figure 6. Spinel $MgAl_2O_4$ grain Q in alumina I; grain P is Al_2O_3. Faults arrowed are in the spinel grain (after [7]).

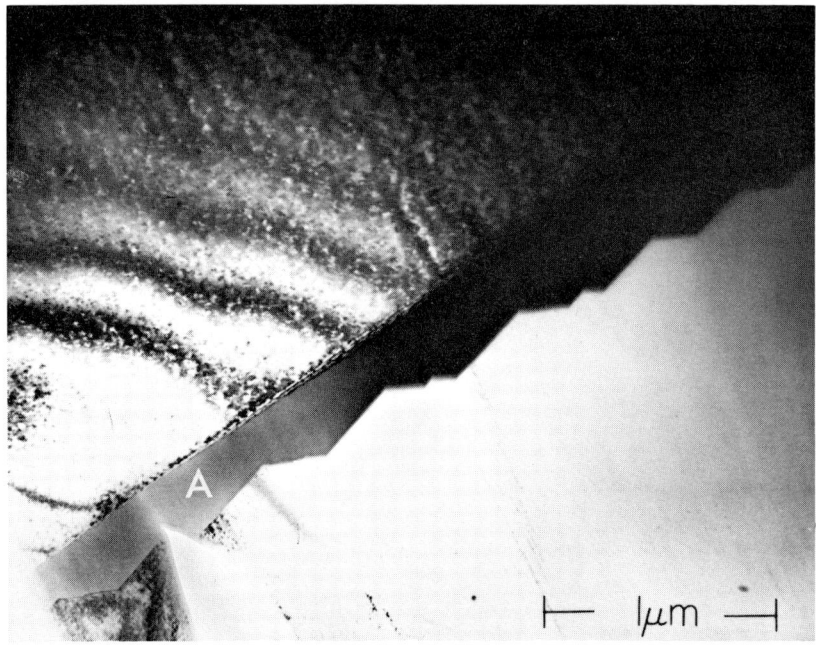

Figure 7. Amorphous phase A between two grains of MgO in magnesia I.

115

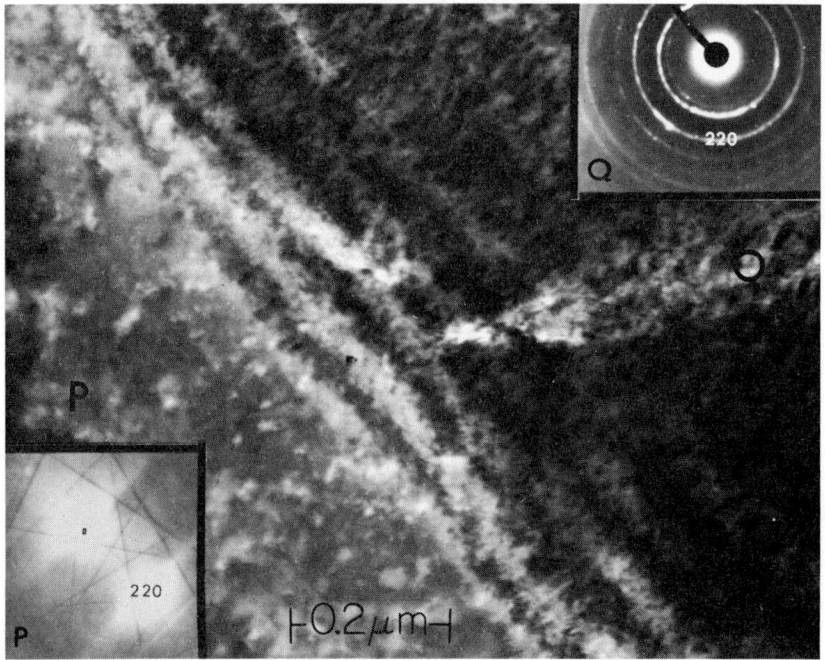

Figure 8. Region in magnesia I showing fine-particle MgO precipitate in a glassy matrix P and an MgO grain Q. Thickness fringes are distorted by diffraction from the MgO precipitates.

brick. Region P has a fine precipitate of MgO contained in a glass matrix. The thickness fringes along the boundary are in poor diffraction contrast even though the grain is tilted to a strong Bragg reflection.

Sub-Boundaries

Small-angle grain boundaries or sub-boundaries were found in most of the specimens. Such boundaries are composed of dislocations, but they often exhibit complex fringe patterns [21, 22]. The dislocations forming the sub-boundaries can be revealed after tilting the specimen with respect to the electron beam to obtain different diffraction conditions. The fine-grain alumina specimens provided many good examples of such sub-boundaries. The composite micrograph of Figure 9 shows a region in alumina IV with the various grains in different diffraction contrast. An example of an extensive dislocation network forming a sub-boundary in alumina IV is shown in Figure 10. In this figure, the hexagonal

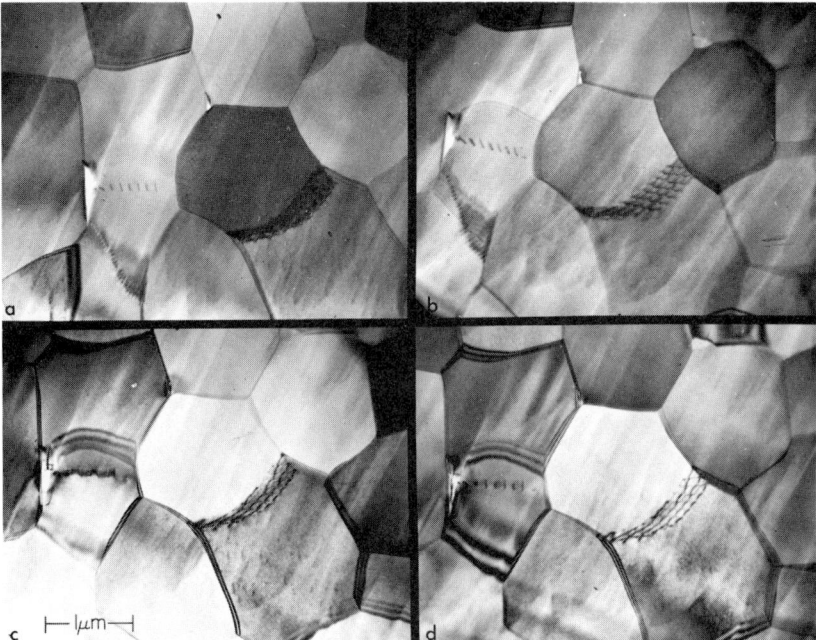

Figure 9. Grains in hot pressed alumina IV. The contrast of grain boundaries and sub-grain boundaries has changed as a result of tilting the specimen with respect to the electron beam.

network of dislocations can be seen; and, in some portions of the boundary a moiré fringe pattern [3, 22] is visible.

One of the sub-boundaries found in the quartzite (SiO_2 I) is shown in Figure 11. Here the sub-boundary has both dislocation contrast and periodic wedge fringe contrast.

Pores

A number of pore configurations contribute to the total porosity of ceramics. These include: (1) pores the size of grains or larger; (2) pores at grain boundary junctions or along grain boundaries; and (3) pores within grains.

Pores the size of grains are frequently interconnected, and can completely penetrate a 45 μm thin section. Holes are found in materials of low porosity such as alumina I, as well as in very porous materials such as magnesia I.

When the specimen thickness is reduced to a few tenths of a micrometer, a small pore will appear as a hole. Thus the triangular holes at the triple junctions

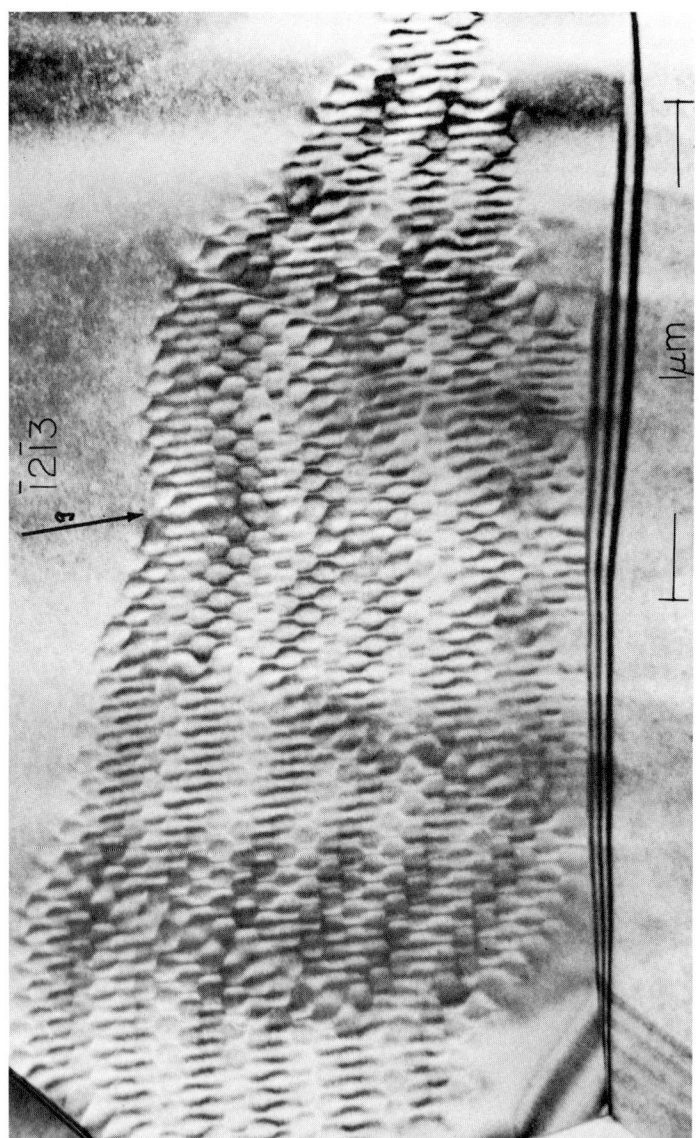

Figure 10. Twist boundary in alumina IV with portions of the dislocation network in dislocation contrast, and also showing moiré-type fringe contrast.

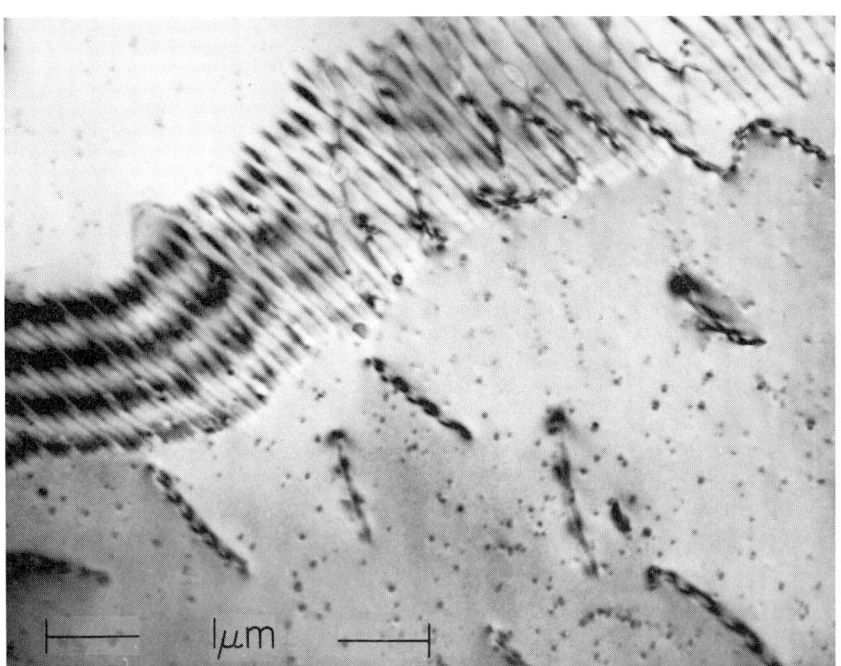

Figure 11. Sub-boundary in quartzite showing dislocations and periodic fringes.

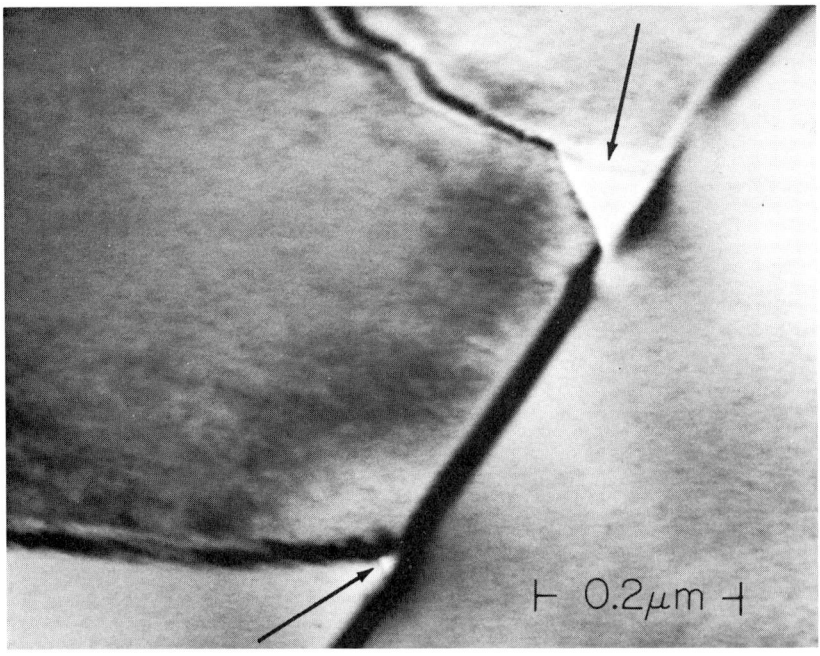

Figure 12. Pores at grain boundary junctions in alumina II.

119

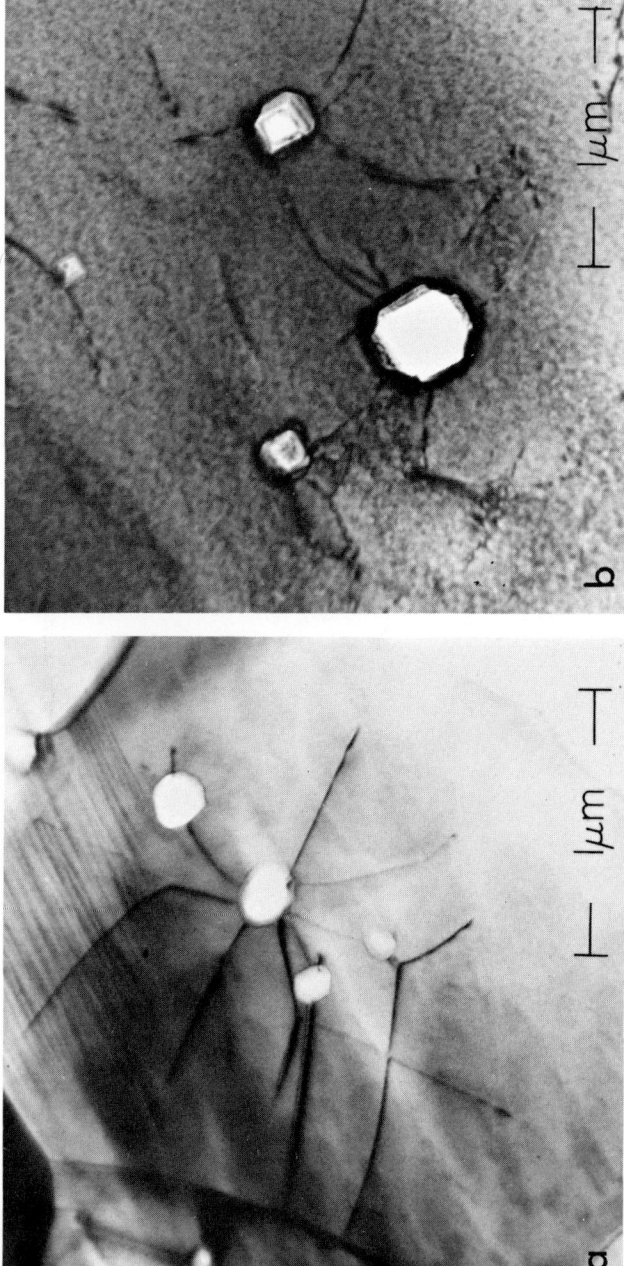

Figure 13. Dislocations associated with pores in: (a) hot pressed alumina III, and (b) extruded magnesia II.

shown in Figure 12 are presumed to have been pores which were not eliminated during sintering of the alumina II sample. Intergranular pores could act as obstacles to dislocation movement during high-temperature processing of the bulk specimen. Two examples of pore-dislocation configurations are shown in Figure 13, where 13a is alumina III and Figure 13b is magnesia II. The micrographs show the polygonal shape of the holes. In Figure 13b, there are thickness fringes at some of the pores. It is possible that some holes mark sites of impurity precipitates which were removed during thinning.

Intergranular pores in the press-forged alumina V have an appearance different from those in the other alumina samples. Several pores can be seen (arrowed) in Figure 14. Each of the small crystallographic holes joins a small area which

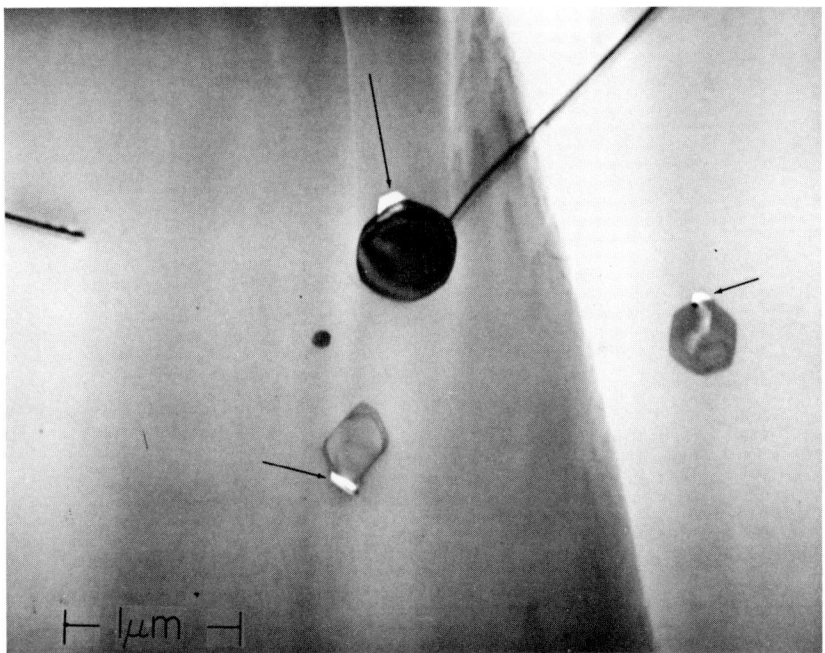

Figure 14. Pores in alumina V.

shows strong fringe contrast, even though they are contained within a single grain. The diffraction conditions can be changed so that the fringes disappear and only the holes remain. This type of diffraction contrast can result from a lenticular void parallel to the foil plane.

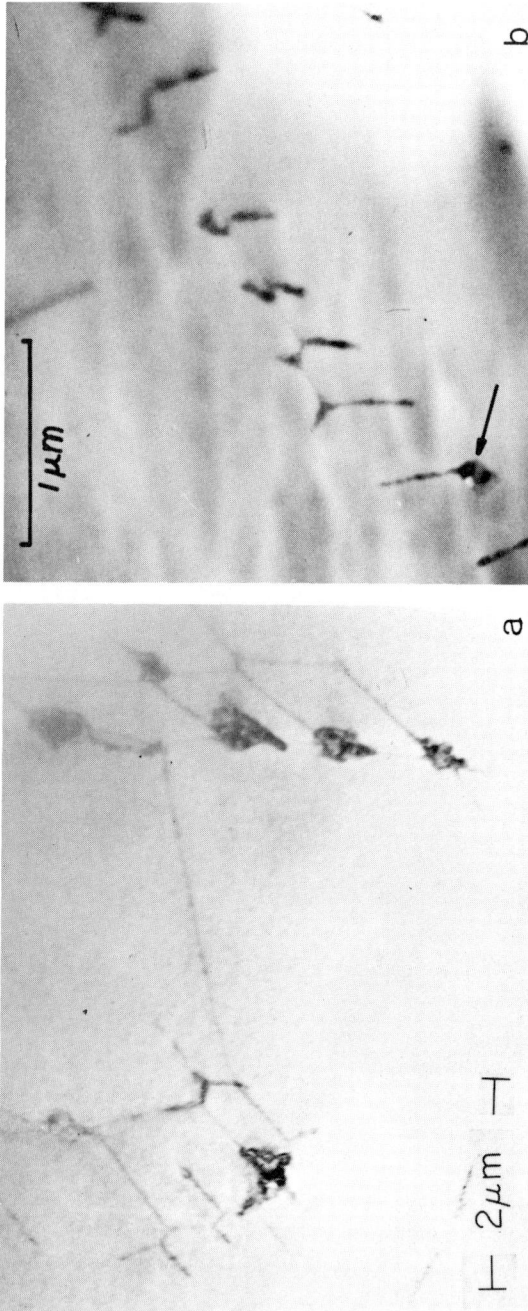

Figure 15. Precipitate particles in magnesia I: (a) at nodes of dislocation network, and (b) at dislocations in sub-boundary.

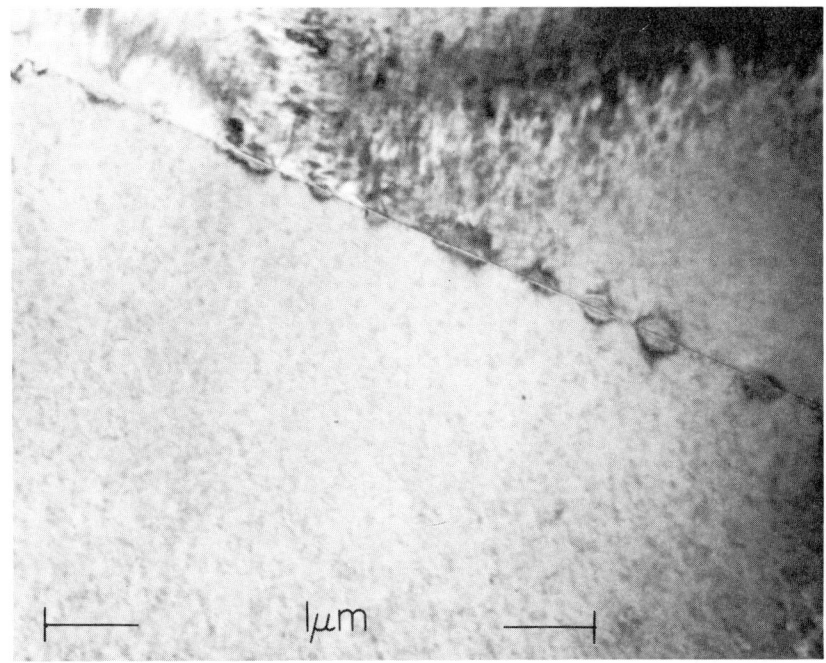

Figure 16. Inclusions along grain boundary in $ZrO_2 + Y_2O_3$.

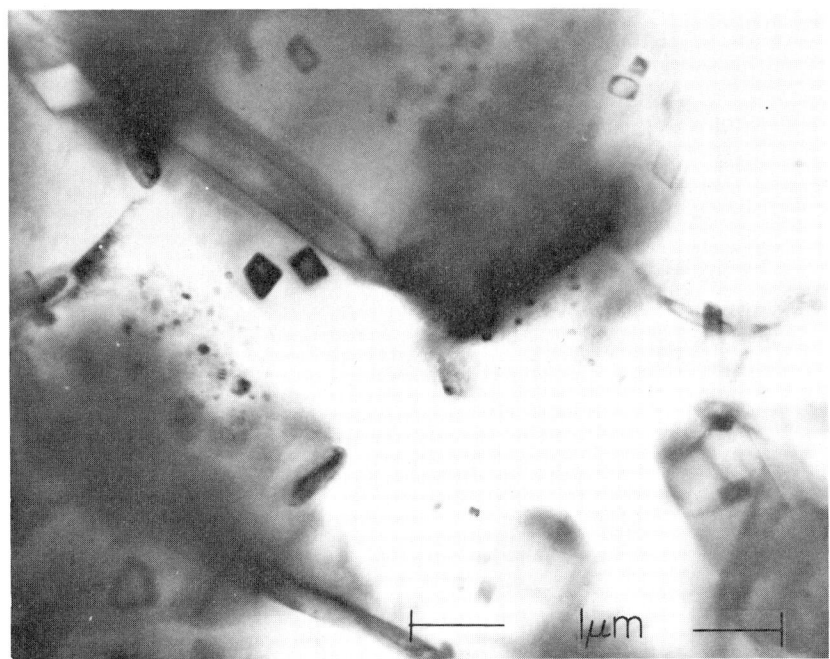

Figure 17. Crystalline and glass phases in a glass-ceramic (courtesy of D. J. Barber).

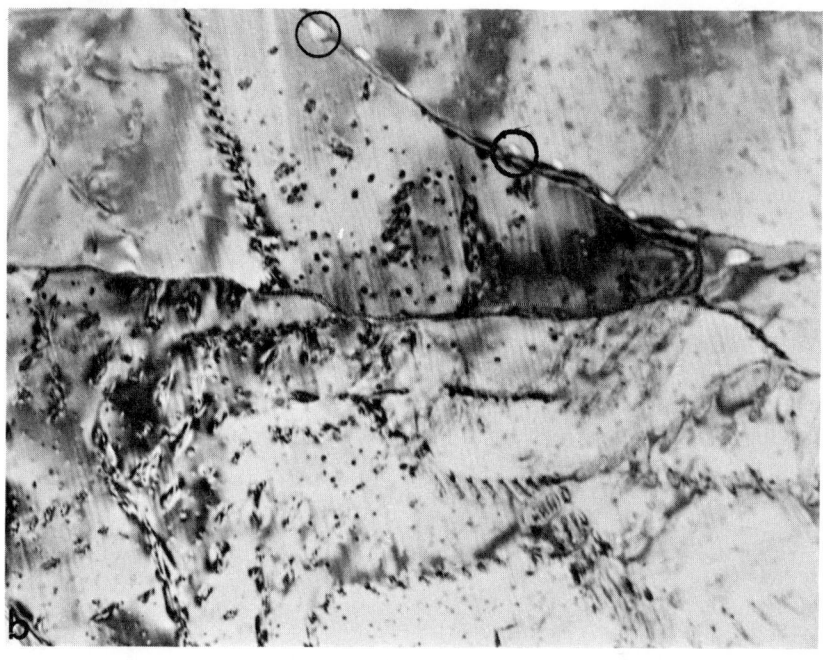

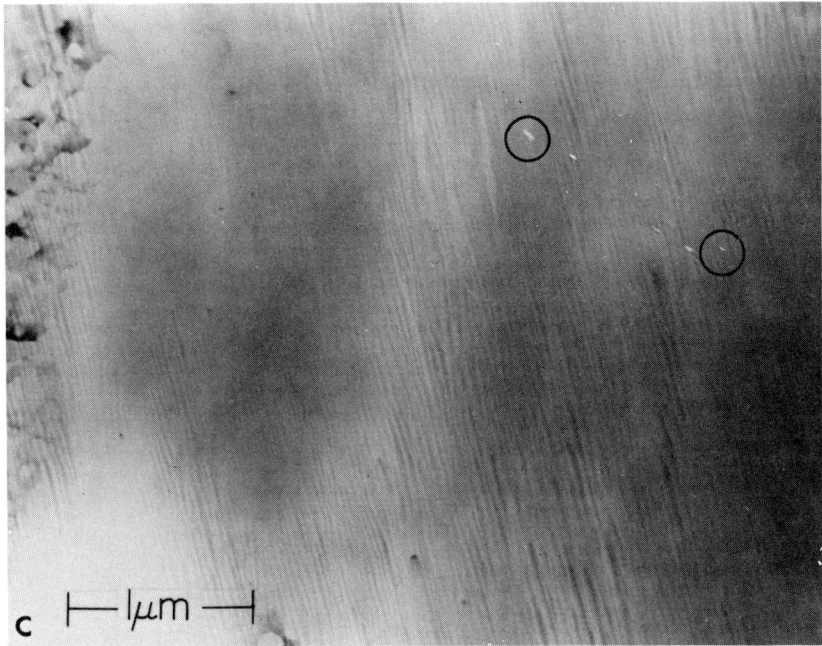

Figure 18. Crystalline (in a,b) to amorphous (in c) phase change in SiO_2 I induced by the electron beam. Micrographs taken after: (a) <30 sec, (b) 1 minute, and (c) ~ 10 minutes in the electron beam.

Second-Phase Materials

Second-phase material usually could be distinguished from the major phase, although it was not always possible to make a positive identification of the phase. Second phases were present in different forms: (1) large single-crystal grains; (2) precipitates within grains; (3) precipitate particles along grain boundaries; (4) non-crystalline regions. Crystalline phases were identified from electron-diffraction patterns by comparing observed interplanar spacings and angles with those of possible phases. The monticellite and spinel grains shown in Figures 5 and 6 were identified from their diffraction patterns in this manner. An amorphous phase and an amorphous phase containing a crystalline precipitate were shown in Figures 7 and 8, respectively.

Examples of impurity precipitation on dislocations are given in Figure 15, and were found in MgO I. Figure 15a shows precipitate particles at the nodes of the dislocation network; 15b shows precipitate particles along the dislocations.

The stabilized zirconia was the only material studied which had a second phase distributed throughout the foil. In Figure 16 the second phase is barely resolved in the grains, but the presence of some particles along the grain boundary is indicated by the fringes seen at intervals along the boundary. These fringes are caused by a strain contrast mechanism [23, 24]. In each grain, the reflection causing the contrast was normal to the grain boundary.

A good example of a multiphase ceramic is the thin foil of a glass-ceramic dinnerware shown in Figure 17 (courtesy of D. J. Barber). The material is composed of crystals of different composition and size contained in a glass matrix. There is considerable variation in intensity and in diffraction contrast from the various components.

Phase Changes

The quartz rock specimens underwent a phase change from crystalline to amorphous (vitrification) while in the electron beam. Micrographs showing vitrification in quartzite (SiO_2 I) are in Figure 18. The small black spots (dislocation loops) in 18a appeared while focusing; 18b was taken after one minute, and 18c after an additional ten minutes. Pores along the sub-boundary in 18a serve to locate the same areas in the sequence. The phase change was rapid, and

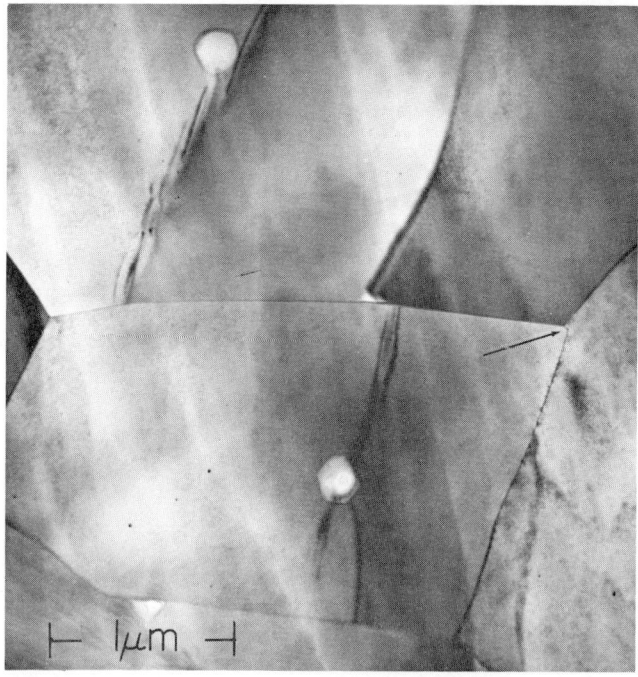

Figure 19. Area in alumina IV deformed 3%, showing grain boundary sliding.

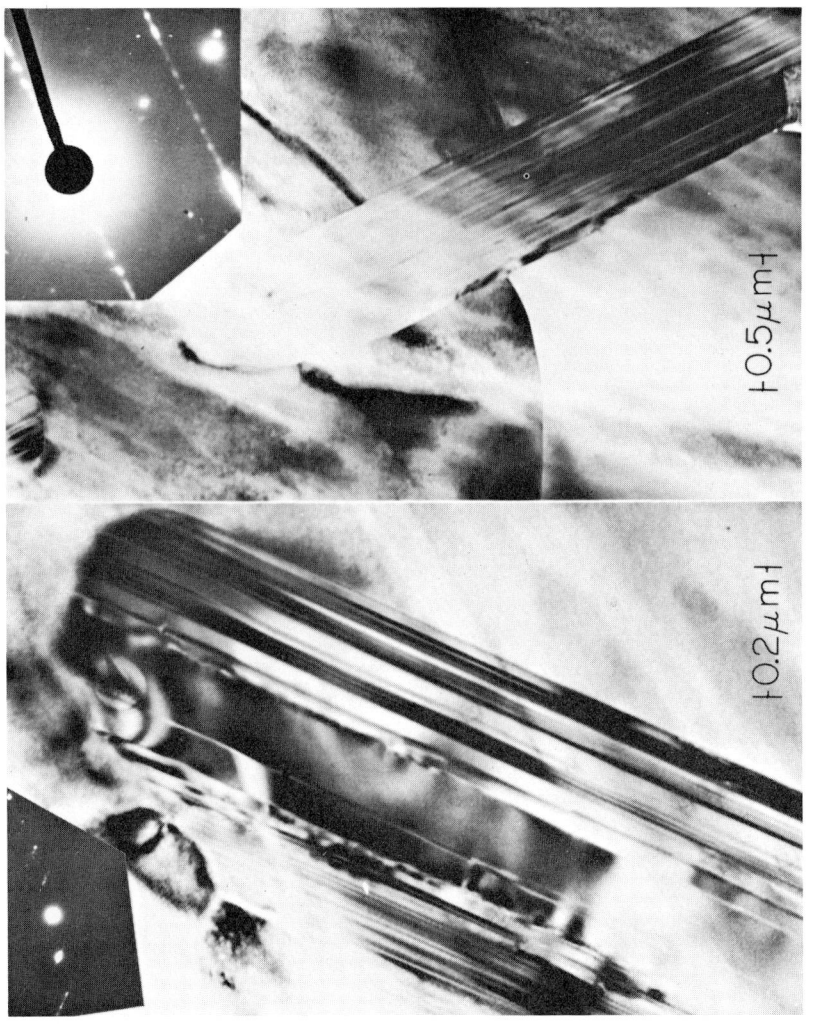

Figure 20. Twinned region in alumina IV.

was confined to the area of the specimen being irradiated by the electron beam. The phase change occurred as a result of damage induced by the electron beam. Such damage and a similarly rapid phase change were observed in specimens of amethyst quartz [25].

Deformation Sub-Structure

The as-received condition of most of the samples included here represented a well-annealed sub-structure in which any dislocations present were arrayed in stable configurations such as sub-boundaries. Some discussion of the dislocation sub-structures in alumina I and alumina II has been presented recently [7]. Sub-boundaries composed of dislocations having Burgers vectors of the types 1/3 $[11\bar{2}0]$ and 1/3 $[10\bar{1}1]$ —c/a = 2.730—were found in alumina II [7] and in sintered alumina containing silica [11].

Only a limited number of observations have been made on additionally deformed ceramic bodies, so that it is not possible to describe completely the dislocation structures corresponding to a given deformation. Deformation in the rock specimens will be discussed separately [26]. Observations on deformed hot pressed alumina showed that grain boundary sliding and twinning occurred, but that the frequency of occurrence of sub-boundaries could not be related directly to additional deformation of this specimen. Plastic deformation of this material is discussed more completely by Heuer *et al.* in a later chapter [27].

Figure 19 shows an area in alumina IV in which grain boundary sliding occurred. There is a small jog (arrowed), and there appear to be dislocations along the adjacent grain boundary. The small triangular holes at triple junctions, which would result from grain boundary sliding, were numerous in this specimen and were not found in alumina III—the undeformed hot pressed sample. Several of these holes can be seen in Figure 9.

The deformation twins, such as those in Figure 20, belong to the rhombohedral mode where twinning occurs on the $\{\bar{1}012\}$ plane in the $\langle 10\bar{1}1\rangle$ direction [29, 30]. Identification of twins in Figure 20 was simplified somewhat by tilting the specimen about an axis parallel to the fringes until the diffraction pattern was streaked in a direction normal to the trace of the twin plane [30].

Magnesia II was heavily deformed by the extrusion process, and within the grains the dislocation density was high, being of the order of 10^8 cm^{-2} (Figure 21). It is apparent in the figure that numerous dislocation interactions have occurred at the grain boundaries. The dislocations in grain P are out of contrast for a 200 reflection, suggesting a $\frac{1}{2}$ [011]-type Burgers vector such as that found in studies in single-crystal MgO [31]. Rice has discussed aspects of plastic deformation of this material [32].

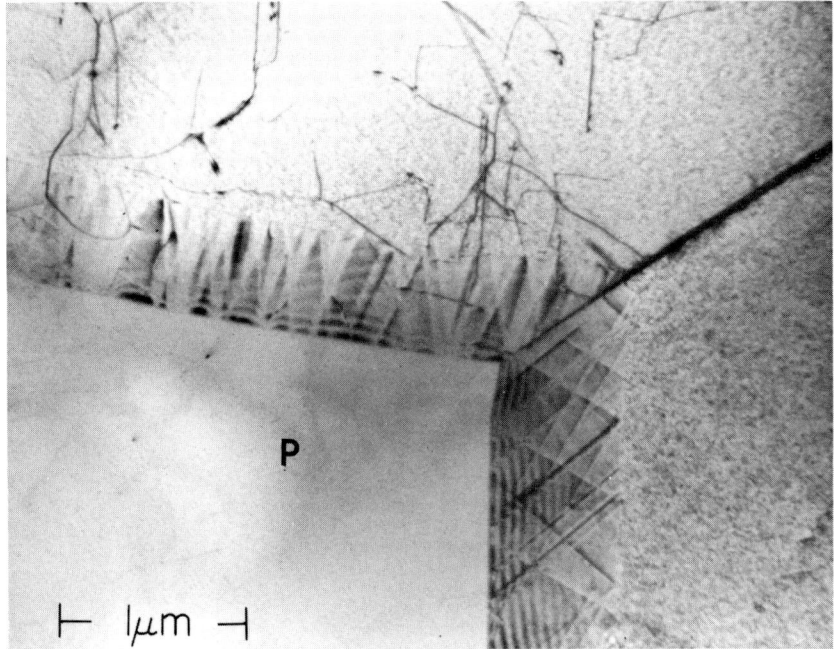

Figure 21. Triple junction in extruded MgO + 2% NiO.

Ion Bombardment Induced Radiation Damage

Radiation damage resulting from argon ion bombardment is visible, under suitable diffraction conditions, as small black-white streaked images which are similar to those produced in neutron-irradiated single-crystal samples [33]. The images are presumably small dislocation loops of vacancy or interstitial character, or both. The loops can be made to grow by pulse heating in the electron beam with the second condenser lens turned off and the condenser aperture removed. Dislocation loops in Figure 22 were produced in a specimen of alumina I by pulse heating. Loops of similar size were produced in neutron-irradiated sapphire by furnace annealing for a few minutes at 1350°C [33].

High-Voltage Microscopy

The micrographs reproduced above were taken with conventional electron microscopes operating at 100 kv. The best resolution and contrast in most of

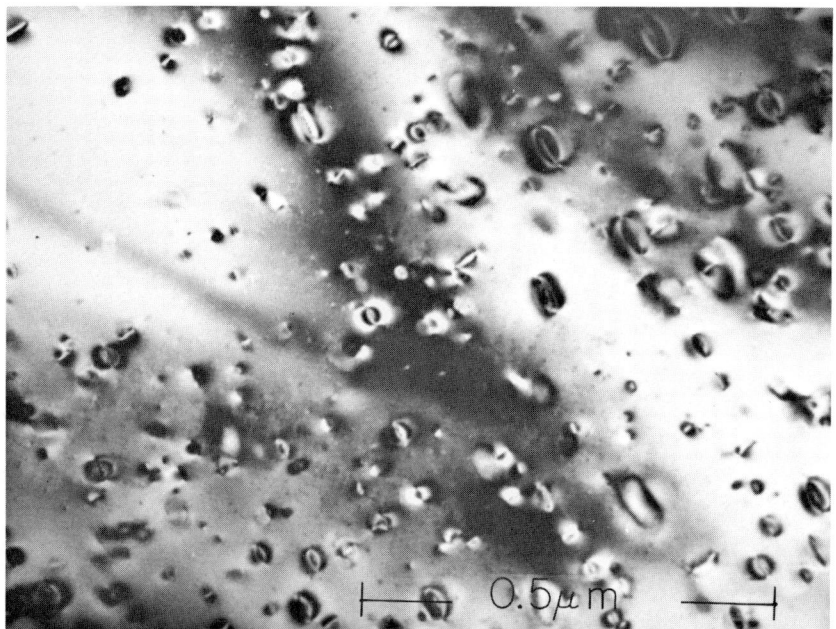

Figure 22. Dislocation loops in alumina I formed after pulse heating in electron beam.

Figure 23. Grains in alumina III, micrograph taken with an electron microscope operating at 1 megv (courtesy of P. R. Swann).

the figures was obtained in areas ~3,000Å thick or less. High-voltage microscopes operating at 750 to 1,200 kv allow resolution and contrast in considerably thicker specimens [34–37]. Figure 23 is a micrograph of alumina III, taken by P. R. Swann with a 1 megv electron microscope. The top and bottom of a grain (arrowed) are imaged, indicating good contrast for at least 1 μm thickness.

Conclusions

The results briefly presented here demonstrate that it is possible to prepare satisfactory electron-microscope specimens from bulk ceramic materials having different composition, grain size, and porosity. Specimens prepared by ion bombardment have sufficiently large thin areas to be representative of the bulk samples. The direct observation by transmission electron-miscoscopy of fine-grain ceramics provides the means for characterizing their microstructure and for studying the dislocation structures of deformed specimens.

Acknowledgments

The author thanks A. H. Heuer, B. J. Hockey, and S. M. Wiederhorn for helpful discussions; D. J. Barber and P. R. Swann for micrographs; and J. M. Christie, A. H. Heuer, J. R. Kreglo, Jr., K. S. Mazdiyasni, and R. W. Rice for samples of: SiO_2, Al_2O_3, MgO, ZrO_2 and MgO+NiO, respectively.

References

1. Kingery, W. D., *Introduction to Ceramics,* John Wiley and Sons, Inc., New York (1960).
2. Insley, H. and Fréchette, V. D., *Microscopy of Ceramics and Cements,* Academic Press, New York (1955).
3. Hirsch, P. B., Howie, A., Nicholson, R. B., Pashley, D. W. and Whelan, M. J., *Electron Microscopy of Thin Crystals,* Plenum Press, New York (1965). Butterworth & Co., Ltd.
4. Thomas, G., *Transmission Electron Microscopy of Metals,* John Wiley and Sons, Inc., New York (1962).
5. Amelinckx, S., *The Direct Observation of Dislocations,* Academic Press, New York (1964).
6. Heidenreich, R. D., *Fundamentals of Transmission Electron Microscopy,* Interscience Publishers, New York (1964).
7. Tighe, N. J. and Hyman, A., "Transmission Electron Microscopy of Alumina Ceramics," in *Anisotropy in Single-Crystal Refractory Compounds*, Plenum Press, New York (1968), 121–36.
8. Tighe, N. J. and Heuer, A. H., "Substructure of Hot Pressed Polycrystalline Al_2O_3," *Bull. Am. Ceram. Soc.,* 47 (1968), 349 (abstract).

9. Tighe, N. J. and Kreglo, J. R., "Microstructure of MgO Brick," *Bull. Am. Ceram. Soc.*, 47 (1968), 426, to be published.

10. Langdon, T. G., "Thinning of Polycrystalline MgO for Transmission Electron Microscopy," *J. Sci. Inst.*, 38 (1967), 125.

11. Gulden, T. D., "Direct Observation of Nonbasal Dislocations in Sintered Alumina," *J. Am. Ceram. Soc.*, 50 (1967), 472-75.

12. Crosby, A., Personal communication.

13. Barber, D. J. and Tighe, N. J., "Observations of Dislocations and Surface Features in Corundum Crystals by Electron Transmission Microscopy," *J. Res. NBS*, 69A (1965), 271-80.

14. Doherty, P. E. and Leombruno, "Transmission Electron Microscopy of Thin Glass Samples," *J. Am. Ceram. Soc.*, 47 (1964), 368-70.

15. Williams, J. P., Carrier, G. B., Holland, H. J. and Farncomb, F. J., "The Determination of the Crystalline Content of Glass-Ceramics," *J. Mat. Sci.*, 2 (1967), 513-20.

16. Christie, J. M., Personal communication.

17. Bauer, T. E. and Beauchamp, R. H., "Mechanical Thinning of Ceramic Materials for Transmission Electron and Optical Microscopy," AEC-BNWL-652 (1968), Battelle Northwest.

18. Ashbee, K. H. G., Lyall, R. and White, D., "Plastic Deformation of a Glass-Ceramic," *Phil. Mag.*, 17 (1968), 225-34.

19. Paulus, M. and Reverchon, F., "Dispositif de Bombardement Ionique pour Préparations Micrographiques," *J. Phys. Radium*, 22 (1961), 103A.

20. _____. "Étude des Parametres du Bombardement Ionique des Ferrites," in *Le Bombardement Ionique*, C. N. R. S. No. 113, Paris (1966), 223-34.

21. Barber, D. J. and Tighe, N. J., "Electron Microscopy and Diffraction of Synthetic Corundum Crystals. II. Dislocations and Grain Boundaries in Impurity-doped Aluminum Oxide," *Phil. Mag.*, 14 (1966), 531-45.

22. Gevers, R., van Landuyt, J. and Amelinckx, S., "On a Simple Derivation of the Amplitudes of the Electron Beams Transmitted and Scattered by a Crystal Containing Planar Interfaces—Images of Subgrain Boundaries," *Phys. Stat. Sol.*, 18 (1966), 325-42.

23. Phillips, V. A. and Livingstone, J. D., "Direct Observation of Coherency Strains in a Copper-Cobalt Alloy," *Phil. Mag.*, 8 (1962), 969-80.

24. Ashby, M. F. and Brown, L. M., "On Diffraction Contrast from Inclusions," *Phil. Mag.*, 8 (1963), 1649-76.

25. McLaren, A. D. and Phakey, P. P., "A Transmission Electron Microscope Study of Amethyst and Citrine," *Aust. J. Phys.*, 18 (1965), 135-41.

26. Christie, J. M. and Tighe, N. J., to be published.

27. Heuer, A. H., Cannon, R. M. and Tighe, N. J., "Plastic Deformation in Fine-Grain Ceramics," this volume.

28. Barber, D. J. and Tighe, N. J., "Electron Microscopy and Diffraction of Synthetic Corundum Crystals. I. Pure Aluminum Oxide Grown by the Verneuil Process," *Phil. Mag.*, 11 (1965), 495-512.

29. Heuer, A. H., "Deformation Twinning in Corundum," *Phil. Mag.*, 13 (1966), 379-93.

30. Swann, P. R. and Warlimont, H., "The Electron Metallography and Crystallography of Copper–Aluminum Martensites," *Acta Met.*, 11 (1963), 511-27.

31. Washburn, J., Groves, G. W., Kelly, A. and Williamson, G. K., "Electron Microscope Observations of Deformed Magnesium Oxide," *Phil. Mag.*, 5 (1960), 991-99.

32. Rice, R. W., "Internal Surfaces of MgO," in *Materials Science Research*, Vol. 3, *The Role of Grain Boundaries and Surfaces in Ceramics*, Plenum Press, New York (1966).

33. Barber, D. J. and Tighe, N. J., "Observations of Neutron Damage in Single Crystal Aluminum Oxide," *J. Am. Ceram. Soc.*, 51 (1968), 611–17.
34. Dupouy, G., and Perrier, F., "Microscope Électronique Fonctionnant sous une Tension d'un Million de Volts," *J. Microscopie*, 1 (1962), 167.
35. Thomas, G., "Electron Microscopy at High Voltages," *Phil. Mag.*, 17 (1968), 1097–1108.
36. Cosslett, V. E., "The Voltage Electron Microscope," *Contemp. Phys.*, 9 (1968), 333.
37. Uyeda, R. and Nonoyama, M., "The Observation of Thick Specimens by High Voltage Electron Microscopy. II. Experiment with Molybdenite Films at 50–1200 kV," *Japan J. Appl. Phys.*, 7 (1968), 200.

SESSION IV

PROCESSING OF ULTRAFINE-GRAIN CERAMICS

MODERATOR: RICHARD M. SPRIGGS
Lehigh University
Bethlehem, Pennsylvania

8. Fine Particulates to Ultrafine-Grain Ceramics

T. VASILOS and W. RHODES
Avco Space Systems Division
Wilmington, Massachusetts

· *ABSTRACT*

The preparation of dense ceramics with ultrafine-grain sizes has, in general, required a coupling of ultrafine particles as starting materials with applied pressure during densification.

The state-of-the-art of a number of ceramic materials so processed is described, with attention given to raw material characteristics and processing variables on final microstructure development.

Introduction

Until recently, the development of fine-grain ceramics from particulates has generally required the simultaneous application of pressure and temperature to lower densification temperatures so that grain growth rates are sufficiently retarded.

However, with the availability of improved particulate materials, advancements have been made in sintering technology which now make it possible to prepare a number of fine-grain bodies without the assistance of applied pressure.

The current discussion, therefore, is concerned with both methods; i.e., sintering and hot pressing of particulates into fine-grain materials. Important to both techniques is the nature of the compositions and particulates employed, and the discussion emphasizes recent findings.

Sintering

General

The sintering of oxides has, in general, resulted in microstructures characterized by 1-7 percent porosity and grain sizes $> 20\,\mu$. However, in the last decade numerous investigators have achieved grain sizes $< 5\,\mu$ and porosities

137

typically of the order of 1-3 percent and occasionally < 1 percent. These results have been obtained through the effective utilization of the following grain growth control mechanisms; pore-grain boundary interaction, grain growth stabilization with discrete second phases, grain growth control with solid solution additions, or the sintering of Ångstrom-size powders. Each of these areas will be discussed with regard to both theory and results. Current work by the authors is concerned with the latter category, thus experience and examples will be weighed in this direction. Some of the pitfalls of general applicability in sintering will be pointed out from this experience.

The theory of sintering has been the subject of a recent review by Coble and Burke [1]. The authors support the view that sintering is a diffusion-controlled process (rather than plastic flow) and the initial shrinkage $\Delta L/L_o$ is represented by an equation of the form

$$\frac{\Delta L}{L_o} = \left(\frac{N\gamma\Omega Dt}{R^3\, kT} \right)^m \tag{1}$$

where N and m are constants, γ the surface energy, Ω is the atomic volume of the diffusing species, D the rate controlling diffusion coefficient, t the time, R the particle radius, and kT have the usual meaning. The power dependence, m, has theoretical values between 0.4 and 0.5 depending on details of the details of the derivation. Thus, the difference in free energy of the neck area and surface of the particle causes a driving force for material transfer from the interparticle boundary to the neck area. The rate of shrinkage is approximately inversely proportional to particle size and constantly decreases with time.

The intermediate and final stages of sintering have been empirically studied by numerous investigators and have been quantified by Coble [2] and Coble and Gupta [3]. For volume diffusion control of a tetrakaidekahedron grain with cylindrical pores along three grain edges, porosity, P, removal is governed by

$$-\frac{dP}{dt} = \frac{N\gamma\Omega D}{(GS)^3\, kT} \tag{2}$$

where GS is the grain size. The fact that grain size continuously changes during the rather large porosity change in the intermediate region has led to some reservations concerning the use of equation (2) [4]. Nevertheless, it has been applied to the sintering of Al_2O_3 [2], Cu [3], BeO [5], and ZnO [6], and found to describe adequately the kinetics of this stage. The application to the Cu sintering was particularly satisfying, as calculated D's were in excellent agreement with tracer measurements.

Pore-Grain Growth Interactions

It has been recognized by Burke [7] that grain growth-pore interactions play a significant role in the outcome of sintering. Pores as the sink for matter transport must remain on or close to grain boundaries (matter source) for sintering to proceed. Pores apparently act as grain growth inhibitors in a manner governed approximately by the equation

$$GS \approx \frac{d}{V_f} \qquad (3)$$

where GS is the limiting grain size, d is the size of the inclusion (pore), and V_f is the volume fraction of inclusions.

Thus, up to some level of density depending on diffusion coefficients, activation energies and relative pore grain boundary surface energies, grain growth will be controlled by the porosity percentage. It was noted by Coble and Gupta [3] in the sintering of Cu that up to 85 percent of theoretical density, grain sizes were constant for a given density independent of sintering temperature if the starting density was constant. Similar conclusions can be drawn from the data of Coble [2] on Al_2O_3, and Clare [8] on BeO. In the oxide cases, densities in the range of 90–97 percent are obtained before the linear relation between density and grain size is destroyed. This is presumably due to the failure of equation (3) from the low V_f and the ability of grain boundaries to sweep past pores. Also, secondary grain growth may be nucleated at this point, in which case the driving force for growth of the secondary grains would be in excess of that for a normal six-sided grain.

If equation (3) is universal, one needs only to choose a small enough starting particle size and the proper green density (the starting density may influence the slope of the grain size/density curve) to obtain the requisites for a slightly porous sub-5μ grain material. Indeed, the data of Coble [2] shows the 5μ level to be reached at 96 percent density in Al_2O_3, and Clare [8] shows BeO to be 5μ at 94 percent density (the Al_2O_3 possessed an MgO grain growth inhibitor which, of course, aids in the obtainment of the desired effect). Bruch [9], in his work on Lucalox, has shown the inter-relations between pores and grain growth, and has derived an equation treating these relations which is represented graphically in Figure 1. This, of course, indicates that one specific green density is required for a given combination of grain size and porosity. The relation leads to the prediction of a sub-5 μ 99$^+$ dense Lucalox body. The "hooker" in the prediction is the requirement for a high green density, and this will be discussed in the section on Sintering Active Powders.

Coble [2] and, more recently, Bruch [9], have shown that pores can actually grow in size as sintering proceeds. This possibility has been discussed from a fundamental standpoint by Kingery and Francois [10], and they have shown that pores also grow in UO_2. Arguments are presented for the migration of pores with grain boundaries and the pore size being governed by a summation of pore growth by coalescence together with their shrinkage due to the matter flux, equation (2).

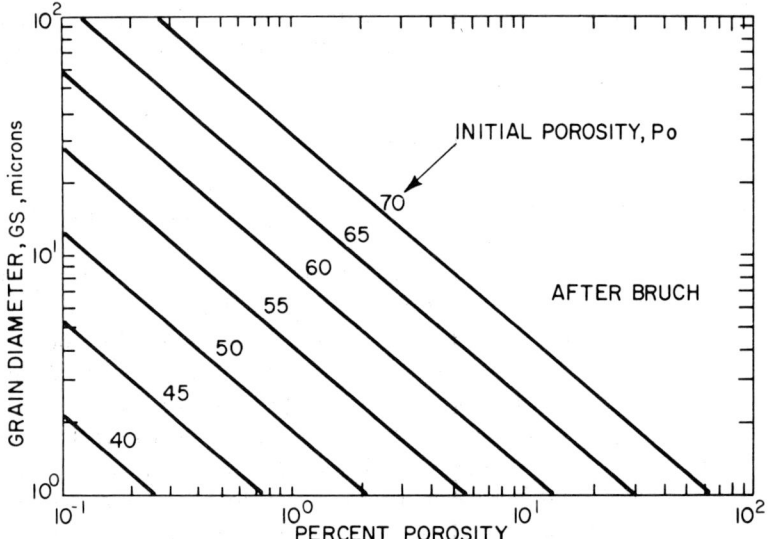

Figure 1. The influence of initial porosity on the grain size and porosity of Linde A alumina with MgO after sintering in hydrogen (after Bruch).

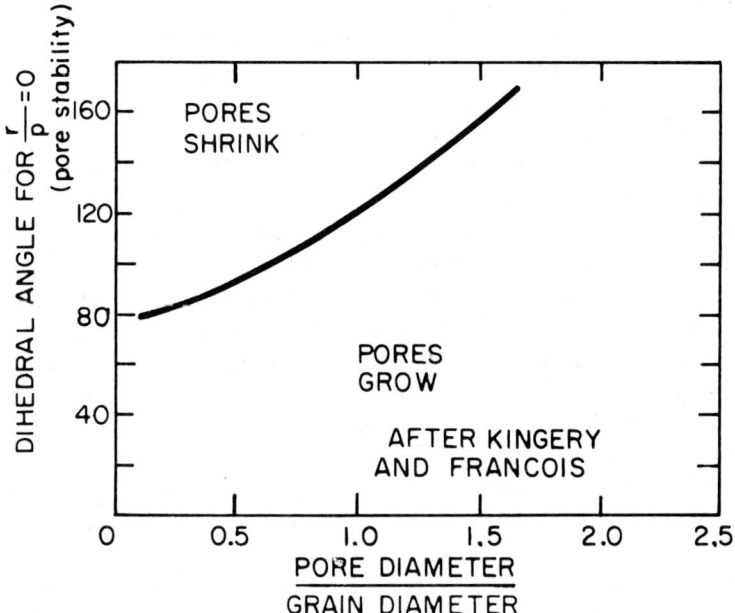

Figure 2. Conditions for pore stability (after Kingery and Francois).

It was reasoned that the surface curvature of a pore, p, was governed by the equilibrium dihedral angle which is given by

$$\cos \frac{\phi}{2} = \frac{\gamma_g}{2\gamma_p} \qquad (4)$$

where γ_g is the grain boundary surface energy and γ_p is the pore surface energy. The condition for pore stability will depend on dihedral angle and the ratio of pore diameter to grain diameter, as shown graphically in Figure 2 (after Kingery and Francois). Thus, material systems having a low dihedral angle have an increased tendency for pore growth. UO_2 has a low ϕ of $92°$, and clear evidence for pore growth was given.

The influence of pores on grain growth is quite significant, but it may be used to advantage for certain property-dependent applications. For example, strength is a property adversely affected by both increasing grain size and porosity; however, the two variables have different dependencies. Figure 3 (after Charles and

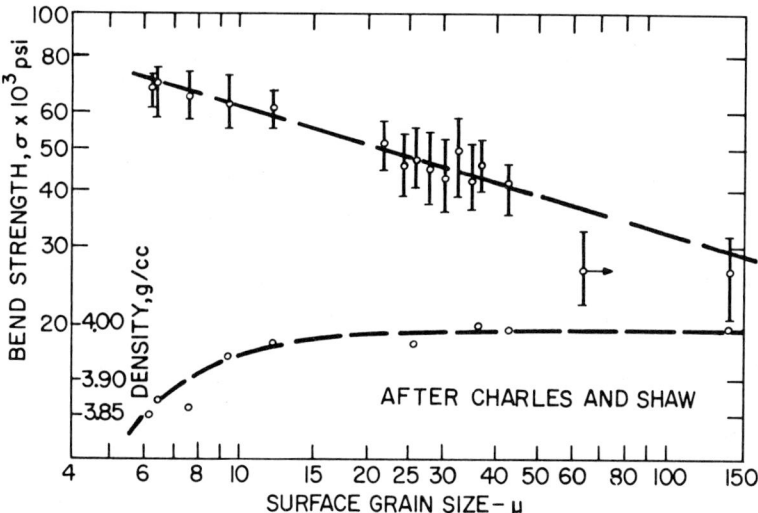

Figure 3. Bend strength and density of Lucalox alumina extruded rods as a function of surface grain size (after Charles and Shaw).

Shaw [11]) demonstrates that improved strength (in 4-point bending) clearly favors the porous fine-grain material. This principal is apparently being used to good advantage in a commercially available [12] sintered Al_2O_3 cutting tools having 2.5 percent porosity and a 4μ grain size.

Effect of Additions on Grain Stabilization

The control of grain growth by chemical additions to the composition is probably the best documented technique in this area. This technique is responsible for the commercial success of a sintered 97 percent dense 3 μ grain size abrasive grain [13] (in practice a bauxite is sintered which results in alumina stabilized by a pseudo-brookite phase). Also, grain stabilization can be accomplished by additives which form solid solutions with the matrix, and this principal has been applied in the sintering of the commercial Al_2O_3 Luca-lox [14] and Coors AD 999 [15]. The former material is available at sub-5 μ grain sizes at 96-88 percent density, while the latter apparently is 99$^+$ percent dense at the 2-3 μ grain size level. The two methods for grain size control will be discussed separately.

Discrete Second Phases

Other than the example cited above, grain refinement in UO_2 has been achieved by the addition of Pt by Elyard et al. [16]. Densities of 97 percent were achieved at a 3 μ grain size. Grain refinement has been achieved in BeO by the addition of ZrO_2 [17], graphite [18], SiC [19], and MgO [19]. Other examples such as SiC in ZrB_2 [20] can be cited for grain refinement in hot pressing.

In general, grain refinement by second-phase additions is governed by equation (3), as was discussed in the section on porosity. Extra work is required to move the migrating grain boundary past the particle and this is governed by the particle size, volume fraction of particles as well as factors such as the boundary-particle surface energy and the geometrical location of the particle (i.e., random, 3 grain edges, or 4 grain corners).

Hillert [21] has analyzed these criteria, and this analysis has been applied to the stabilization of grain size to the sub-5 μ level in the sintering of ceramics by Woolfrey [22]. Equation (3) is modified to give

$$V_f = 0.3 \ \frac{d}{GS} \tag{5}$$

for a random dispersion of particles of size d,

$$V_f = 0.6 \ \frac{d^2}{GS^2} \tag{6}$$

for particles located entirely on grain boundaries,

$$V_f = 1.05 \ \frac{d}{GS} \tag{7}$$

for a random dispersion of particles with effective particles located on 3 grain edges, and

$$V_f = 1.86 \; \frac{d^3}{GS^3} \tag{8}$$

for particles located entirely on 3 grain edges.

The hypothetical situation of particles located at 4 grain corners was analyzed, and it was found that particles three times as large as for the other cases were required for stabilization. These equations are solved for grain stabilization in the 0.25–5 μ regime using 1,000 Å dispersed particles, and graphically shown in Figure 4 (after Woolfrey).

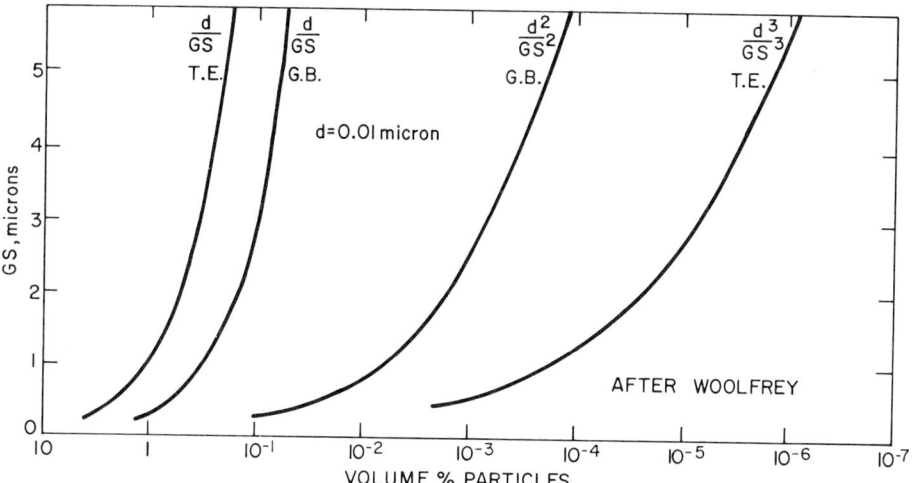

Figure 4.   Volume percentage of 0.01 μ particles required to stabilize grain sizes in the 0.25–5.0 μ region (after Woolfrey).

This approach toward grain refinement has proved successful and it is expected that application of the above relations will be applied to many future systems. This is not to imply that the approach is without difficulties, since factors such as segregation of the particles, particle growth, particle solubility in the matrix, and surface energy between the matrix and dispersed phase will affect the choice and success of the additive.

Solid Solution Effects

The additions of MgO to Al_2O_3 to prevent discontinuous grain growth is probably the best documented example of this category, and has been the subject of research by Coble [2], Jorgensen and Westbrook [23], and Jorgensen [4]. There are other examples of sintering to high density with good grain growth control. One recent result is the sintering of yttrium-iron garnet to 99^+ percent density with a final grain size of 1–3 μ by Paladino and co-workers [21]. In this case, specific additives were not made, but stoichiometry effects were noted; thus, segregation of an ion is probably responsible for the prevention of pore entrapment.

In fact, the segregation of the normal impurities for a given powder may impart significant grain growth control. As is noted in the section on hot pressing, high-purity materials exhibit a high tendency for the onset of exaggerated grain growth. Segregation of impurities or solid solution dopants at grain boundaries is expected from thermodynamic considerations, as expressed by the well-known Gibbs adsorption equation

$$d\gamma = - \Gamma_2 \, d\mu_2 \qquad (9)$$

where μ_2 is the chemical potential and Γ_2 is the excess amount of component 2 per unit area relative to component 1. If Raoult's law applies it can be shown that upon substitution and integration equation (9) takes the form of

$$\Delta\gamma = -\Gamma_2 \, kT \, \ln\frac{\gamma_2}{x_2{}^o} \qquad (10)$$

where $x_2{}^o$ is the concentration of the solute. A positive absorption decreases the interfacial energy, as noted by equation (10). However, the surface tension will not be reduced to zero, so we have the condition that

$$0 < \gamma_{(x_2)} = \gamma_{(x_2{}^o)} - \int_{x_2{}^o}^{x_2} \frac{\Gamma_2 \, kT \, dX}{x} \qquad (11)$$

where $\gamma_{(x_2)}$ is the surface tension at a concentration of solute, x_2, where no adsorption takes place (or where $\gamma_{x_2}{}^o$ is known).

The application of equation (11) generally leads to the conclusion that a monolayer of impurities are adsorbed on grain boundaries. However, hardness measurements [23] across grain boundaries, auto-radiography [23], and grain boundary diffusion measurements generally lead to the conclusion that the affected layer may be hundreds of Ångstroms, and in some cases microns, wide. If this is true, the layer may be a non-equilibrium structure, and perhaps due to a vacancies gradient arising from the normal sink-source relations or from quenching.

In the case of Al_2O_3 + MgO it appears clear that the solid solution substitution creates either Schottky or Frenkel defects that reduce both the rate of grain growth and sintering. Thus the rate controlling diffusion process is reduced.

Pores and grain boundaries remain in close proximity, allowing complete pore removal.

This mechanism of grain growth control will certainly receive increasing attention. It has been reported that Y_2O_3-ThO_2 [25] sinters to full density apparently by this process, although the final grain size in this case (as with full-density Lucalox) is greater than 5 μ.

Sintering Active Powders

There are several centers of activity concerned with the sintering of powders $< 1,000$ Å in ultimate crystallite size. ThO_2 and ThO_2-UO_2 solid solutions are produced on a fairly large scale by the "sol-gel" process. 80 Å particles are sintered to $99^+\%$ dense $-6+16$ mesh aggregates for vibratory compaction in a metal tube and an ultimate nuclear application [26]. A typical grain size after sintering at $1150°C$ is 0.5 μ. Morgan and co-workers have studied initial stage sintering [27] and creep [28] in bulk samples of ThO_2 and ThO_2-CaO solid solutions prepared in a similar manner. Their reported grain sizes were 5–20 μ for ThO_2, and 4–8 μ for ThO_2-CaO. Densities corresponding to specific grain sizes were not listed, but it is inferred from the higher temperature of sintering ($1800°C$ for 2 hours) that difficulties were encountered in translating the technology of sintering fine-grain aggregates to the sintering of bulk samples. Mazdiyasni, Lynch, and Smith [29] have been engaged in producing 100 Å-size powders of yttria-stabilized zirconia from the decomposition of metal alcoholates. They demonstrated that under optimum sintering conditions densities of 99^+ percent and grain sizes of 2–5 μ could be achieved with a $1450°C$, 16-hour cycle. This is $500°$-$700°C$ lower than normally employed with micron particle size powders, and difficulties are encountered when sintering zirconia to high density [30] because of the low cation diffusivities. The authors [31] have been concerned with the sintering and hot pressing behavior of this yttria-stabilized zirconia (Zyttrite) for the past year, thus the emphasis in the discussion will be on this material.

Consideration of equation (1) leads to the explanation for the decreased sintering cycle for Ångstrom-size powders. That is, for two powders, R_1 and R_2 the following relation should hold for a given fraction of initial stage shrinkage

$$\left(\frac{D_1}{R_1^3 \, T} \right)^m = \left(\frac{D_2}{R_2^3 \, T} \right)^m \tag{12}$$

where m is approximately 0.5. Thus, a factor of 10 to 100 in R will result in a large change in the required T and its corresponding D. This change will depend on the activation energy for sintering, and the lower the activation energy the lower the temperature for equivalent shrinkage.

One of the principal factors that must be considered when sintering a bulk sample from fine powders is the attainment of sufficient green density to (1) allow normal sintering to proceed, and (2) minimize shrinkage. Some aspects of pore-grain growth interactions have been discussed above; however, in this case we are concerned with attaining a high particle-to-particle coordination number to permit homogeneous shrinkage. Low green density, of course, results in high shrinkage and a danger of crack development due to setter drag frictional effects.

The effect of pressing pressure on green (unsintered) density is shown in Figure 5 and compared with data obtained by Bruch on various Al_2O_3 powders.

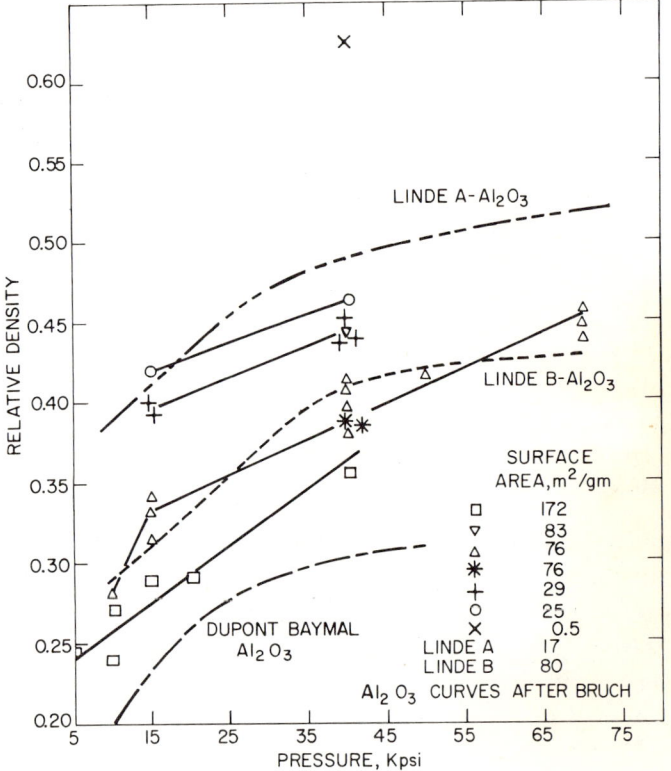

Figure 5. Influence of pressure on green density of yttria-stabilized zirconia powders of varying surface area, and comparison with various Al_2O_3 powders.

All of these pressings were accomplished on as-calcined powders. (It will be shown in a subsequent paragraph that the powder treatment can significantly affect the result.) The results for the two materials are quite similar, and at low pressures < 35 kpsi the green density at a given pressure is inversely proportional to particle size. Green density is also approximately directly proportional to pressure in this same interval. There is a vast amount of void space in this regime, thus particles are free to slide under increased pressure to a denser packing arrangement. The higher density resulting with increased particle size is what would be expected for the decreased frictional effects from larger particles and fewer total contact surfaces.

The effect of green density on the progression of normal sintering is illustrated for two temperatures in Figure 6. This discontinuous relation is thought

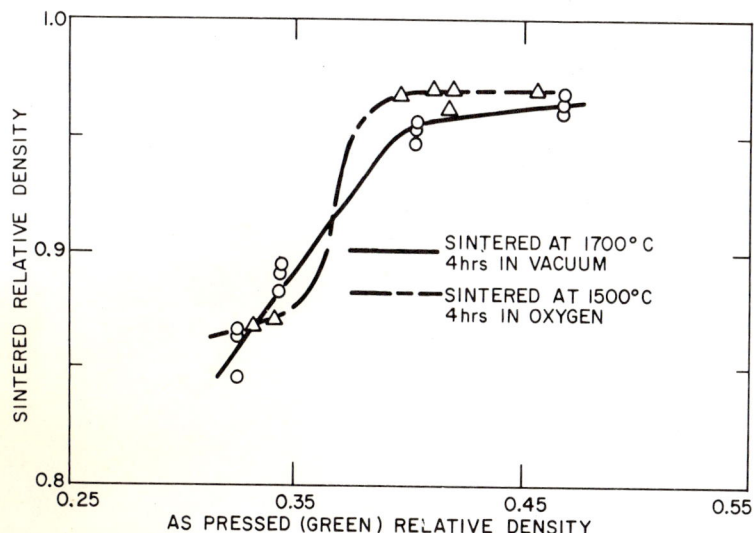

Figure 6. Relation of sintered density to green density for yttria-stabilized zirconia.

to be caused by the low coordination number for the packing of agglomerates in these compacts. Slight variations in density, or perhaps grain boundary sliding due to the relaxation of stresses generated in the initial shrinkage, cause fissures to open. An example of this in zirconia is shown in Figure 7. It will be noted that these fissures are surrounded by areas of near-theoretical density. Increased thermal cycles usually resulted in slight density increases, but maximum density in these cases was approximately 90 percent. Similar structures have been reported by Mansour and White [32] for UO_2, and by Livey et al. for BeO [33]. In the latter case, preheating at a temperature below that required for normal shrinkage alleviated much of the effect, presumably due to an improvement in the coordination and/or bonding (significant surface diffusion can occur prior to shrinkage) [34].

Bruch [9] noted the occurrence of subnormal sintering in his studies of the Lucalox composition $(Al_2O_3 + MgO)$. His structures did not exhibit fissures, but

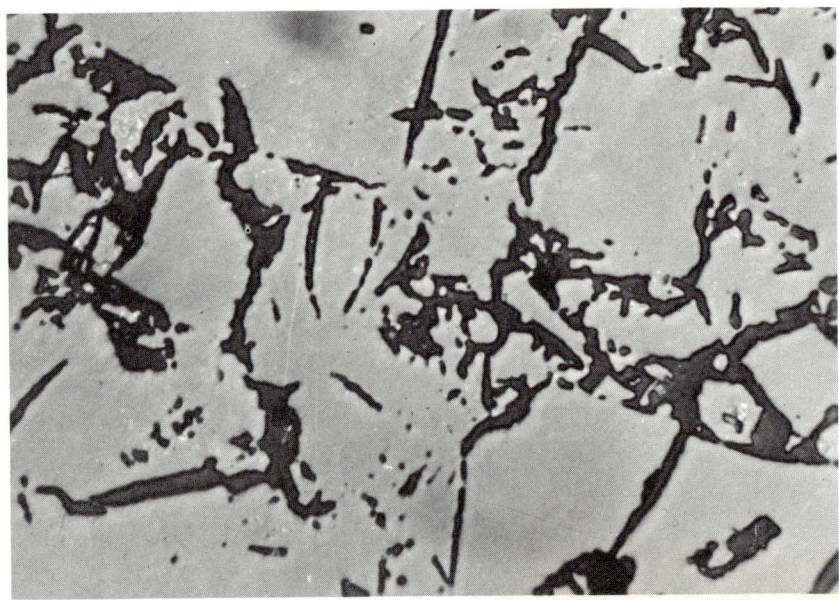

Figure 7. Evidence for pore growth in as-polished microstructure of Zyttrite, sintered at 1600°C for 940 minutes. Green density 29% and sintered density 84%. (Dark grey phase in plastic vacuum impregnated to fill pore structure.) 750X.

were characterized by large pores and slow densification kinetics. He attributed this to pore growth and coalescence. It was expected from surface energy considerations that pores would become less stable at higher temperatures, thus the green density required to achieve normal sintering would decrease with increasing temperature. This was observed by Bruch for alumina, and has been experimentally verified for yttria-stabilized zirconia, as shown in Figure 8. The main difference between the two systems in the subnormal range is the pore structure.

All processers of sub-micron powders give considerable attention to powder agglomerates. There are undoubtedly systems that are more sensitive to agglomerate structure than others. Systems employing chemical additions to control grain growth may exhibit a low sensitivity to agglomerate structure. It is the authors' belief that Ångstrom-size powders are more sensitive to agglomerate structure than 0.5 μ material, for example. It is important to obtain a uniform pore spacing and size distribution, and it is generally conceded that this can be accomplished either by reducing the existing agglomerate structure to the basic crystallite size or by forming agglomerates with uniform pore structure. Success in the latter technique would require that the agglomerates be packed to a high density and cold pressed to a packing geometry approaching that within the agglomerate itself. The results of several agglomeration and de-agglomeration treatments for zirconia are given in Table I.

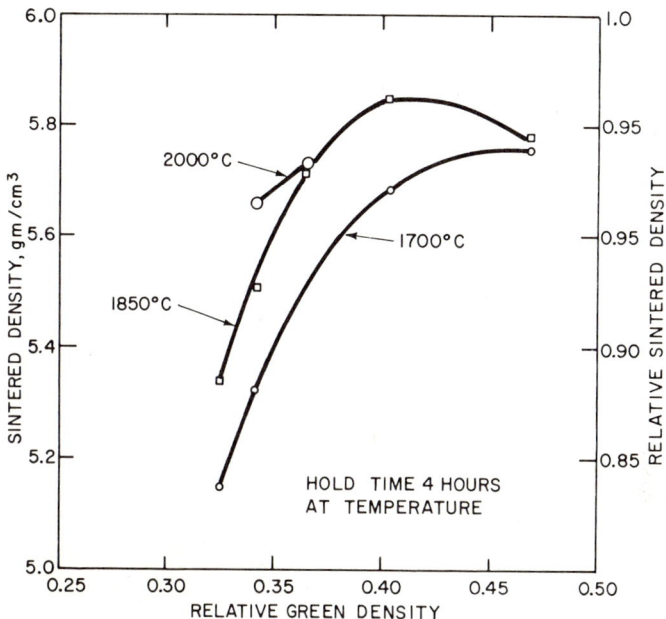

Figure 8. Sintered density as a function of green density at various temperatures for yttria-stabilized zirconia.

The most remarkable results were obtained by separating agglomerates from a slip so that only the basic crystallites remained. A disk was centrifugally cast from this remaining suspension. Upon drying, the green density was measured to be 72 percent of theoretical (close packing of uniform spheres results in a theoretical density of 74 percent). This type of sample sintered to 94.1 percent at 950°C, 99.5 percent at 1100°C and 99.9 percent at 1300°C. The highest tem-

TABLE I

Effect of Powder Treatments* for Various Lots of
Yttria-Stabilized Zirconia

Treatment	Increase in Relative Green Density,%	Increase in Relative Sintered Density,%
Dry Ball Mill	4.5	4.6
Fluid Energy Mill	2.0	0.8
Centrifugally Cast†	28.3	7.6
Slip Cast and Pressed	6.8	1.6
Mortar and Pestle	1.0	8.7

*Under comparable pressing pressure and sintering cycles.
†As-cast, no pressure–compared with 70 kpsi as-calcined powder.

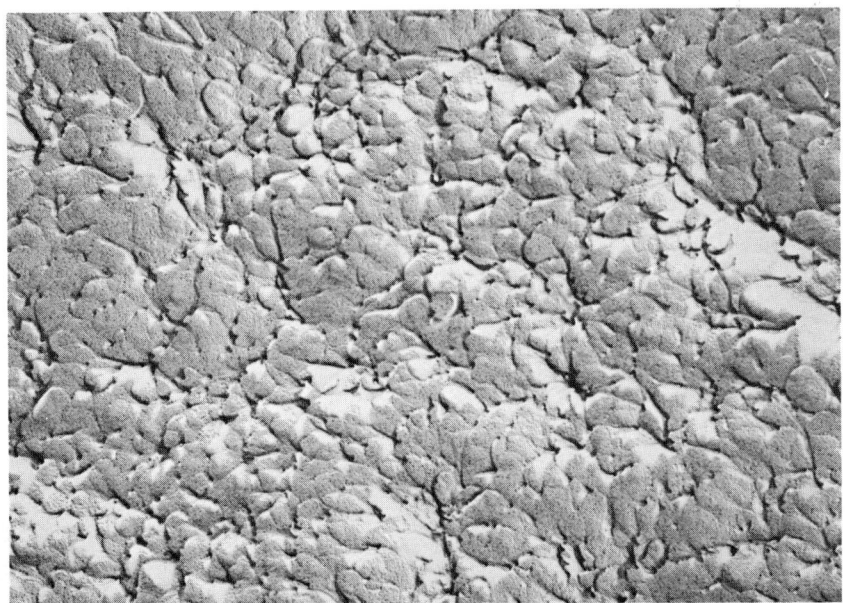

Figure 9.   Yttria-stabilized zirconia sintered at 1300°C to 99.9% theoretical density. 30,000X.

perature used is 200°C lower than that utilized for the sintering of as-calcined powder, which results in samples 95–99 percent dense. The microstructure of the 1100°C sample is illustrated in Figure 9 and has an average grain size of 0.2 μ, making it the finest grain high-density sintered body known to the authors.

Direct evidence for the influence of agglomerate structure on microstructure is shown in Figure 10. This particular lot of yttria-stabilized zirconia possessed 75 percent +400 mesh agglomerates (an unusually large percentage). The non-uniform pore distribution of Figure 10a is spaced at 100–150 μ intervals which would correspond to a mesh size of from 160 to 100, and indeed this sample was made from the as-calcined powder. However, when only –325 mesh agglomerates were used, the microstructure in Figure 10b was obtained and the large pore spacing was reduced to 20–60 μ, which corresponds closely with the 325 screen size. This effect is not as pronounced with the normal as-calcined agglomerate size distribution.

It is emphasized that for the achievement of the full potential from fine particles, as indicated by equations (1) and (12), optimum conditions must be realized in the preparation of the green compact; that is, high green density and uniform pore size distribution. This condition has perhaps been achieved by many other investigators, but the most recent example that has come to the authors' attention is the sintering of Lucalox by R. Coble [34] to 99[+] percent

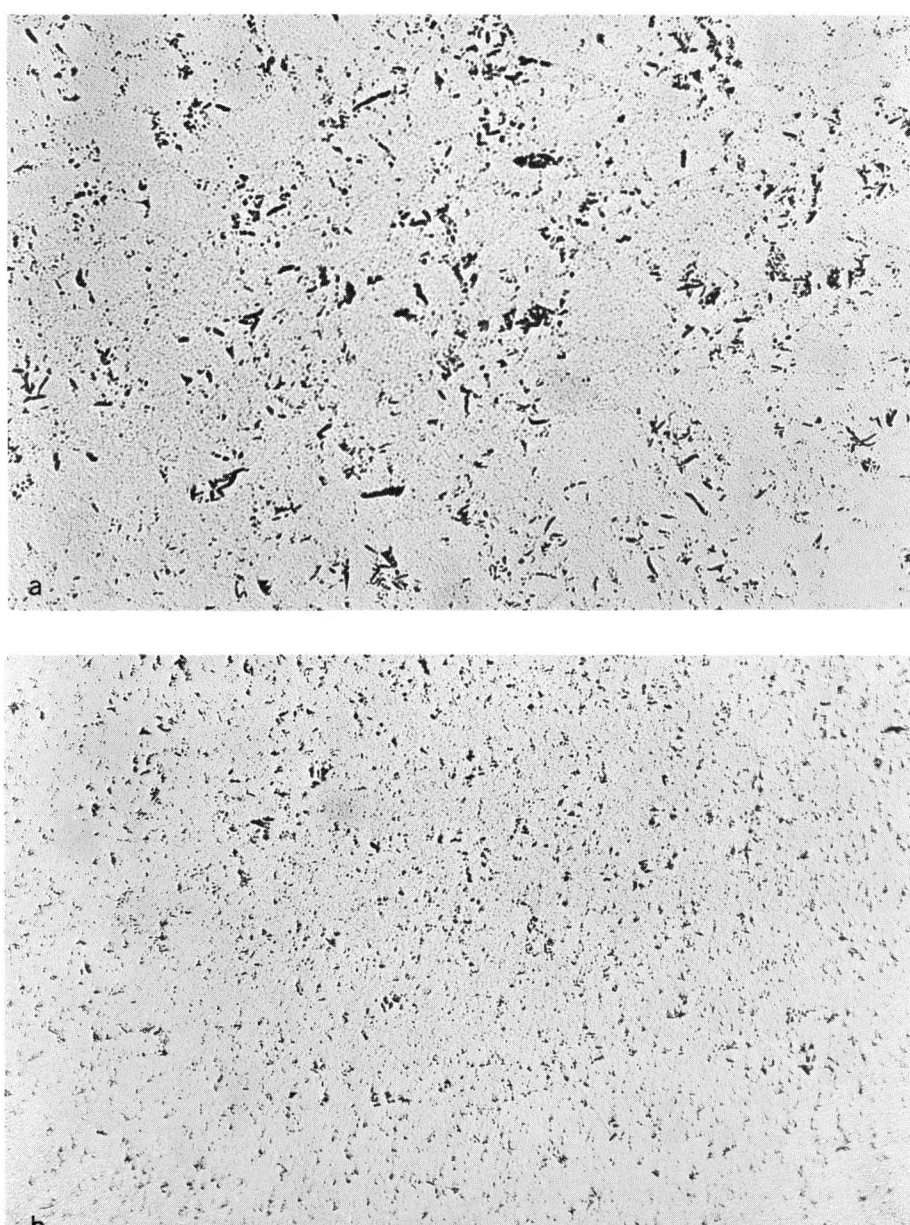

Figure 10. Pore distribution in pressing made from (a) as-received zirconia powder, and
(b) screened powder with agglomerate less than $^-325$ Mesh. Both as-polished. 100X.

dense, with a final grain size of approximately 0.5 μ, at a temperature several hundred degrees lower than normally employed. In this case, the development of a uniform agglomerate structure was the successful approach.

· Hot Pressing

· General

The hot pressing of a number of oxide and non-oxide compositions has been given particular attention in recent years with respect to achieving dense fine-grain microstructure. The process, with its relatively short time cycle or lower equivalent densification temperature, has been used to fabricate fine-grain micro-structures (with or without the use of additives, etc.) to control or restrict grain growth—as is normally the case with pressureless sintering.

As in the case of sintering, particulate characteristics play an important role in determining microstructure development, and the parameters of pressure, tem-perature, and time are selected and modified in accordance with densification behavior of the particulates under study.

There is a diversity of opinion with respect to the mechanisms by which densi-fication proceeds during hot pressing of particulates.

Some investigators have supported plastic flow models [35], while others have cited particle rearrangement by grain boundary sliding and fragmentation, as im-portant densification mechanisms—particularly during early compaction stages and at very high pressures with hard materials such as carbides [36] and borides [37]. Recent investigations of hot pressing have recognized the probable occur-rence of more than one mechanism, with considerable attention given to the sig-nificance of diffusion. Further, attempts have been made explicitly to derive diffusion models. Vasilos and Spriggs [38], for example, concluded that the process of densification during conventional graphite-die hot pressing of most ceramics (i.e., for nominal pressures of 1,000 to about 10,000 psi and at temper-atures up to 2500°C) was essentially diffusion-controlled. Rossi and Fulrath [39] have also demonstrated that the densification of alumina by hot pressing pro-ceeds by a diffusional process. A densification model employed by these inves-tigators to account for the densification mechanism is a modification of the Nabarro–Herring creep expression. The strain rate, $\dot{\epsilon}$, in the relation is approxi-mated by use of isothermal rate of change of porosity or linear shrinkage rate, e.g.,

$$\dot{\epsilon} \approx \frac{13.3\, D\, \sigma_e\, \Omega_o}{kT\, (GS)^2} \approx -\frac{dL}{Ldt} \qquad (13)$$

where: σ_e = the effective stress

$\dot{\epsilon}$ = strain rate $(\sec^{-1})$

$\dfrac{dL}{Ldt}$ = linear shrinkage rate

The observed consistency in the calculated diffusion coefficient values for hot pressing, as well as the reasonable comparison between activation energy values, has suggested that for the low to moderate pressures employed, densification beyond the initial stages is a diffusion-controlled process.

Coble [40] recognized that hot pressing by a diffusional process should also incorporate a term due to the vacancy gradient present from surface energy and pore curvature effects, as in normal sintering. For the final stage of pore shrinkage the deviation in vacancy concentration is

$$\Delta C = \frac{2\,Co\,\gamma\,\Omega}{kTr_p} \qquad (14)$$

where r_p is the pore radius. Combined into a flux equation for the effect of pressure, one obtains

$$J = \frac{4\,\pi\,DCo\,\Omega}{kT}\left(\frac{2\gamma}{r_p} + \sigma_e\right) \qquad (15)$$

and an effective creep rate of

$$\dot{\epsilon} = \frac{13.3\,D\,\Omega}{kT(GS)^2}\left(\frac{2\gamma}{r_p} + \sigma_e\right) \qquad (16)$$

This equation has not been tested explicitly; however, if the data of Vasilos and Spriggs [38] is corrected for the driving force of porosity, calculated diffusion coefficients for Al_2O_3 hot pressing agree with direct aluminum ion diffusion measurements within a factor of two.

The work of Coble and Ellis [41] on neck growth between single crystal spheres of alumina under stress is consistent with the conclusion that diffusion is a major transport mechanism in hot pressing. They found that the contribution of plastic flow to alumina densification at the pressures normally used in hot pressing (e.g., less than 10,000 psi) was small. They concluded that the final stage of densification of alumina during normal hot pressing occurs by enhanced diffusion under the influence of stress. The contribution of plastic flow at higher pressures (over 10,000 psi) and with "softer" materials (such as MgO and Ni) was not specifically evaluated by Coble and Ellis, but it was recognized that for materials with higher symmetry and multiple primary slip planes, the contribution of plastic flow should be greater. It was suggested that the creep behavior of given materials should serve to elucidate probable mechanisms; those which undergo creep by plastic flow should also undergo final densification by a diffusion mechanism. Although creep data of this type at the stresses and grain sizes of the studies referenced are not yet available, it is of interest to note that creep measurements of fine-grain $(1-3\ \mu)$ dense (98–99.5 percent), pure magnesia and alumina suggest a diffusional mechanism at lower stress levels (up to 4,000 psi) for both materials.

Based on the available information, it is believed that the principal mechanism by which densification of ceramics proceeds during hot pressing at pressures up to about 10,000 psi is essentially pressure-directed diffusion with a number of other operative and perhaps overlapping mechanisms, as outlined below:

1. The first step is similar to one of cold compaction where the degree of densification is proportional to log applied pressure and covers a densification range up to about 70 percent of theoretical, depending on particle characteristics. Fragmentation as a coincident step has not been found to be significant in the case of sub-micron size particles.

2. The second step is probably one of sliding and rearrangement wherein the degree of densification is further increased and is proportional to the ratio of applied pressure to the powder particle size.

3. A third step involving plastic flow at grain contacts occurs rapidly (within minutes) and may occur in combination or at the same time as step 2, above. However, the possible contribution of plastic flow to densification is apparently limited to less than 84 percent of theoretical, and is dependent on the ratio of applied stress in the die to the stress at particle contact points, i.e.,

$$Ps = \frac{Pso}{(1 - 2\sigma_A/N\sigma_\rho)^3} \tag{17}$$

where:

$$\sigma_A = \text{stress applied to die}$$
$$\sigma_\rho = \text{contact stress between particles}$$
$$N = \text{coordination number of particles}$$
$$\text{(approximately 12 for ideal packing)}$$
$$Pso = \text{initial packed density}$$
$$Ps = \text{final density}$$

4. The final stage of densification (up to theoretical) occurs by diffusion under the influence of stress, the densification rate being greater than the pressureless sintering case by a factor essentially proportional to $L/R\gamma$, where L is applied load, R is grain radius, and γ is surface energy; see equation (16).

Ultimate densification is limited by pore entrapment. Over-all densification mechanisms at the higher pressures are not clearly established, although it is generally conceded that the plastic deformation characteristics of the higher symmetry materials are important. Further effort is needed to describe the densification kinetics in these areas.

Process Characteristics

The reader is referred to various available reviews [42, 43] on the subject, and a general description of process is described below.

In hot pressing, the pressure can be applied uniaxially, biaxially or isostatically. With the exception of recent hot isostatic pressing and ultrahigh-pressure work, uniaxial hot pressing represents the "standard" technique and the one about which the most information is available.

While there are no sharp distinctions in the pressure ranges employed, it has been possible to identify arbitrary separations for various mold materials employed, as seen in Table II.

TABLE II

Ranges in Hot Pressing Fabrication

Pressure Range and Application	Mold Materials	Size Limits for Fabricated Parts
1,000 to 5,000 psi uniaxial	Graphites	Up to 2 ft. in diameter (smaller sizes as pressures increase above 2,000 psi)
5,000 to 20,000 psi uniaxial	Special graphites	
10,000 to 50,000 psi uniaxial	Ceramics (e.g., Al_2O_3, SiC, TiB_2, ZrB_2)	Less than $3''$ in diameter
10,000 to 50,000 psi isostatic	Gas pressure contained in steel	Increased lengths (up to $20''$) with diameters $< 3''$
50,000 to 300,000 psi uniaxial	High-strength steel and cemented tungsten carbide at the higher temperatures	Generally $1''$ in diameter tending to $3''$ as limit
300,000 to 750,000 psi	Cemented tungsten carbide	$\frac{1}{4}''$ to $\frac{1}{2}''$ in diameter

The available configurations within these listed ranges provides decreasing specimen size capacity as the operating pressure increases. The size limits shown in each range are generally intended as a guide, and actual diametral limits are dependent on specimen thickness, densification temperature, and desired microstructure.

The relatively low strength levels for conventional graphite die parts have been known to limit the minimum attainable grain size at full density for particular ceramic raw materials; however, the recent availability of high-strength graphite and filament-wound graphite has made it possible to attain a level of 10,000–15,000 psi in a laboratory scale.

As a result, increasing attention has been given to the use of dies and plungers constructed from ceramic materials such as alumina, silicon carbide, and refractory borides.

Microstructure Development

The kind of microstructure that is obtained for a particular composition under specific processing conditions can be significantly altered as a result of variations in starting particle characteristics. In this connection, in hot pressing as in sintering, developed microstructure in bodies is known to depend not only on particulate size and distribution and impurity chemistry distribution, but also on the character of agglomerates. Even in hot pressing, segregation in agglomerate size can lead to a gradient in densification and resulting microstructure.

Oxide Compositions

A number of oxides are currently prepared in fine-grain form by hot pressing routinely at less than 10,000 psi. In particular, considerable effort over the past few years has been expended on fine microstructure development in alumina, magnesia, and other oxide materials in the ~ 99.9 percent pure (0.1 percent cation impurity) and > 99.9 percent pure single-phase composition ranges. Some variations in microstructure have been observed as a function of impurity levels and other particulate characteristics, and these are described below.

Alumina: Various grades of fine-particle pure alumina are currently available today, and considerable effort has been expended in fabricating fine-grain bodies from them in recent years.

Much of the earlier effort has been concerned with alumina ~ 99.9 percent pure (~ 0.1 percent cation impurity). Linde A alumina (average particle size, 0.3μ) is representative of such a composition class, and despite some difficulties in impurity segregation, duplex phase, and particle structure, one can obtain fairly uniform, dense, fine-grain alumina microstructure via hot pressing below $1500°C$ for short periods of time. By way of example, Figure 11 shows an electron-microphotograph of a dense alumina sample so prepared. The grain size distribution is generally $1-2 \mu$, with some grains above the average. The reproducibility of such microstructures is made difficult by raw material lot variability. Duplex grain structure in densified bodies is not uncommon and has been attributed to possible impurity segregation as well as significant agglomerate structure variation.

As in the case of sintering, the use of additives has also been found beneficial in damping out effects of raw material lot variation which lead to exaggerated grain growth. Magnesia has been found to be a useful additive here, and Figure 12 shows an electron-micrograph of hot pressed Al_2O_3 containing MgO additive, revealing dense uniform grain structure (average, 0.9μ). In this instance, the additive appears to have interdiffused in the alumina structure and does not exist as a discrete phase in the grain boundaries. In other cases, i.e., where Cr_2O_3 and V_2O_3 have been used as additives to alumina, these compounds

Figure 11. Electron-micrograph of dense Al_2O_3, 1–2 μ grain size. 17,000X.

Figure 12. Electron-micrograph of hot pressed Al_2O_3 with additive showing dense uniform grain structure, 0.9 μ. 15,000X.

Figure 13. Electron-micrograph of chrome addition to Al_2O_3. 30,000X.

are seen as boundary phases, as shown by the micrographs in Figures 13 and 14. The interdiffusion rates for these compounds with Al_2O_3 are sufficiently slow so that during the hot pressing cycle only partial diffusion occurred. In all cases, the grain growth rates have been effectively retarded by the additives chosen.

Various high-purity (> 99.9 percent pure) grades of fine particle size alumina are now currently available, albeit in limited lot size, and a number of these are undergoing study in terms of composition effects on resulting microstructure and densification behavior [44].

Figure 15 shows electron micrographs of a number of such high-purity alumina powders. A qualitative description of the characteristics of the powders is as follows:

1. *Linde Laser Grade*—This powder has a very fine particle size and a uniform distribution ranging 0.02–0.04 μ in diameter of γ Al_2O_3.

2. *Cominco HPM*—Particles of three groupings were found; 0.05 μ, a range of particles 0.05–0.1 μ, and large flakes ranging from 1–7 μ.

3. *Johnson–Matthey*—Very fine particles are predominantly gamma Al_2O_3 while smooth-sided particles are alpha Al_2O_3. Quantitative X-ray analysis has shown this powder to contain 58 w/o α alumina with the remainder being transitional phases, predominantly γ Al_2O_3.

Hot pressed samples made from either the Johnson–Matthey or Linde Laser

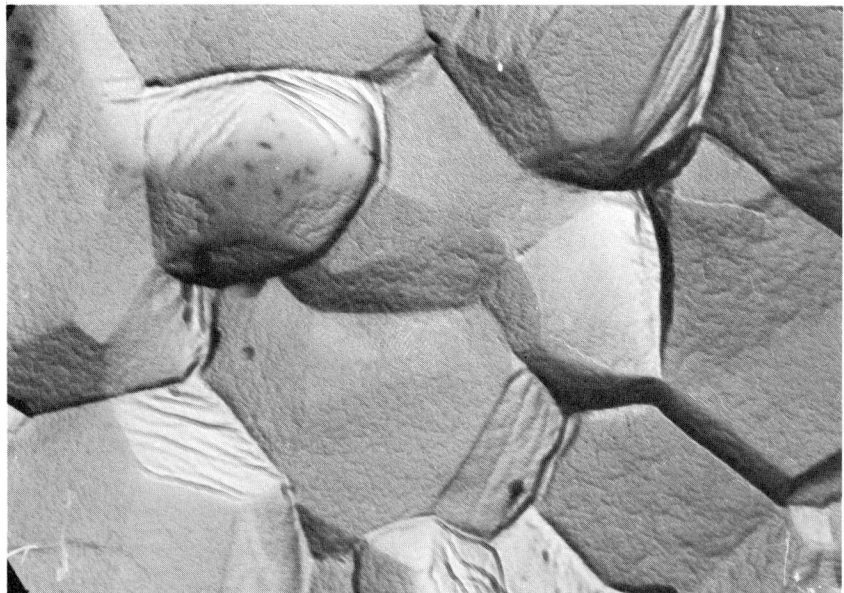

Figure 14. Electron-micrograph of vanadia addition to Al_2O_3. 30,000X.

Grade aluminas have generally exhibited uniform grain structure and relatively high density under normal Al_2O_3 hot pressing conditions, despite γ-phase transition and duplex particle structures. Figure 16 shows an electron micrograph of hot pressed Johnson–Matthey material to illustrate such a grain structure.

On the other hand, the Cominco powder, when hot pressed, shows an interesting variation in developed grain structure, as shown in Figure 17. This structure shows one population of grains at 1 μ and a second population at 9 μ average diameter. This range is far outside the normal grain size variation and is thought to be related to the starting powder. As shown in Figure 15, the Cominco powder had a wide variation in particle size, and it is believed that the micron and greater size particles act as centers for secondary grain growth at the expense of the smaller population and actually halt their rapid growth only when they meet a neighboring grain growing by the same mechanism. Because only a limited number of nuclei for secondary growth are available, regions of small grains remain in the final structure illustrated.

Besides variability in grain structure, hot pressed samples of such starting material were also found to exhibit porosity nests in an otherwise translucent-appearing body. A 10 X macrophoto of such a sample (1/8″ thick), taken in transmitted light, is shown in Figure 18 and reveals a significant area distribution of porosity—showing up as dark regions in the photo. These observed "nests" of

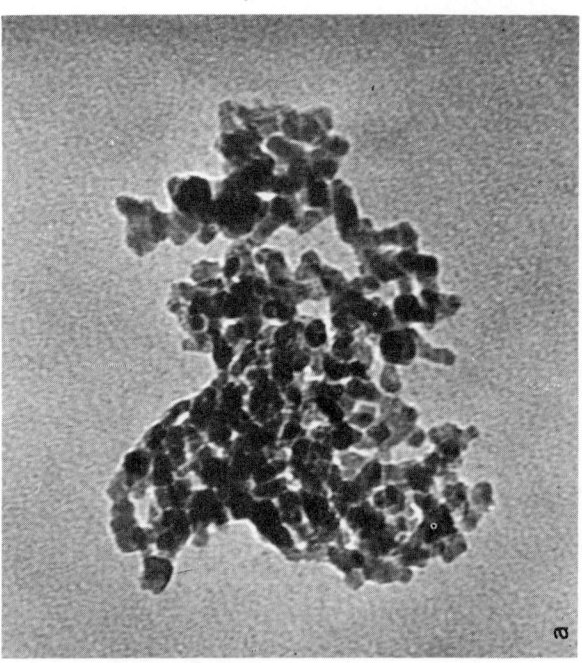

Figure 15. Electron-micrographs of alumina powders: (a) Linde Laser Grade, 320,000X; (b) Johnson-Matthey, 30,000X; and (c) Cominco, 30,000X.

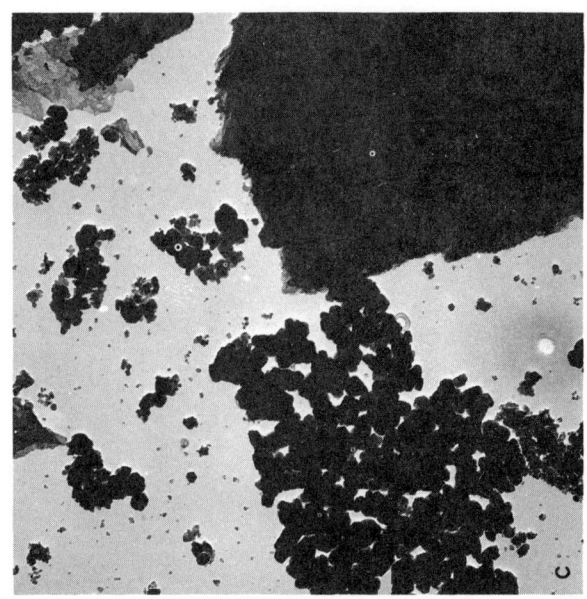

Figure 16. Electron-micrograph of Al_2O_3 made from Johnson-Matthey powder. 1,500X.

Figure 17. Electron-micrograph of Al_2O_3 made from Cominco powder, exhibiting duplex structure. 2,000X.

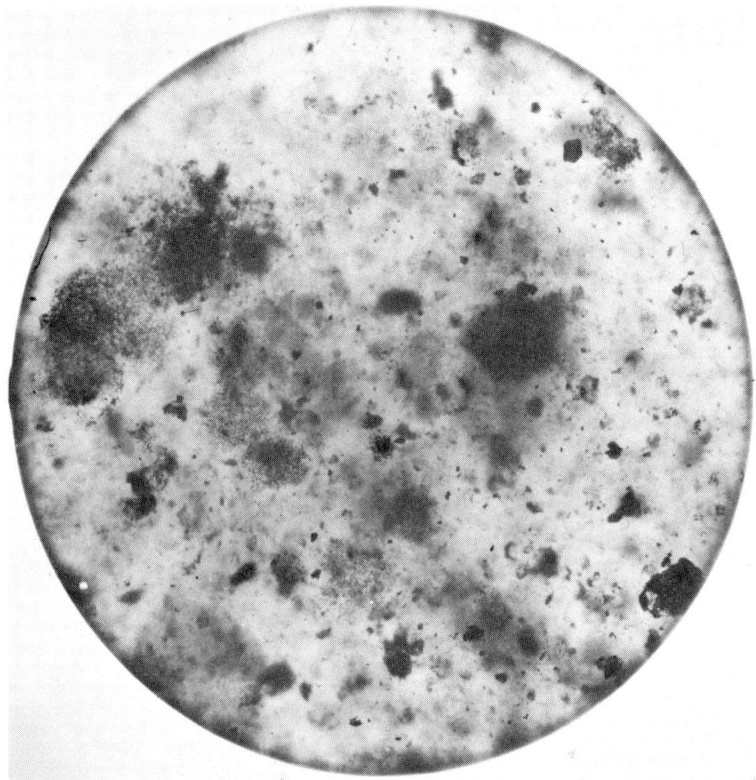

Figure 18. Light transmission through dense Al_2O_3 showing porosity nests due to agglomerate structure. 10X.

porosity are thought to be associated with the original segregated agglomerate structure of the raw material. Such nesting has also been observed in the Johnson–Matthey powder.

Magnesia: This material is also available for a number of sources as $< 0.1\ \mu$ and ~ 99.5 percent pure compound, and at this purity level has been the subject of a number of fabrication and property evaluation programs [38].

Full-density samples with grain sizes ranging from $0.1\ \mu$ to several microns (as shown in Figures 19, 20, and 21) have been prepared by varying hot pressing pressures from as high as 250,000 psi to as low as 4,000 psi, and respective temperatures from as low as 800°C to as high as 1550°C.

A comparison of densification kinetics between such a starting material (99.5 percent pure) and > 99.9 percent pure MgO reveals that somewhat higher temperatures were required for the purer material to attain full density despite its finer particulate size $< 0.1\ \mu$ [44]. Figure 22 shows densification plots for

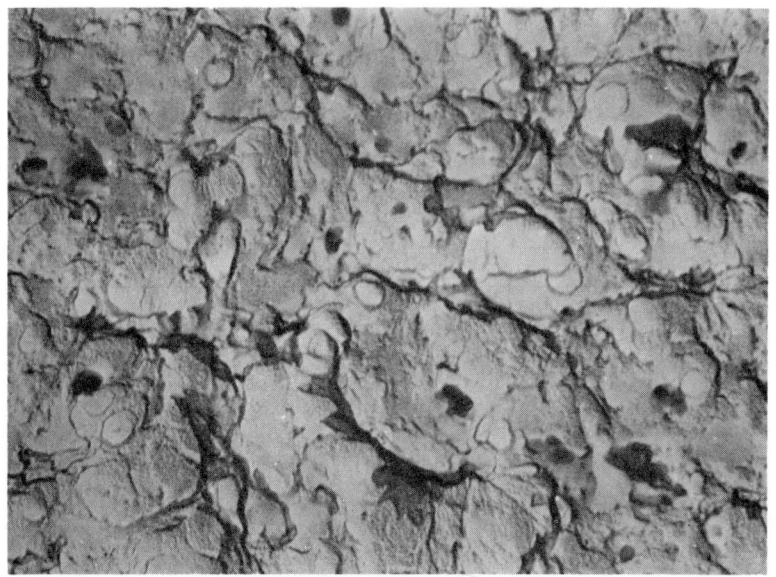

Figure 19. Electron-micrograph of MgO, hot pressed at 250,000 psi at 920°C. 28,000X.

Figure 20. Electron-micrograph of MgO, hot pressed at 10,000 psi at 1120°C. 28,000X.

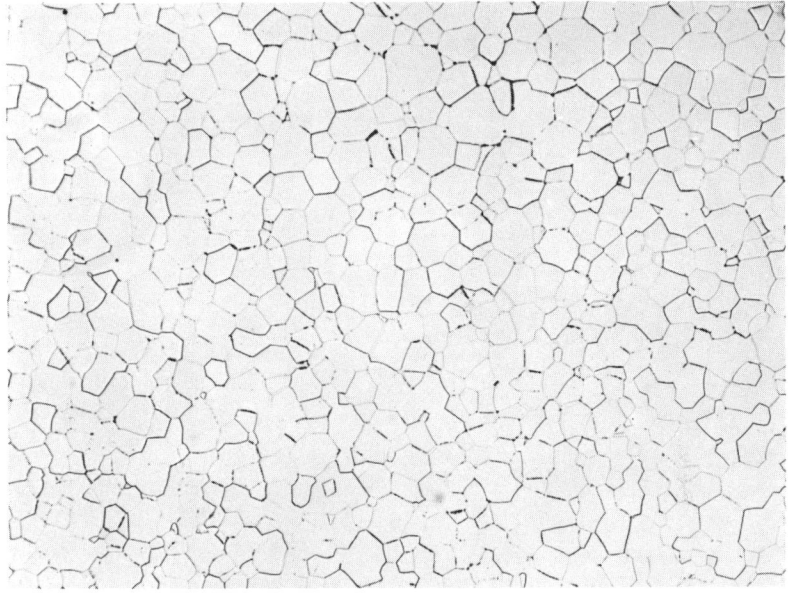

Figure 21. Optical micrograph of MgO, hot pressed at 4,000 psi at 1550°C. 250X.

the two powders pressed at equal pressures but at a temperature 250°C lower in the case of the less pure powder. In spite of the temperature disadvantage, the less pure MgO has reached full density, whereas the pure MgO sample had not reached theoretical density after 100 minutes. This result cannot be explained by a particle–size relation as they are both in the 200–400 Å range. It is thought, rather, that this is related to the impurity differences between the materials.

Figure 23 shows an electron micrograph of a dense sample of the high-purity MgO, possessing a uniform 9 μ grain size (fabricated at 1200°C and 10 ksi for 70 minutes). Finer sizes are attainable by reducing dwell time. Grain size data gathered from a number of MgO billets hot pressed in the 1400°–1500°C range reveal that as long as some porosity remains, the high-purity MgO experiences approximately equivalent grain growth to the 99.4 percent pure MgO. However, when pores are eliminated, the high-purity MgO grows at a rate that appears to be several times faster than less pure material. Spriggs *et al.* [45] noted that grain growth for full-density 99.4 percent MgO was faster than was reported in earlier studies on 3 percent porous material, and this was attributed to the pinning of grain boundaries by pores. Leipold [46] similarly has observed "excessively" high grain growth rates in high-purity hot pressed MgO.

Other Oxides: A number of other refractory oxides with dense fine-grain

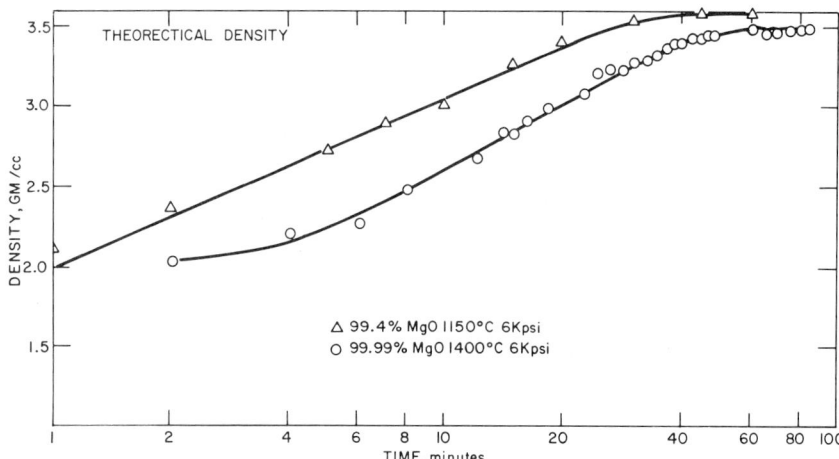

Figure 22. Densification curves for high- and low-purity MgO.

microstructures have been prepared by hot pressing. These include magnesium aluminate–spinel, $MgAl_2O_4$ [47]; nickel oxide, NiO; chromium oxide, Cr_2O_3; zinc oxide, ZnO [48]; yttrium oxide, Y_2O_3, etc.; as well as a number of ferromagnetic and ferroelectric oxide compositions. Where the compositions are sus-

Figure 23. Electron-micrograph of dense high-purity MgO. 1,500X.

ceptible to reduction, oxide dies in an oxidizing environment have been employed for densification purposes. This is particularly true for such compounds as NiO, ZnO, Fe_2O_3, etc.

In addition, the microstructure that develops is a strong function of the character of the particulates comprising the starting material composition. As in the cases mentioned earlier for MgO and Al_2O_3, fine crystallite size starting materials (i.e., $1\,\mu$) are required in order to realize preparation of dense, fine-grain bodies. The availability or lack of such fine-size starting materials is often dominant in determining whether particular oxides can be prepared with dense fine-grain microstructure.

Non-Oxide Compositions

A number of carbide, nitride, and boride compounds are hot pressed to high density at pressures less than 10,000 psi (temperatures approaching 2000°C are usually required for these compounds, which dictates the requirement for graphite-die components). Very few of these compounds are available in sub-micron powder, thus the starting particulate is usually in the $1\text{--}5\,\mu$ range. By careful control of temperature and time, grain sizes in the $3\text{--}5\,\mu$ range have been fabricated in 98^+ percent dense SiC and B_4C. However, a grain size of from

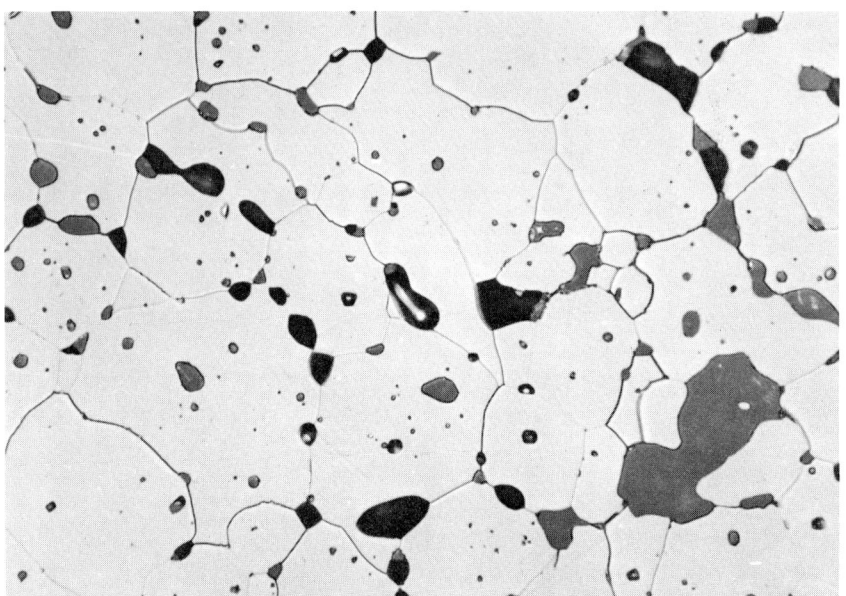

Figure 24. ZrB_2 with ZrO_2 impurity phase. 500X.

5-20 μ is more common, especially when the last 2 percent of porosity is removed.

Research on the refractory diborides illustrates the effectiveness of discrete second-phase additions as grain growth inhibitors in hot pressing [20]. Commercially available ZrB_2 typically possesses a 15-30 μ grain size when hot pressed to full density, as illustrated in Figure 24. A second impurity phase (ZrO_2) provides some grain boundary pinning, as shown by the large percentage of second-phase particles on the boundaries. However, when the grain size was small and a high driving force was present the boundaries effectively swept past the impurity phase, as demonstrated by the presence of many ZrO_2 particles within grains. The addition of SiC accomplishes two significant structural modifications, as shown in Figure 25. Grain growth is retarded so that a typical final grain size of

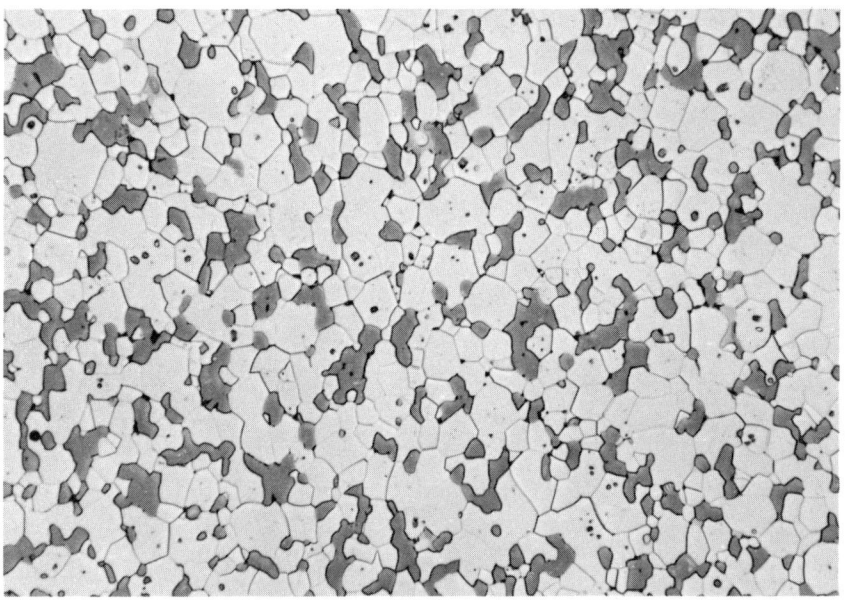

Figure 25. ZrB_2 with SiC addition to limit grain growth and refine impurities. 500X.

8 μ is achieved, and a chemical reaction reduces the ZrO_2 content to approximately 1 volume percent. The addition of C plus SiC results in further grain refinement, as illustrated in the fully dense 4 μ grain size structure shown in Figure 26. As well as enhancing strength due to the grain size refinement, these additions have proved beneficial to other physical properties such as oxidation resistance.

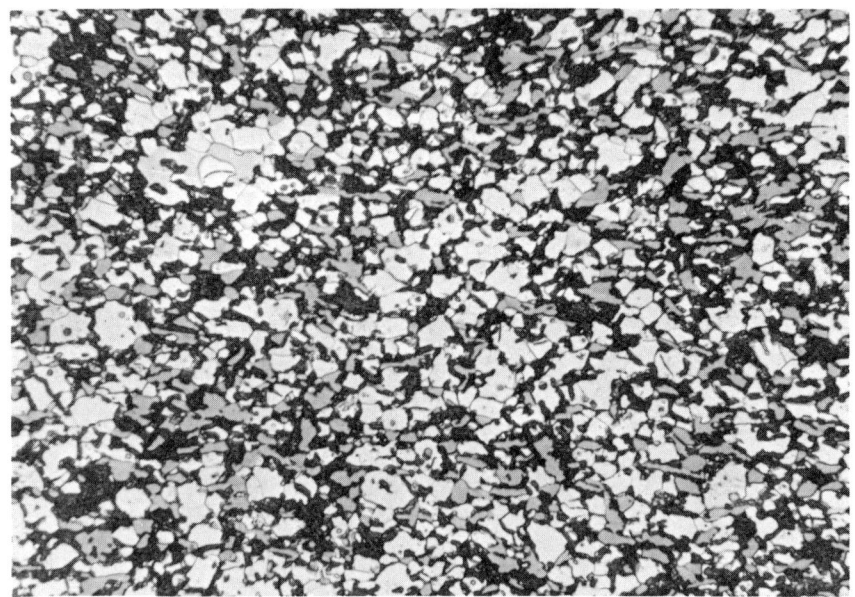

Figure 26. ZrB_2 + SiC + C addition which limits grain growth to $< 5 \mu$. 500X.

Conclusions

The discussion has shown that by utilizing one of four grain growth control mechanisms, sub-5 μ grain size bodies may be fabricated by sintering. In most cases where such a grain size has been obtained, several percent or more porosity remain. However, several notable exceptions are the sintering of Al_2O_3 and Y_2O_3-ZrO_2 to 99^+ percent dense sub-5 μ grain size body.

There are many examples of 99^+ percent dense fine-grain hot pressed oxide structures. The discussion showed that particulate and agglomerate structure and size were extremely important in developing a uniform microstructure. Several examples of fine-grain non-oxide hot pressed material were discussed, but examples are limited by the non-availability of sub-micron size powders. Grain growth control by discrete particles is particularly effective for these materials.

Acknowledgment

The authors would like to thank their many colleagues who contributed to the research summarized in this paper. They also gratefully acknowledge the support of the U.S. Army, the U.S. Navy, and the U.S. Air Force on completed programs which contributed to this paper. Specifically, they call attention to three current programs which were abstracted extensively; the U. S. Air Force programs, "Ultra High Purity Fine Particulate Stabilized Zirconia," F33615–67–C–1963, and "Research and Development of Refractory Oxidation Resistant Diborides," AF 33(615)–3671 (Subcontract from ManLabs, Inc.), and the U.S. Navy-sponsored program, "Microstructure Studies of Polycrystalline Refractory Oxides," N000 19–68–C–0108.

References

1. Coble R. L. and Burke, J. E. "Sintering in Ceramics," *Prog. Cer. Sci.,* Vol. 3, Macmillan Co., New York (1963), 197.
2. Coble, R. L., "Sintering Crystalline Solids. I. Intermediate and Final State Diffusion Models," *J. Appl. Phys.,* 32 (1961), 787. See also: Coble, R. L., "Sintering Crystalline Solids. II. Experimental Tests of Diffusion Models in Powder Compacts," *ibid.,* 793.
3. Coble R. L. and Gupta, T. K., "Intermediate Stage Sintering," in *Sintering and Related Phenomena,* G. C. Kuczynski *et al.,* eds., Gordon and Breach, New York (1967), 423.
4. Jorgensen, P. J., "Modification of Sintering Kinetics by Solute Segregation in Al_2O_3," *J. Am. Ceram. Soc.,* 48 (1965), 207.
5. Bannister, M. J., "Interdependence of Pore Removal and Grain Growth During Later Stages of Sintering in Beryllium Oxide," in *Sintering and Related Phenomena,* G. C. Kuczynski *et al.,* eds., Gordon and Breach, New York (1967), 581.
6. Gupta, T. K., "Sintering of Zinc Oxide," Sc.D. Thesis, Massachusetts Institute of Technology (December 1966).
7. Burke, J. E., "Role of Grain Boundaries in Sintering," *J. Am. Ceram. Soc.,* 40 (1957), 80–85.
8. Clare, T. E., "Sintering Kinetic of Beryllia," *J. Am. Ceram. Soc.,* 49 (1966), 159.
9. Bruch, C. A., "Sintering Kinetics for the High Density Alumina Process," *Bull. Am. Ceram. Soc.,* 41 (1962), 799.
10. Kingery W. D. and Francois, B., "Sintering of Crystalline Oxides. I. Interactions Between Grain Boundaries and Pores, and II. Densification and Microstructure Development in UO_2," in *Sintering and Related Phenomena,* G. C. Kuczynski *et al.,* eds., Gordon and Breach, New York (1967), 471.
11. Charles R. J. and Shaw, R. R., "Delayed Failure of Polycrystalline and Single Crystal Alumina," General Electric Co. Report No. 62–RL–3081M.
12. Harrisville Tool Co., distributors for Feldmuhle, Germany.
13. Ueltz, H. F. G., Abrasive Grain, U.S. Patent 3,079,243.
14. Coble, R. L., Transparent Alumina and Method of Preparation, U.S. Patent 3,026,210.
15. Wolkodoff, V. W. and Weaver, R. E., Alumina Ceramic, U.S. Patent 3,377,176.
16. Elyard, C. A., Gibbs, B. M. and Rawson, H., "Effect on Uranium Oxide of Controlling Grain Growth during Sintering," *Trans. Brit. Ceram. Soc.,* 62 (1963), 127.
17. Duderstadt, E. C. and White, J. E., "Sintering BeO to Variable Densities and Grain Sizes," *Bull. Am. Ceram. Soc.,* 44 (1965), 907.
18. Langrod, K., "Graphite as Grain Growth Inhibitor in Hot-Pressed Beryllium Oxide," *J. Am. Ceram. Soc.,* 48 (1965), 110.
19. Hill, N. A., O'Neil, J. S. and Livey, D. T., AERE R5056 (1966).
20. Clougherty, E., Kalish, D. and Peters, E. T., "Research and Development of Oxidation Resistant Diborides," AFML–TR–68–190.

21. Hillert, M., "On the Theory of Normal and Abnormal Grain Growth," *Acta Met.*, 13 (1965), 227.
22. Woolfrey, J. L., "Feasibility of Dispersed Phase Grain Refinement in Ceramics," AAEC/ E170 (February 1967).
23. Jorgensen P. J. and Westbrook, J. H., "Role of Solute Segregation at Grain Boundaries During Final Stage Sintering of Al_2O_3," *J. Am. Ceram. Soc.*, 47 (1964), 332.
24. Paladino, A. E., Private communication.
25. Jorgensen P. J. and Anderson, R. C., "Grain Boundary Segregation and Final Stage Sintering of Y_2O_3," *J. Am. Ceram. Soc.*, 50 (1967). 553-58.
26. Ferguson, D. E., Dean O. C. and Douglas, D. A., "The Sol-Gel Process for the Remote Preparation and Fabrication of Recycle Fuels," presented at Third International Geneva Conference on the Peaceful Uses of Atomic Energy (September 1964).
27. Morgan C. S. and Yust, C. S., "Material Transport During Sintering of Materials with Fluorite Structure," *J. Nucl. Mat.*, 10 (1963), 182. See also: Morgan, C. S., McHargue, C. J. and Yust, C. S., "Material Transport in Sintering," *Proc. Brit. Ceram. Soc.*, 3 (1965), 177.
28. Morgan C. S. and Hall, L. L., "The Creep of ThO_2 and ThO_2–CaO Solid Solutions," *Proc. Brit. Ceram. Soc.*, 3 (1965), 177.
29. Mazdiyasni, K. S., Lynch C. T. and Smith, J. S., "Cubic Phase Stabilization of Translucent Yttria–Zirconia at Very Low Temperatures," *J. Am. Ceram. Soc.*, 50 (1967), 532.
30. Rhodes, W. H., "Sintering Characteristics of Stabilized Zirconium Oxide," General Electric Co. Report No. 61-RL-2703M.
31. Rhodes W. H. and Haag, R. M., "Ultra-High Purity Fine Particulate Stabilized Zirconia," Contract F 33615-67-C-1693, Progresss Reports 1-4 (1967-68).
32. Mansour N. A. and White, J., *Powder Metallurgy*, 12 (1963), 108.
33. Livey, D. T., Brett, N. H., Denton I. and Murray, P., AERE Report M/R, 2794 (1959).
34. Coble, R. L., Private communications.
35. Murray, P., Rodgers, E. P. and Williams, A. E., "Practical and Theoretical Aspects of Hot Pressing of Refractory Oxides," *Trans. Brit. Ceram. Soc.*, 53 (1954), 471-510. See also: Jackson J. S. and Palmer, P. F., "Hot Pressing Refractory Hard Materials," Chap. 18 in *Special Ceramics*, P. Popper, ed., Heywood, London (1960), 305-28; and McClelland, J. D., "A Plastic Flow Model of Hot Pressing," *J. Am. Ceram. Soc.*, 44 (1961), 526.
36. Chang R. and Rhodes, C. G., "High Pressure Hot Pressing of Uranium Carbide Powders and Mechanism of Sintering of Refractory Bodies," *J. Am. Ceram. Soc.*, 45 (1962), 379-82.
37. Clougherty E. and Kalish, D., Private communications.
38. Vasilos T. and Spriggs, R. M., "Pressure Sintering: Mechanisms and Microstructures for Alumina and Magnesia," *J. Am. Ceram. Soc.*, 46 (1963), 493-96.
39. Rossi R. C. and Fulrath, R. M., "Final Stage Densification in Vacuum Hot Pressings of Alumina," *J. Am. Ceram. Soc.*, 48 (1965), 558-64.
40. Coble, R. L., "Mechanisms of Densification During Hot Pressing," in *Sintering and Related Phenomena*, G. C. Kuczynski *et al.*, eds., Gordon and Breach, New York, 329.
41. Coble R. L. and Ellis, J. S., "Hot Pressing Alumina–Mechanisms of Material Transport," *J. Am. Ceram. Soc.*, 46 (1963), 438-41.
42. Vasilos T. and Spriggs, R. M., "Pressure Sintering of Ceramics," Chap. 2 in *Progress in Ceramic Science*, Vol. 4, J. Burke, ed., Pergamon Press (1966).
43. Fulrath, R. M., "Hot Forming Processes," in *Critical Compilation of Ceramic Forming Methods*, Air Force Materials Lab. Tech. Doc. Report No. RTD-TDR-63-4069 (January 1964), 33-43.

44. Rhodes, W. H., Sellers, D. J., Cannon, R. M., Heuer, A. H., Mitchell W. R. and Burnett, P. L., "Microstructure Studies of Polycrystalline Refractory Oxides," Summary Report, U.S. Naval Air Systems Command, Contract NOw 19–67–C–0336 (May 1968).

45. Spriggs, R. M., Brissette L. A. and Vasilos, T., "Grain Growth in Fully Dense Magnesia," *J. Am. Ceram. Soc.,* 47 (1964), 447.

46. Leipold, M. H., "Impurity Distribution in MgO," *J. Am. Ceram. Soc.,* 49 (1966), 478.

47. Kriegel, W. W., Palmour, Hayne III and Choi, D. M., "The Preparation and Mechanical Properties of Spinel," Chap. 12 in *Special Ceramics, Proceedings of a Symposium held by British Ceramic Research Association in 1964,* P. Popper, ed., Academic Press (1965), 167.

48. Spriggs, R. M., Brissette, L. A., Rossetti, M. and Vasilos, T., "Hot Pressing Ceramics in Alumina Dies," *Bull. Am. Ceram. Soc.,* 42 (1963), 477.

9. Solid-State Sintering

D. LYNN JOHNSON
Northwestern University
Evanston, Illinois

ABSTRACT

The theory of solid-state sintering is reviewed, with particular emphasis on the effects of several concurrent mechanisms on sintering behavior. Computer synthesis of both non-isothermal and isothermal sintering data has been used to predict initial sintering behavior under a wide variety of conditions. This approach has been successful in describing the sintering of compacts of spherical particles of Fe, Cu, Fe_2O_3, Ag, and LiF. The relative importance of surface, grain boundary, and volume diffusion and vapor transport are discussed. Experimentally observed sintering behavior is discussed in terms of the theoretical model.

Introduction

Although the important phenomenon of sintering has been the subject of many studies in recent years, considerable uncertainty still exists concerning the mechanisms of material transport operative during the sintering of any given material. The possible mechanisms have been well discussed, but the interpretation of the data in terms of these mechanisms has frequently led to conflicting opinions. Among the reasons for this state of affairs are the continued reliance on the early order-of-magnitude models and/or the assumption of a single mechanism being responsible for the observed behavior.

As an example, consider the case of silver. Kuczynski [1, 2] interpreted his neck-growth measurements on sintered wires and spheres of silver in terms of his volume diffusion model, and computed volume diffusion coefficients which are somewhat greater than the values determined by tracer analysis. Johnson and Clarke [3] studied shrinkage of compacts of sized spherical particles. Their data were compatible with Kuczynski's volume diffusion model, as modified by Kingery and Berg [4] to describe shrinkage, again yielding volume diffusion coefficients which were too high. However, their own model, which assumed that volume and grain boundary diffusion could occur simultaneously, but that surface diffusion was insignificant, resulted in calculated volume and grain boundary diffusion coefficients which were within the scatter bands of the tracer values. Finally, Nichols [5] re-evaluated Kuczynski's data for sintered wires and concluded that surface diffusion was the principal mechanism of material transport.

Similar ambiguities for many materials have caused some to despair of a method to describe sintering phenomena reliably. On the other hand, many have been content to name volume diffusion as the sintering mechanism whenever the time-dependence of shrinkage or neck growth is approximately equal to that predicted by Kuczynski [1] and Kingery and Berg [4], which it frequently is. Thümmler and Thömma [6] have reviewed most sintering work.

Recently, the status of volume diffusion has been challenged as the chief sintering mechanism. The calculations of Seigle [7], and the surface diffusion models of Nichols [5] and Nichols and Mullins [8], using the surface diffusion coefficients calculated from scratch healing and grain boundary grooving experiments, predict that surface diffusion should predominate in most cases where diffusion coefficients are known. Wilson and Shewmon [9] demonstrated that this was so for copper by measuring both shrinkage and neck growth. However, compacts of most materials shrink during sintering and therefore, since surface diffusion cannot cause shrinkage, volume and/or grain boundary diffusion or plastic flow must be operating as well as surface diffusion. It is quite likely that the sintering of most materials is caused by more than one mechanism of material transport.

The present author has proposed a method of studying sintering which permits a considerably more reliable determination of the various mechanisms which may be effecting the sintering [10]. All material transport mechanisms which are significant can be determined qualitatively and, if vapor transport and plastic flow are found to be negligible, the volume, grain boundary, and surface diffusion coefficients can be calculated. Preliminary data for Fe, Cu, and LiF have given diffusion coefficients which are in close agreement with the published coefficients. This method will now be discussed in greater detail.

Experimental Procedures and Data Analysis

The primary experimental feature of this method is that the shrinkage, shrinkage rate, and neck size must be measured simultaneously. The two sintering particles in contact must have either a common axis of rotation or a common mirror plane. The basic sintering equations are

$$\frac{x^3 \rho}{x + \rho \cos \alpha} \dot{\delta} = \frac{4\gamma\Omega D_V}{\pi kT} \cdot \frac{A_V}{x} + \frac{8\gamma\Omega b D_B}{kT} \tag{1}$$

for two bodies of revolution possessing a common axis of rotation (two spheres, sphere on a plane, cone on a plane, etc.) and

$$\frac{x^3 \rho}{x + \rho \cos \alpha} \; \dot{\delta} = \frac{3\gamma\Omega D_V}{kT} \cdot A + \frac{3\gamma\Omega b D_B}{kT} \qquad (2)$$

for two bodies with a common mirror plane (two cylinders, cylinder on plane, wedge on plane, etc),
where

x = neck radius or half-width
ρ = neck surface radius of curvature
A_V = neck surface area (for circular necks)
A = neck surface area per unit length (for rectangular necks)
α = complement of half the grain boundary groove angle
γ = surface tension
Ω = atomic volume
D_V = volume diffusion coefficient
b = grain boundary width
D_B = grain boundary diffusion coefficient
k = Boltzmann's constant
T = absolute temperature
δ = interpenetration of the original surfaces of the bodies
$\dot{\delta}$ = interpenetration rate

These equations are independent of any material transport from the particle surfaces to the neck surface, provided the geometry parameters x, ρ, and A_V are known. If these parameters and the interpenetration rate (shrinkage) are determined for a series of times at a fixed temperature, then a plot of the left-hand side versus A_V/x or A, whichever is appropriate, will yield a straight line. The two diffusion coefficients can be calculated from the slope and intercept of this line.

For two identical spheres, or a compact of identical spheres in which the number of interparticle contacts is independent of shrinkage, equation (2) can be written as

$$\frac{X^3 R}{X + R \cos \alpha} \; \dot{y} = \frac{2\gamma\Omega D_V}{\pi k T a^3} \cdot \frac{A_V}{ax} + \frac{4\gamma\Omega b D_B}{kT a^4} \qquad (3)$$

where

$X = x/a$
$R = \rho/a$
a = sphere radius
$y = \delta/(2a) = \Delta L/L_o$ = fractional shrinkage
$\dot{y}$ = shrinkage rate

Furthermore, the other geometric parameters can be computed if both y and X are known:

$$\rho = \frac{a(y^2 - 2y + X^2)}{2[1 - (1 - y)\sin\alpha - X\cos\alpha]} \tag{4}$$

$$A_V = 4\pi\rho \left[(\theta - \alpha)(a + \rho)\cos\theta - \rho(\sin\theta - \sin\alpha) \right] \tag{5}$$

where

$$\theta = \sin^{-1}\left[\frac{a - ya + \rho\sin\alpha}{a + \rho} \right] \tag{6}$$

Thus, by measuring y, $\dot{y}$, and x, the volume and grain boundary diffusion coefficients can be calculated, provided plastic flow is unimportant. (Plastic flow would cause the plot to be concave downwards, and could thus be identified.) In addition, the surface diffusion coefficient can be estimated if vapor transport is negligible.

The existence of significant surface diffusion or vapor transport can be identified by comparing the measured neck size with that predicted for no mass transport from the sphere surfaces. A computer program has been written which synthesizes sintering data for any combination of diffusion coefficients. The surface diffusion coefficient is obtained by using the values of D_V and bD_B obtained from equations (1), (2), or (3), and a guessed value of D_S. The calculated neck size at a given shrinkage is compared with the measured size, and a new estimate of D_S is made on the basis of this comparison. By successive approximations, a value of D_S is chosen which results in a calculated neck size equal to the measured size. The synthesized and measured data may then be compared. Figures 1 and 2 show the results for 12.3 μ radius iron spheres sintered at 875°C in dry hydrogen [10]. If vapor transport had contributed to neck growth, the experimental points in Figure 1 would have been on a steeper line than that calculated [11].

Each of the three diffusion coefficients calculated from the data shown in Figures 1 and 2 agrees with the published values to within a factor of 1.5 to 3.5. Analysis of the two isotherms each of Kingery and Berg [4] and Wilson and Shewmon [9], where neck growth and shrinkage were measured simultaneously, gave similar comparisons with the published volume diffusion coefficient and only slightly less close agreement with the surface diffusion coefficient. Data for LiF agreed with the published Li volume diffusion coefficient [12]. Although the amount of data is quite limited, the results are very satisfactory and indicate that the above approach is useful for studying sintering kinetics.

The principal drawbacks of this method are that the particle geometry must be well characterized, i.e., two particles of known shape or a compact of spheres of narrow size distribution, and there are limitations in the measurement of neck

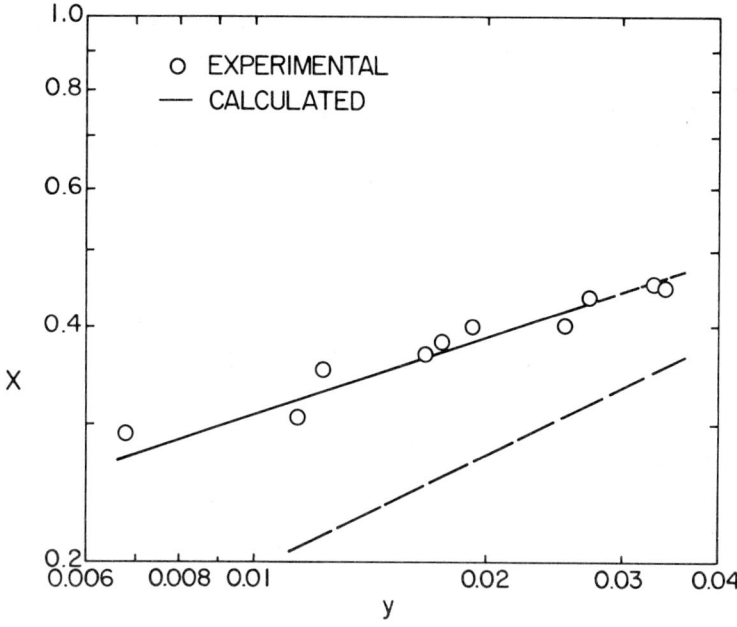

Figure 1. Neck size versus shrinkage for compacts of 12.3 μ radius iron spheres sintered at 875°C in hydrogen. The broken line represents the expected results for no surface diffusion.

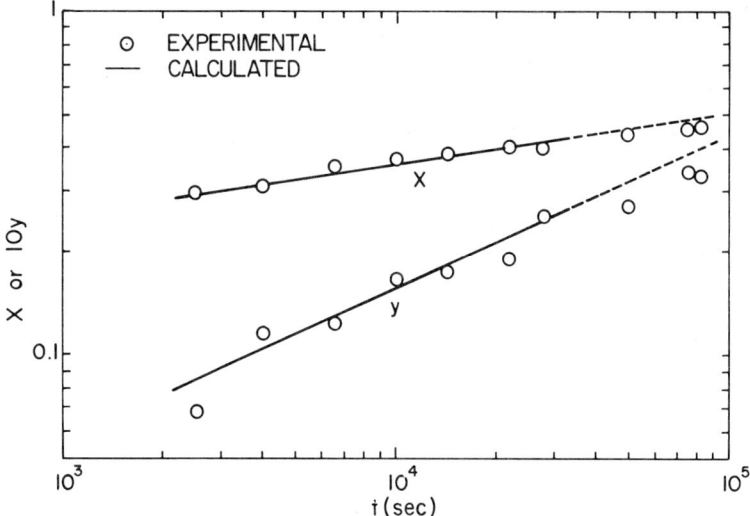

Figure 2. Neck size and shrinkage versus time for 12.3 μ radius iron spheres. Close-packed model breakdown occurs at the ends of the solid lines.

size and shrinkage. Neck size can be measured more readily on two particle systems (sphere on plane, for instance), while shrinkage is most easily measured on compacts. Compacts of spherical particles of narrow size distribution appear to be the best compromise, but neck measurement on sectioned compacts becomes difficult for particle radii of less than 10 μ. Nevertheless, the fact that this method removes so much of the uncertainty as to the sintering mechanisms makes it worth the effort.

Sintering of Ideal Compacts

The theoretical sintering behavior of ideal compacts by combinations of volume, grain boundary, and surface diffusion will be considered in this section. An ideal compact will be defined as an array of spheres of uniform diameter in which the number of interparticle contacts is independent of shrinkage. The fractional shrinkage of the compact as a whole will then be the same as that of any two of the particles.

The relative importance of each of the mechanisms is a function of the particle size, the temperature, and the instantaneous geometry [7]. A measure of the relative magnitudes of volume and grain boundary diffusion is given by the ratio of the slope to the intercept of equation (3), namely

$$\frac{\text{slope}}{\text{intercept}} = \frac{aD_V}{2\pi bD_B} \tag{7}$$

Thus volume diffusion is favored at higher temperatures and for larger particles. The particle-size dependence of surface diffusion and grain boundary diffusion is the same, but the relative importance of these two will be a function of the temperature. The volume diffusion to grain boundary diffusion and grain boundary diffusion to surface diffusion flux ratios will increase with sintering time at a given temperature and particle size. These generalizations will become apparent in the discussion below.

In order to investigate theoretical sintering kinetics, a computer program was written which constructs sintering data for combinations of volume, grain boundary, and surface diffusion, based on the model discussed above and presented in more detail elsewhere [10]. This program calculates the fluxes of atoms from the grain boundary between the spheres to the neck via volume and grain boundary diffusion, and from the sphere surfaces to the neck via volume and surface diffusion. As the geometry evolves, the program calculates the geometric parameters, the shrinkage, shrinkage rate, and time. The temperature can be made to increase linearly with time to a certain point and then follow an exponential decay curve to the isothermal temperature. The model applies only to the initial stage of sintering; that is, when the individual necks on any given

sphere are separate from each other. Two model breakdown points were
calculated, one for close-packed spheres and one for simple cubic packing. A
compact of random-packed spheres would begin to deviate from the theoretical
curve at the close-packed model breakdown point, since there would be regions
of close packing within the compact.

The logarithm shrinkage/logarithm time plot will be used to demonstrate the
effects of surface diffusion. An ideal compact heated instantaneously to the
sintering temperature will give a straight line on this plot with a slope of about
0.52 for pure volume diffusion, or about 0.33 for pure grain boundary diffusion,
or anything in between for both mechanisms occurring simultaneously. If
surface diffusion is added, it will displace the line downward and increase its
slope. Curve A on Figure 3 is for pure grain boundary diffusion with instantaneous heat-up.

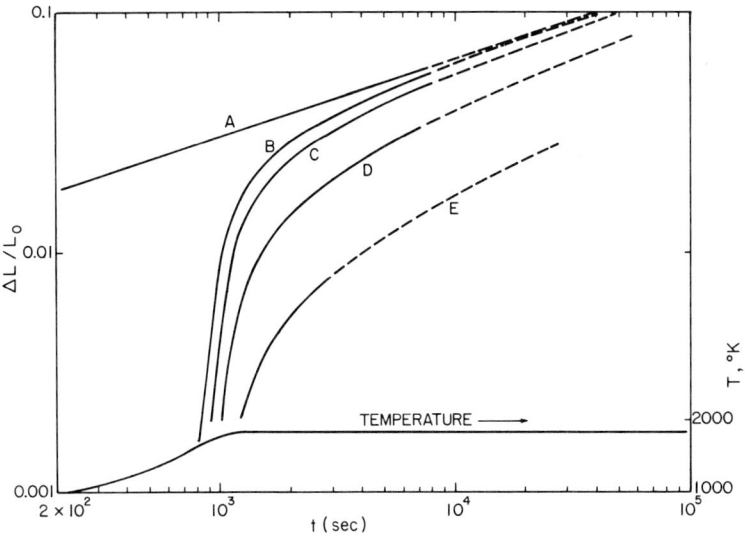

Figure 3. Calculated shrinkage curves. Heating rate = $1°C$/sec, particle radius = 1.0 μ,
surface tension = 950, atomic volume = 1.4 $\times$ 10^{-23}. bD_B = 0.27 exp $(-133,000/RT)$.
Curve A, D_{OS} = 0, instantaneous heating to $1800°K$. Curve B, $D_{OS} \leqslant 10^{-3}$. Curve C,
D_{OS} = 0.1. Curve D, D_{OS} = 1.0. Curve E, D_{OS} = 10.

Next, consider the case of gradual heating to the sintering temperature. For
these examples the heating rate was $1.0°$ per second from $773°K$ to $1700°K$,
followed by an exponential decay to $1800°K$. The zero of time was taken at
$773°K$. Curve B on Figure 3 is for no surface diffusion, and curves C through E
are for increasing amounts of surface diffusion. Curve B can be brought into
coincidence with curve A by subtracting 1,200 seconds from each time. At 1,200

seconds the temperature is $1794°K$, $6°$ below the isothermal temperature. Subtracting $1200°K$ from the other curves straightens them out also, with resulting slopes of 0.35, 0.41, and 0.48 for curves C, D, and E, respectively.

The close-packed model breakdown points are indicated on Figure 3 by the ends of the solid lines, while the simple cubic model breakdown points are at the ends of the broken lines. The shrinkage of a random-packed compact would fall below the broken lines, giving a lower slope to that portion of the curves.

The effect of heating rate is shown in Figure 4, using the surface diffusion

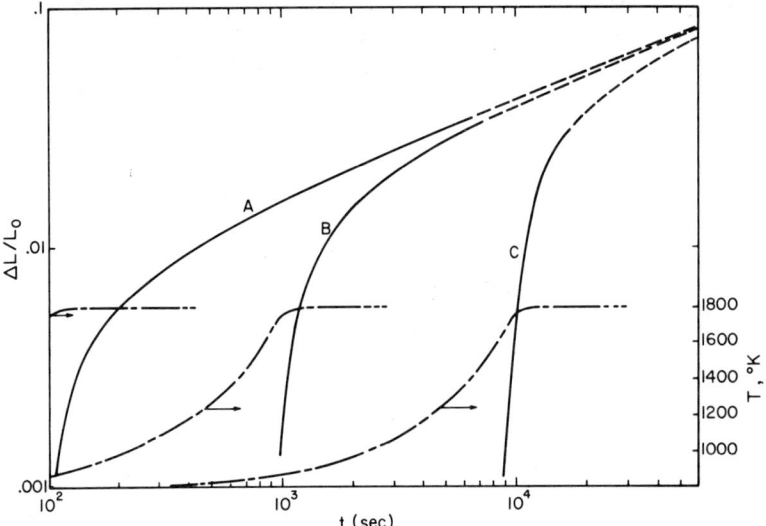

Figure 4. Calculated shrinkage curves. Data same as Figure 3 except D_{OS} = 1.0. Curve A, rate = $10°C/sec$ to $1700°K$. Curve B, rate = $1.0°C/sec$ to $1700°K$. Curve C, rate = $0.1°C/sec$ to $1700°K$. All curves isothermal at $1800°K$.

coefficient of curve D of Figure 3. Note that the slower heating rate results in a model breakdown at lower shrinkages. This is because surface diffusion is relatively more important at lower temperatures, and the neck will grow more by surface diffusion at lower temperatures. If these curves are corrected for heat-up time, the slope is greater for the slower heating rate.

The effect of particle size is indicated in Figure 5. Here, also, the surface diffusion coefficient of curve D of Figure 3 was used. Model breakdown occurs at lower shrinkage for smaller particles, again because surface diffusion is more important at lower temperatures, and more of the sintering takes place at lower temperatures for smaller particles.

The results for volume diffusion plus surface diffusion are similar, except that the particle size effect is more pronounced since surface diffusion is relatively more important than volume diffusion for smaller particle sizes. The time-

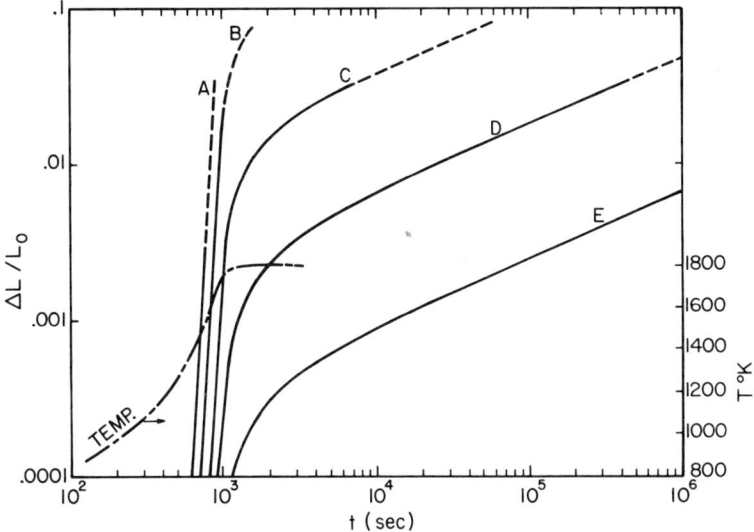

Figure 5. Calculated shrinkage curves. Data same as Figure 3 except D_{OS} = 1.0. Rate = 1.0°C/sec to 1700°K, isothermal at 1800°K. Sphere radii (μ): curve A, 0.1; curve B, 0.3; curve C, 1.0; curve D, 3.0; curve E, 10.0.

corrected logarithm shrinkage/logarithm time plots can have slopes in excess of 0.60.

One of the most important conclusions that can be drawn from the foregoing discussion is that the slope of the logarithm shrinkage/logarithm time plot is a completely unreliable indicator of the sintering mechanism. The slope on time-corrected plots can vary from 0.33 upward before model breakdown, and any slope in excess of 0.33 could be the result of surface diffusion or volume diffusion or both, plus grain boundary diffusion. After model break, slopes will be flatter, and measurements of shrinkage alone cannot determine the model breakdown point. The isothermal shrinkage plots of compacts represented by curves A and B of Figure 5 would undoubtedly have low slopes or be continuously curved. The entire range of slopes reported in the literature can thus be explained, but the actual mechanisms for any case cannot be. The logarithm neck size/logarithm time plots for the synthesized data show a similar degree of variability. It must be concluded that the sintering mechanism cannot be determined by measuring only shrinkage or neck growth.

There have been some attempts to determine sintering mechanisms by comparing the sintering activation energy with that of the published diffusion coefficients. However, synthesized data show that surface diffusion will alter the activation energy. Smart [13] found that for α-Fe_2O_3 the sintering activation energy, based on shrinkage measurements only, was 78 kcal/mole, in agreement with the activation energy for volume diffusion of oxygen. By employing the method outlined in the previous section, he found a grain boundary plus surface diffusion combination, with a grain boundary diffusion activation energy of 30 kcal/mole and a low surface diffusion activation energy.

It thus appears that sintering mechanisms cannot be determined with confidence using the usual techniques. It is quite likely that in no case published thus far has the sintering mechanism of a ceramic material been correctly determined, simply because there is no information about surface diffusion and model breakdown. The method described above and in reference [10] has proved successful in those few cases where it has been applied, and is perhaps a useful approach to the study of sintering kinetics.

· Conclusions

The classical methods of determining mechanisms of material transport during sintering are unreliable. Combinations of diffusion mechanisms have been shown to affect the observed sintering behavior in such a way that the sintering mechanisms cannot be inferred from the time-dependence of the shrinkage or neck growth, or from the apparent activation energy. A method of investigation is discussed which will permit qualitative identification of whatever mechanisms may be operative. Under conditions when no plastic flow or vapor transport occurs, it is possible to calculate the volume, grain boundary, and surface diffusion coefficients.

Acknowledgment

The research upon which this work is based was supported by the Advanced Research Projects Agency of the Department of Defense through the Materials Research Center, Northwestern University.

References

1. Kuczynski, G. C., "Self-Diffusion in Sintering Metal Powders," *Trans. Amer. Inst. Min. (Metall.) Engrs.*, 185 (1949), 169.
2. Kuczynski, G. C., "Measurement of Self-Diffusion of Silver Without Radioactive Tracers," *J. Appl. Phys.*, 21 (1950), 632.
3. Johnson, D. L. and Clarke, T. M., "Grain Boundary and Volume Diffusion in the Sintering of Silver," *Acta Met.*, 12 (1964), 1173.

4. Kingery, W. D. and Berg, M., "Study of the Initial Stages of Sintering Solids by Viscous Flow, Evaporation-Condensation, and Self-Diffusion," *J. Appl. Phys.*, 26 (1955), 1205.

5. Nichols, F. A., "Theory of Sintering of Wires by Surface Diffusion," *Acta Met.*, 16 (1968), 103.

6. Thümmler, F. and Thömma, W., "The Sintering Process," *Metallurgical Reviews*, 12 (1967), 69.

7. Seigle, L. L., "Atom Movements During Solid State Sintering," *Progress in Powder Metallurgy*, Vol. 20, Metal Powder Industries Federation, New York (1964), 221.

8. Nichols, F. A. and Mullins, W. W., "Morphological Changes of a Surface of Revolution Due to Capillarity-Induced Surface Diffusion," *J. Appl. Phys.*, 36 (1965), 1826.

9. Wilson, T. L. and Shewmon, P. G., "The Role of Interfacial Diffusion in the Sintering of Copper," *Trans. Amer. Inst. Min. (Metall.) Engrs.*, 236 (1966), 48.

10. Johnson, D. L., "New Method of Obtaining Volume, Grain Boundary, and Surface Diffusion Coefficients from Sintering Data," accepted for publication in *J. Appl. Phys.*

11. Johnson, D. L., Shingu, P. H. and Smart, J. S. III, unpublished.

12. Moore, D. J., "Initial Sintering Kinetics of Lithium Fluoride in an Inert Atmosphere," M.S. Thesis, Northwestern University (1968).

13. Smart, J. S. III, "Initial Solid State Sintering of Hematite," Ph.D. Dissertation, Northwestern University (1968)

10. Determination of an Effective Viscosity of Powders as a Function of Temperature

WAYNE S. YOUNG, STEPHEN T. RASMUSSEN, and IVAN B. CUTLER
University of Utah
Salt Lake City, Utah

ABSTRACT

A method is described for determining the effective viscosity of powder compacts during densification as a function of temperature. The method utilizes constant rate of heating and is easily adapted to determining the sintering temperature of a powder compact as well as measuring the effective viscosity. The method can be used for crystalline as well as glassy (amorphous) systems. Values of the coefficient of viscosity or diffusion can be determined for powders of simple particle geometry and known particle size distribution. For practical applications, an effective viscosity is proposed as a better measure of densification characteristics. Illustration of the method is made by examining shrinkage characteristics of various ceramic powder compacts at constant rates of heating.

Introduction

Since the pioneering work of Frenkel [1] describing sintering in terms of simple particle geometry and atomic movement, there have been many important contributions to the theory of sintering with additional refinements to cover all of the methods of atomic transport occurring during sintering. Johnson has reviewed the current understanding of sintering theory as it pertains to crystalline materials in this volume [2]. In spite of these important developments concerning the theory of sintering, there has been little translation of theory to practical application. It is still impossible to predict the densification characteristics of real powders during a sintering cycle. This does not signal the failure of sintering theory, but rather, the problems of translating theoretical information into practical terms.

The theoretical equations have been developed relating isothermal shrinkage of powder compacts and/or isothermal neck growth between spherical particles, knowing particle size, surface tension, coefficients of atomic diffusion, and temperature. It becomes very difficult to take into account particles of non-spherical shape that are of practical importance. For example, the shape of angular particles may change with time toward a more spherical particle shape. Very little is known at the present time about the anisotropic character of surface tension. Even though an average surface tension could be measured or assumed, it is doubtful that this term can be known within 10 percent accuracy.

Particle size is another very difficult term to define with any degree of accuracy. There are many problems associated with particle size distribution which have so far defied analytical description. For the majority of ceramic materials, diffusivity will be extrinsic. It is very sensitive to the presence of impurities. In practical cases of sintering of powders, the coefficient of diffusion will be many orders of magnitude different than the coefficients that are available in the literature describing the intrinsic or low-impurity extrinsic diffusion.

In the case of aluminum oxide, not only is the diffusion coefficient very sensitive to the type and amount of impurity present, but the mechanism of sintering evidently changes from grain boundary to volume diffusion or volume diffusion to grain boundary diffusion, depending upon the impurity type and content [3]. These factors and others make it very difficult to obtain a practical application of sintering theory to real systems. The prospect of being able to predict the onset of shrinkage or the firing temperature from chemical composition, particle size, shape, and distribution, appears to be remote indeed.

Theoretical treatment of sintering to date has described isothermal shrinkage and neck growth. Isothermal experiments are difficult to perform, because the maximum rate of shrinkage or neck growth should be obtained at zero time just at the instant the sample is inserted into its isothermal environment. It is impossible to generate entirely isothermal conditions for a sample, because it is impossible to raise the temperature of a sample to its isothermal temperature and maintain uniform temperature throughout the sample all on an instantaneous basis. During isothermal experiments the rate of shrinkage and the rate of neck growth decrease rapidly with time. As these processes become slower and slower, the opportunity for grain growth with accompanying loss of grain boundary sinks for shrinkage minimizes the application and effectiveness of the isothermal treatment.

The type of heating program that can be treated analytically and which approaches practical application is a constant rate of heating. Constant rate of heating (CRH) experiments offer an opportunity to simplify sintering theory, particularly for practical applications with real powders. The onset of shrinkage and the initial sintering of powders can be expertly treated through the use of CRH experiments. If these experiments are continued beyond the initial stages of sintering, maximum shrinkage can be obtained and a good description of the firing temperature is made available. With CRH experiments, it is not easy to describe with certainty the mechanism of sintering and it is no easier to obtain diffusion coefficients by this technique than by isothermal measurements, but, fortunately, for practical situations this is not a vital piece of information to obtain. With CRH experiments it is possible to describe an effective activation energy for sintering and effective parameters which describe densification of powder compacts and allow differentiation between various alternative processing techniques and impurity additions in practical sintering situations.

Equations for CRH Sintering

Frenkel [1] derived isothermal sintering equations that have been successfully used for viscous materials such as glass. He predicted that the rate of shrinkage of a compact of spheres should follow the equation

$$d(\Delta L/Lo)/dt = \gamma/2a\eta \qquad (1)$$

where $\Delta L/Lo = (Lo - L)/Lo$ = the fraction of shrinkage compared to the original length of the powder compact, Lo, t is the time, γ is the surface free energy, a is the average spherical particle radius, and η is the viscosity. Over a limited region of temperature, the viscosity can be represented by the equation

$$\eta = A \exp(Q/RT) \qquad (2)$$

where A is a pre-exponential term, Q is the activation energy, R is the gas constant, and T is the temperature on an absolute scale. By heating at a constant rate (CRH) according to the equation

$$T = ct \qquad (3)$$

the exponential character of sintering is easily seen by combining these equations into

$$d(\Delta L/Lo)/dT = (\gamma/2acA) \exp(-Q/RT) \qquad (4)$$

If the exponential term is sufficiently larger (over 10), an approximate integral form of the differential equation can be obtained as

$$\Delta L/(LoT^2) = (\gamma R)/(2acAQ) \exp(-Q/RT) \qquad (5)$$

Equations similar to (4) and (5) can be obtained in the case of the shrinkage of crystalline powders by movement of vacancies, as described by Johnson [2]. For diffusion through the volume of the material

$$Td(\Delta L/Lo)\,dT = [(5.34\,\gamma\,\Omega RTD_l')/(ka^3 Qc)]^{1/2}\, Q/2RT \exp(-Q/2RT) \qquad (6)$$

where k is Boltzmann's constant, Ω is the vacancy volume, and D_l' is the pre-exponential term for volume or lattice diffusion. Johnson has used the exponent 0.49, but the difference between 0.49 and 0.50 is not justified on a practical basis for sintering. The integrated form of this equation is

$$\Delta L/(Lo\sqrt{T}) = [(5.34\,\gamma\,\Omega D_l' R)/(ka^3 cQ)]^{1/2} \exp(-Q/2RT) \qquad (7)$$

Similarly, for grain boundary diffusion

$$Td(\Delta L/Lo)/dT = [(2.14\,\gamma\,\Omega RTbD_g')/(a^4 Qck)]^{1/3}\;Q/3RT\;\exp(-Q/3RT) \quad (8)$$

where all of the terms have been previously defined except b and D_g' which represent the effective grain boundary width and the pre-exponential term for the grain boundary diffusion coefficient. The integrated form of this equation gives

$$\Delta L/(Lo\,\sqrt[3]{T}) = [(2.14\,\gamma\,\Omega bD_g'R)/(ka^4 cQ)]^{1/3}\;\exp(-Q/3RT) \quad (9)$$

The ratio of equation (4) to equation (5) will remain constant as given by

$$\frac{d(\Delta L/Lo)/dT}{\Delta L/(LoT^2)} = Q/R \quad (10)$$

Also, the ratio of (6) to (7) will be a constant as given by

$$\frac{(T)^{3/2}\,d(\Delta L/Lo)/dT}{\Delta L/(Lo\,\sqrt{T})} = Q/2R \quad (11)$$

and the ratio of (8) to (9) is

$$\frac{T^{1.67}\,d(\Delta L/Lo)/dT}{\Delta L/(Lo\sqrt[3]{T})} = Q/3R \quad (12)$$

The most important feature of equations (4) through (9) is the predominant exponential function of temperature which contains the activation energy. This exponential function overshadows all other temperature terms in the equations. The activation energy for sintering of ceramic materials ranges from 50 kcal/mole to 200 kcal/mole. To illustrate the exponential features of the equations that have been listed above, Figure 1 has been drawn for activation energies of 50, 75, and 150 kcal/mole. The smaller the activation energy, the longer will be the temperature range during which shrinkage takes place. In effect, this series of activation energies compares a viscous flow mechanism to a volume diffusion mechanism and to grain boundary diffusion mechanism (150, 75, and 50 kcal/mole).

For densification by shrinkage, the effective activation energy controls the temperature-sensitive portion of equations (4) through (9). This is the exponential term,

$$\exp(-Q/nRT) = \exp(-E/RT) \quad (13)$$

where E is the effective activation energy and $n = 1$, 2, or 3, depending on the mechanism of densification (viscous flow, volume diffusion, or grain boundary diffusion.)

Notice that in equations (5), (7), and (9) the essential differences, other than differing effective activation energies, are in the function of temperature in the

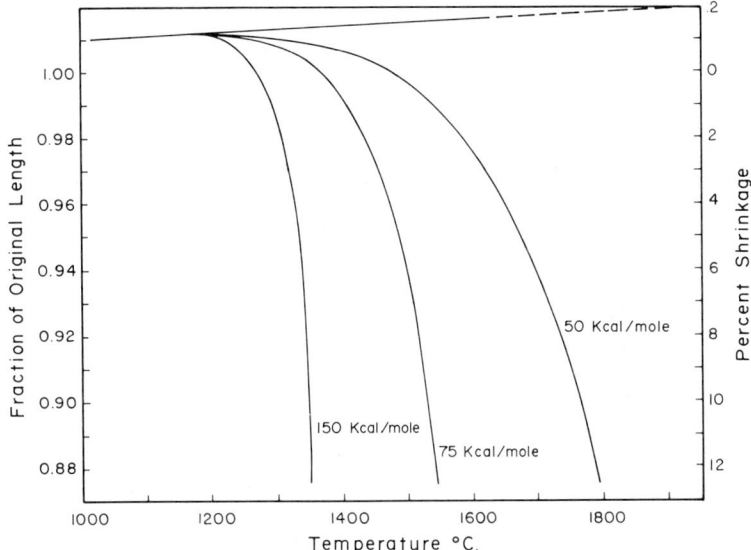

Figure 1. Shrinkage of a powder compact showing the effect of activation energy on the exponential increase in shrinkage as a function of temperature during a constant rate of heating.

pre-exponential term. Figure 2 shows a plot in which the solid lines are based on equation (5) and the dotted lines neglect the T^2 term that falls outside of the exponential relationship. Neglecting the T^2 term may give an error of as much as 15 percent in the activation energy for small effective activation energies, but ordinarily, rate experiments are not performed carefully enough to distinguish an error of less than about 10 percent. Thus it is possible to formulate an empirical equation that would reasonably represent the densification by shrinkage of all types of powders. This equation is presented as

$$\Delta L/(L_o T) = K/\eta' = K' \exp(-E/RT) \tag{14}$$

where K and K' are constants that depend on particle size, shape, and distribution, surface tension, heating rate, activation energy, etc., and η' is an effective viscosity for crystalline and non-crystalline materials.

The shrinkage data at constant rates of heating for almost all powders will fit equation (14), thereby producing an effective activation energy and a parameter, K', which is sensitive to the nature of the powder and method of manufacturing of the powder. K' would ordinarily be most sensitive to particle size, shape, and distribution parameters that could ordinarily be altered by the processing procedure for the ceramic powder. The value of the effective activation energy, E, is most important. In addition to giving an indication of the initial sintering mechanism involved, E displays the temperature-sensitivity of the shrinkage

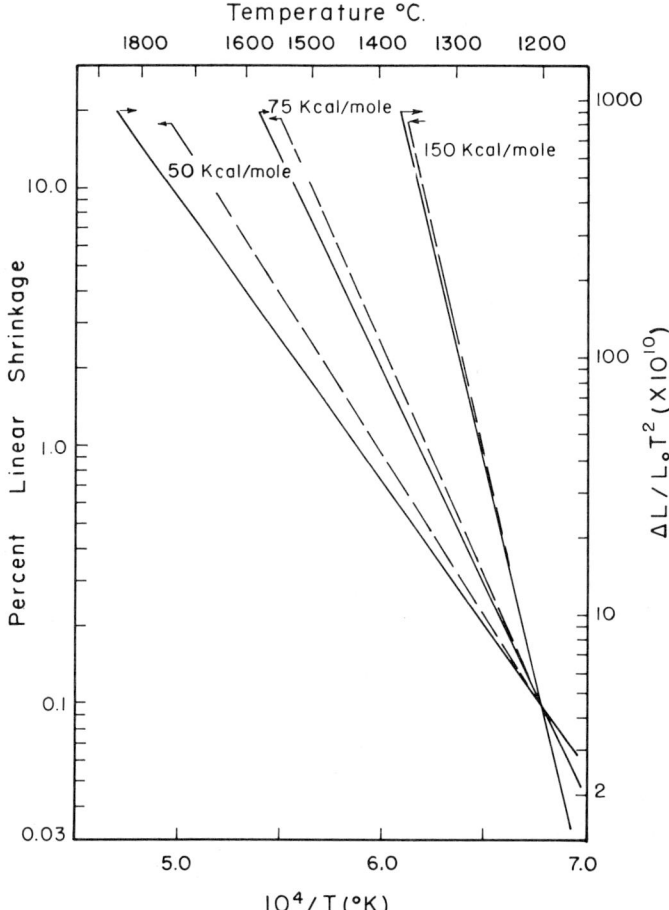

Figure 2. Semi-logarithm plots of the shrinkage of a powder compact, assuming viscous sintering according to equation (5) (right-hand ordinate). If the pre-exponential T^2 term is neglected, a slight error in activation energy would be encountered as indicated by the dotted line. This error is especially important at low activation energies.

process and signals to the investigator the degree of control that must be exercised over the sintering process when being concerned about warpage, shrinkage cracks, over-firing, etc. For powders that sinter without liquid (100 percent crystalline), the effective activation energy will be one third or one half the value measured by other techniques such as ionic conductivity, tracer diffusion, etc.

Of course, if it were possible to control particle size and geometry of the particles sufficiently well, an actual viscosity or diffusion coefficient could be obtained. It would then be valuable to solve the appropriate equation for the viscosity or diffusion coefficient. For most practical systems, it will be difficult to describe particle size, particle shape, and the size distribution. For this reason, these parameters have been lumped into k', which might well be described as a reactivity coefficient which will increase approximately linearly with the reciprocal of the particle size.

The very minor amount of experience gained with impurity effects on sintering suggest that the effective activation energy may or may not be changed with the addition of an impurity; however, for the most part, it would be expected that the pre-exponential term or reactivity coefficient might be increased by several orders of magnitude with the incorporation of a small amount of impurity.

It may not be wise to use such a generalized equation as described as equation (14) in systems that are entirely crystalline and that would be expected to follow an equation such as (7) or (9). On the other hand, unless there is a change of mechanism during the CRH experiment, it may not be possible to distinguish between (7) and (9). Equations (11) and (12) may be of some assistance, and the size of the effective activation energy may also give an indication as to which equation is the correct one to use. If it is deemed desirable to know the mechanism of diffusion, it may be necessary to perform isothermal sintering experiments to further substantiate the information gained from CRH experiments.

Experimental Procedure

The equipment assembled for measuring the shrinkage at constant rates of heating is similar to that previously described for isothermal shrinkage measurements [4]. The essential differences consist of using a programmable temperature controller instead of an isothermal temperature controller, and an XY recorder in the place of an XT recorder. The heart of the equipment is the sensitive transducer or dilatometer that measures the change in length of a powder compact. An LVDT (linear variable differential transformer) has served admirably for this use. The transformer core and following rod are simply counter-balanced in order to minimize external forces on the powder compact. Several different furnaces and experimental set-ups have been used to gather the data reported in this paper. One of these is shown schematically in Figure 3. The data are easily obtained regardless of the experimental apparatus used, and readily interpretable in terms of temperature.

The data for equations (4), (6), (8), (10), (11), and (12) required the differential of shrinkage or length with respect to temperature. These data were obtained by processing the length–temperature data through a computer program designed to yield the differential of shrinkage with respect to temperature. The data for equations (5), (7), (9), and (14) are easily obtained directly from the recorded

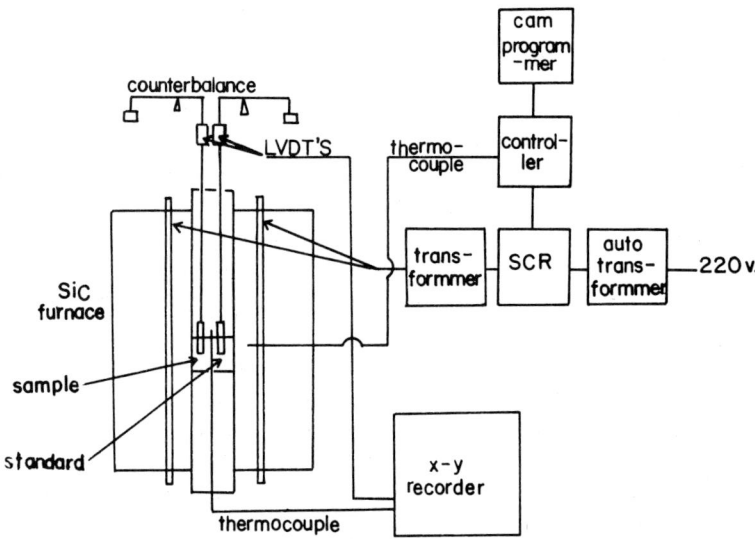

Figure 3. Schematic diagram of experimental apparatus assembled to obtain CRH data.

chart, and although it might be convenient to use, a computer program is certainly not necessary.

CRH experiments were performed with various powders. Spherical glass powder from Cataphote Corporation, available as Unisphere No. 4000, was used to obtain the data reported in Figure 4. This powder was from about $15-25\,\mu$ in diameter. The chemical composition was that of a soda-lime silica glass [6]. An atmosphere of O_2 and water vapor (230 mm partial pressure) was maintained by flowing oxygen through a constant temperature bath.

The alumina powder used in these experiments was reactor-grade Alcoa A–14 which had been dry-ground for 12 hours in a vibratory mill using 1 percent methanol as a grinding aid. The ground alumina was elutriated in a pH 4 HCl solution to give approximately 5–8, 2–5, and $1-2\,\mu$ particle size distributions. Powder compacts were prepared in the form of $1/8''$ and $1-1/4''$ disks by pressing in a steel die at 30,000 psi, using 1 percent PVP solution as a temporary binder. All samples were run with the LVDT output plotted against a similar output from a previously sintered sample to nullify any thermal expansion effects.

The TiO_2 used in these experiments was high-purity material obtained from H. U. Anderson. It was calcined for 10 minutes at $1000°C$, according to the procedure previously reported by Anderson [7]. The calcined powder was pressed into disks in a manner similar to that described for alumina powder. Shrinkage data were obtained with reference to an alumina standard as described previously. By checking the recorder output on heating and cooling, it was determined that the correction for the difference in thermal expansion of the titania compared to

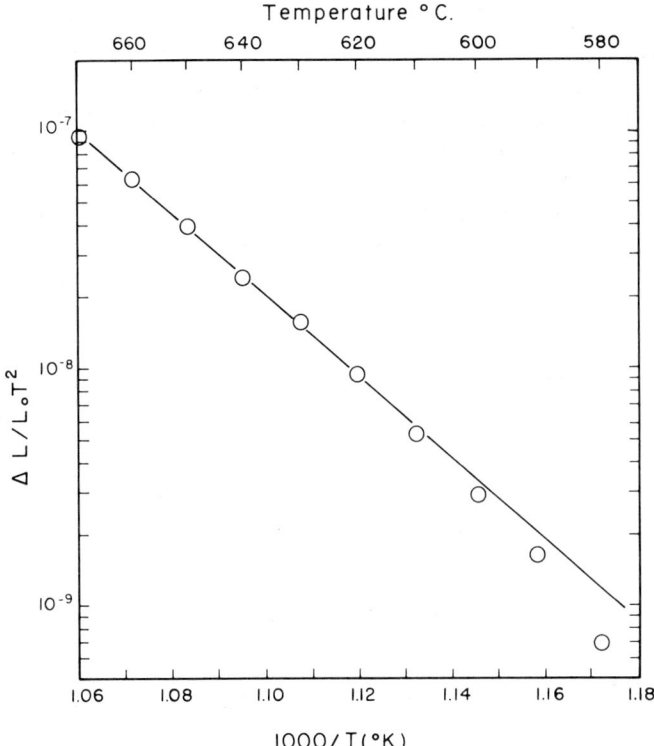

Figure 4. Sintering of glass spheres according to equation (5) at a heating rate of $1.22°C/$ min and an atmosphere of 230 mm of water vapor partial pressure.

the alumina was negligible. The zirconia powder was obtained as "Zyttrite" (yttria-stabilized zirconia) from K. S. Mazdiyasni of Air Force Materials Laboratory. The reported crystallite size was 70–100 Å. It was used as received, with similar pressing and sintering techniques to those used with the titania.

A potassium feldspar was used as one of the powders for investigation. It was ground and sieved to an average particle size of about $40\,\mu$ diameter and compacted into a disk, as previously described.

Results and Discussion

Experiments with spherical glass powders have already been reported [6] testing the validity of equations (4), (5), and (10). Agreement with theory was found to be very good in the case of spherical particles and not satisfactory for angular glass particles. One of many experiments performed with glass powders is shown in Figure 4 and illustrates agreement with equation (5). Because the viscosity of glass does not strictly follow equation (2), the linear portion of Figure 4 is somewhat limited. Deviation from linearity is toward a high activation energy at lower temperatures, and a lower activation energy at high temperatures, which is in accord with experience with viscosity measurements on glass.

Deviation from linearity, however, is not entirely due to non-exponential behavior of viscosity. At high temperature, at between 8 and 12 percent linear shrinkage, the model equations no longer describe shrinkage of glass powder compacts. The rate of shrinkage decreases as pores between spherical particles close and disappear.

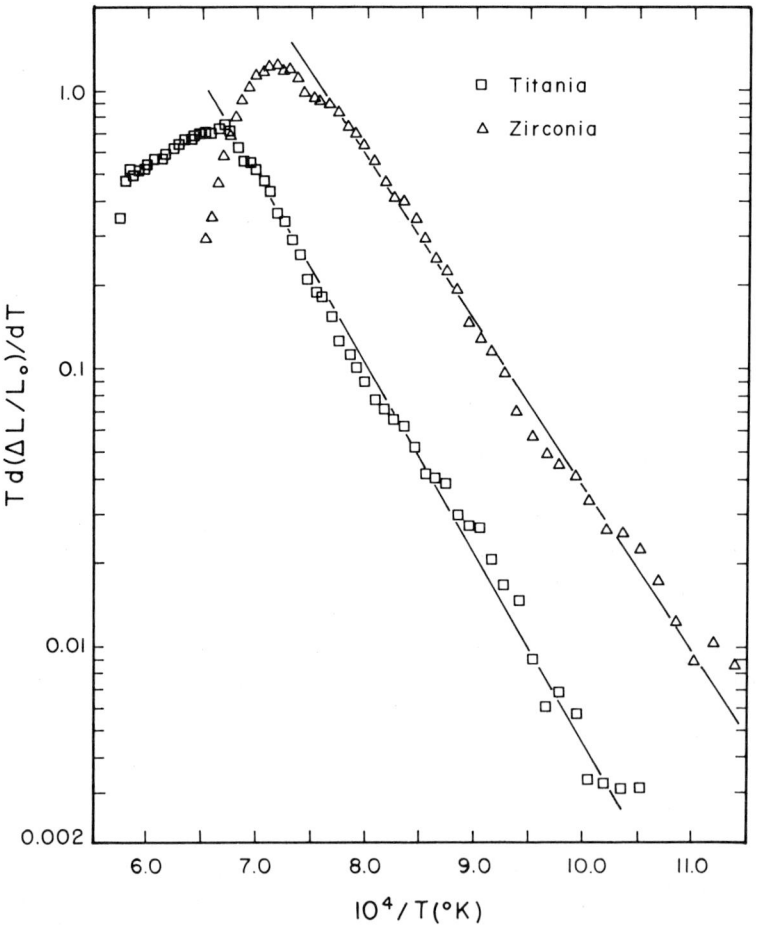

Figure 5. Constant Rate of Heating (CRH) experiments showing the shrinkage of titania and yttria-stabilized zirconia at a heating rate of 4.0°C/min. These data are plotted according to equations (6) and (8).

At low temperatures, there are experimental difficulties to making a distinc-
tion between shrinkage and thermal expansion of the glass. A small error in
judging the thermal expansion of the glass can amount to a large error in ΔL. As
a result, at low temperatures there will always be a tendency toward scatter in
the data, with errors of judgment resulting from evaluating small amounts of
shrinkage.

The CRH experiments for titania and zirconia are shown in Figures 5 and 6.
The calcined titania sintered 16 percent, with deviation from linearity occurring
at about 6.5 percent shrinkage. The effective activation energy, E, was 25 to 30

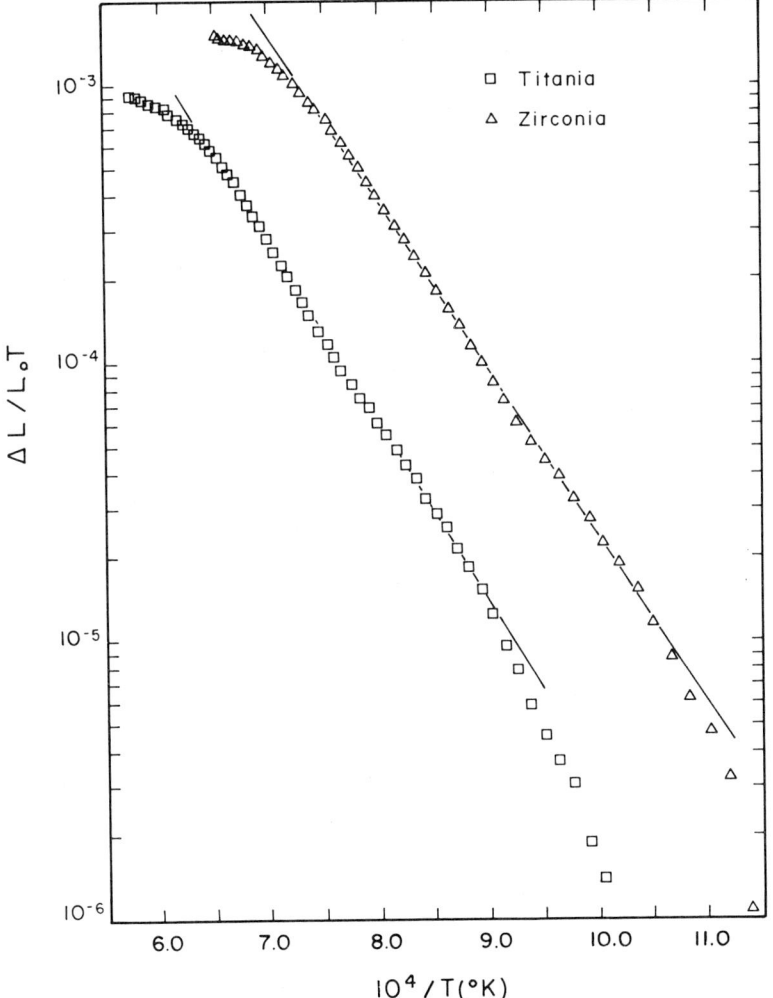

Figure 6. The data in Figure 5 plotted according to equation (14).

kcal/mole. The compacts apparently follow grain boundary diffusion over the sintering temperature range of 700–1500°C, as has been previously reported for isothermal studies [7]. On the basis of grain boundary diffusion the activation energy, Q, should be 75–90 kcal/mole, compared to 78 kcal/mole reported by Anderson. A powder compact of uncalcined TiO_2 was sintered under CRH conditions that deviated from straight-line behavior at low temperatures, perhaps due to a wider particle size distribution.

The zirconia (zyttrite) shrank 23 percent, and exponential behavior was observed up to 10 percent linear shrinkage. The effective activation energy, E, was 29 kcal/mole. If volume diffusion kinetics predominate as reported by Jorgensen [8], this will give an activation energy, Q, of 58 kcal/mole compared to the previously reported value of 92.5 kcal/mole. Grain boundary diffusion would give better agreement with 87 kcal/mole. The shrinkage range for these experiments was 600–1250°C.

Three different CRH experiments are shown in Figures 7 and 8. The alumina used in these experiments was dry-ground and elutriated to three size ranges. The 5–8 μ material follows grain boundary diffusion kinetics over the entire sintering range (1100–1500°C), with an effective activation energy, E, of 40 kcal/mole and 1.8 percent linear shrinkage. The 1–2 μ material gave an E of 97 kcal/mole up to 1200°C and 0.24 percent shrinkage, and then broke to an E of 37.5 kcal/mole for the remainder of the 4 percent shrinkage. The 2–5 μ alumina which sintered to 2.9 percent appears to be an intermediate stage between the 1–2 μ and 2–5 μ size ranges. This alumina sintered with an E of 67.5 kcal/mole up to 1225°C and 0.23 percent shrinkage finished with an E of 34 kcal/mole. It is expected that the exponential relationships would hold to much longer shrinkages than shown in Figures 7 and 8, but the CRH experiments were terminated at the upper limit of the furnace.

The effect of particle size of the sintering of alumina has been seen in experiments with several different aluminas and has given results similar to those shown in Figures 7 and 8. The influence of size is further illustrated in Figure 9 for a CRH experiment with dry-ground, unsized alumina. This material which had a large particle size distribution (10–0.1 μ) sintered with an E of 63 kcal/mole up to 1200°C and 1 percent shrinkage. The remainder of the 9.6 percent shrinkage was accomplished with an E of 39 kcal/mole.

The effect of surface diffusion was studied by annealing a pressed compact for 100 hours at 830°C and sintering under the regular CRH conditions. The results are also shown in Figure 9. The sample sintered 15 percent, with the first 1.5 percent occurring up to 1150°C and giving an E of 93 kcal/mole, and the remainder showing an E of 47 kcal/mole. This is in contrast to an unannealed sample in which the initial and final effective activation energies, E, were 75 kcal/mole and 40 kcal/mole, respectively.

Although there is good evidence that surface diffusion interferes with shrinkage and gives rise to a superficially high activation energy as predicted by John-

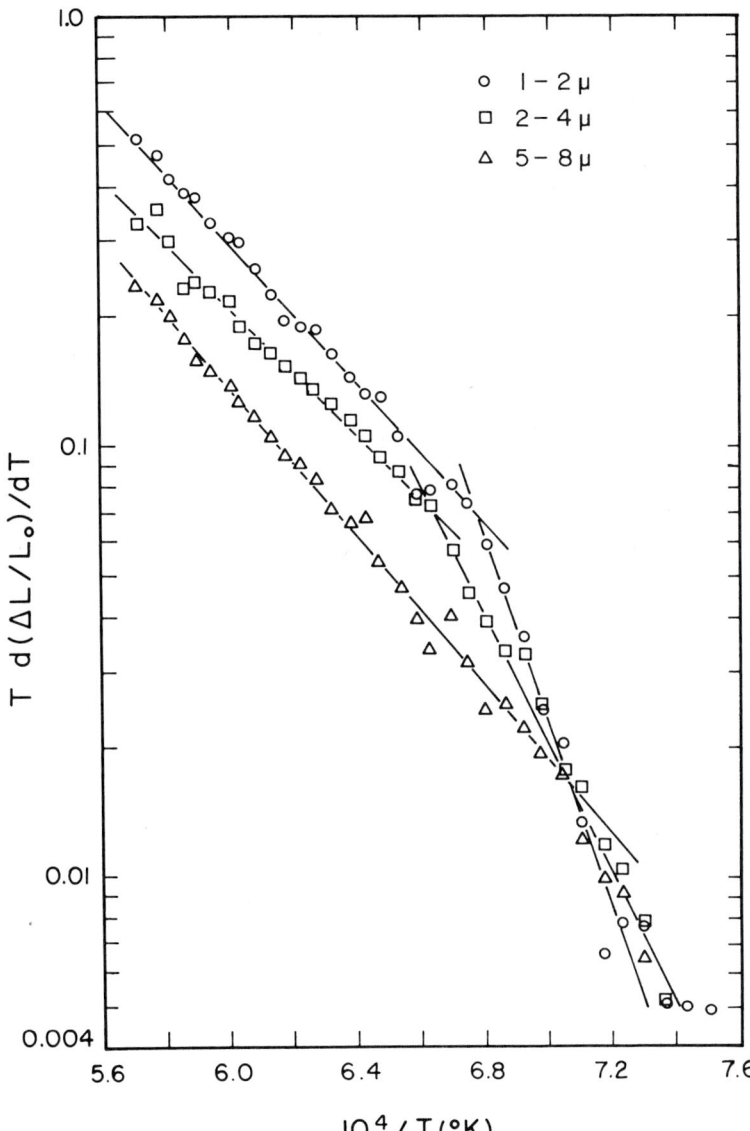

Figure 7. CRH experiments of Al_2O_3 sized 1–2 μ, 2–5 μ, and 5–8 μ, at a constant rate of 4.0°C/min, according to equations (6) and (8).

son [2], there is also a possibility that smaller particles may sinter by volume diffusion. Further research in this area is well justified.

The results for the sintering of feldspar are shown in Figure 10, where the data have been plotted according to equations (4) and (5). The sintering range was

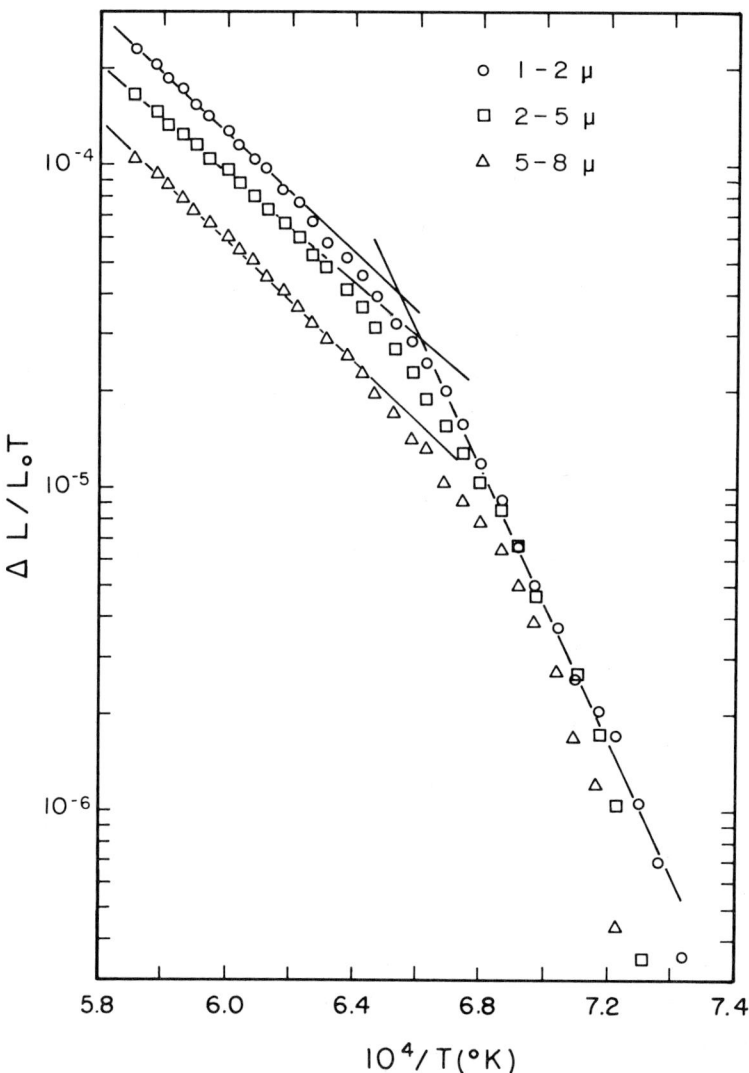

Figure 8. The same data, but plotted according to equation (14).

1050–1250°C and a total of 11 percent linear shrinkage was recorded. The effective activation energy was approximately 73 kcal/mole. The sample showed some warping, which may have resulted in abnormally high values of shrinkage and $d(\Delta L/Lo)/dT$ at the highest temperatures.

In the case of sintering of glass spheres, either by isothermal shrinkage or CRH experiments, agreement with theory is far better than would be expected on the

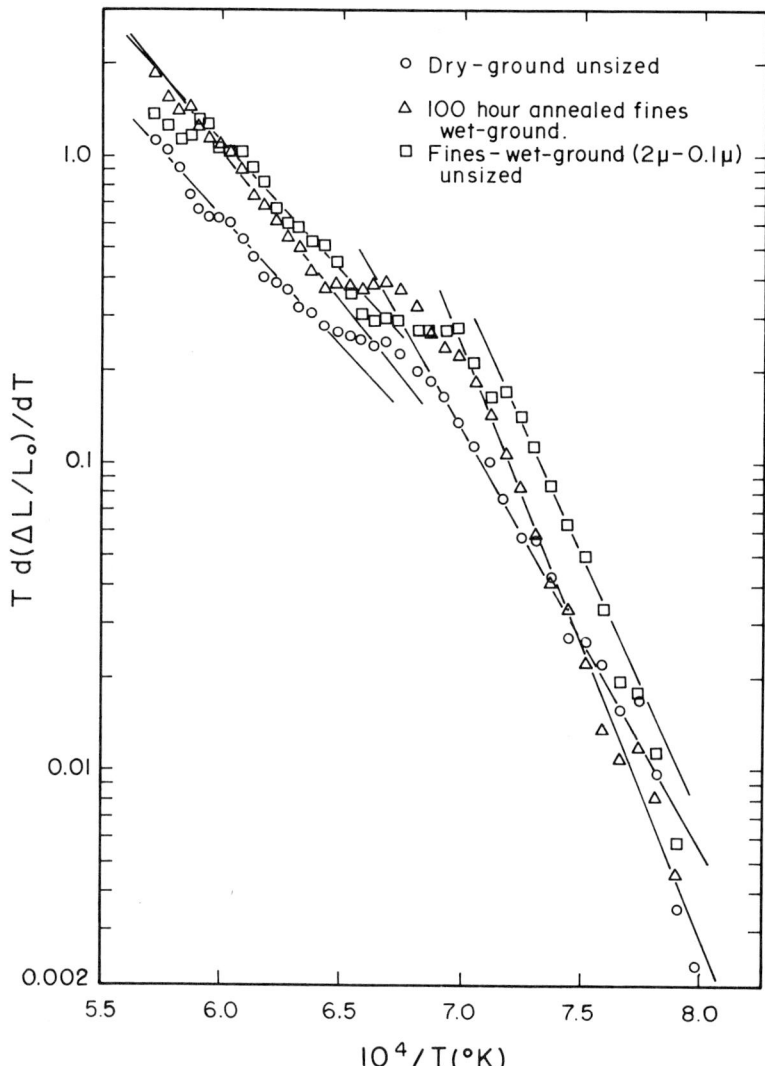

Figure 9. Shrinkage data indicating the effect of surface diffusion on sintering of Al_2O_3. One powder compact was annealed for 100 hours at $830^\circ C$, where surface diffusion was thought to be active, and then heated at a CRH while monitoring shrinkage. This is compared to a CRH experiment with no pre-annealing.

basis of the simple geometrical assumptions, upon which the theory of initial sintering kinetics is based. Johnson has noted that the equations are quite accurate up to 3 percent shrinkage. This would suggest that agreement with theory at 5 or 10 percent shrinkage would be fortuitous. In the case of spherical glass

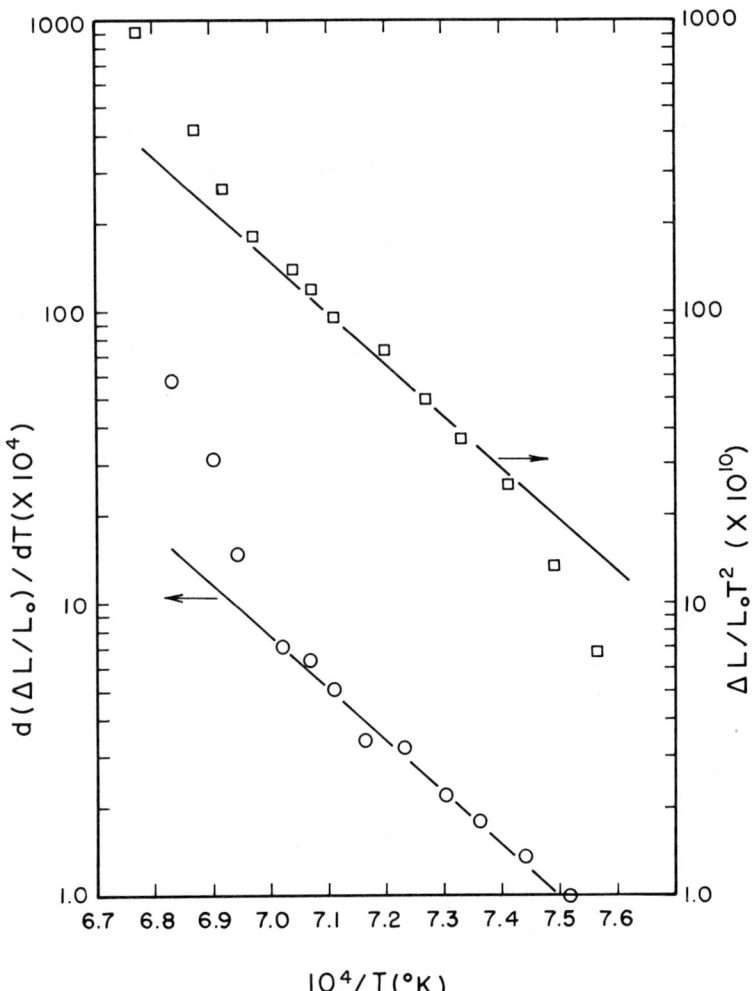

Figure 10. CRH data for a potassium feldspar heated at $2°C/min$ and plotted according to equations (4) and (5).

particles, good agreement with theory is obtained up to 12 percent linear shrinkage. Since pores shrink at the same rate that particles shrink, this suggests that the rate of shrinkage of pores may be a better model to describe initial sintering kinetics from 5–10 percent linear shrinkage than the shrinkage of spherical particles in contact with each other.

In the case of crystalline materials, agreement with theory is surprisingly good. From isothermal shrinkage measurements, it was believed that perhaps 5 or 6 percent linear shrinkage would be the very best that could be expected from equations of this nature. It is obvious from the data that 10–12 percent linear shrinkage is not unreasonable for the range of shrinkage during which the initial sintering kinetic equations describe the shrinkage of crystalline materials. Once again, it is supposed that deviation from the theory occurs when the number of pores begins to decrease through pore closure and shrinkage to the point where pore annihilation takes place.

During isothermal shrinkage measurements, the rate of shrinkage decreases with time. Over the substantial time period involved in isothermal experiments, there is an opportunity for grain boundary migration and the resultant loss of grain boundary sinks for vacancy condensation. It is supposed that the reason CRH experiments agree with the theory to greater shrinkages than isothermal experiments is because of the decreased opportunity for grain boundary migration during the short interval of time involved in the CRH experiments. During CRH experiments the rate of shrinkage, instead of decreasing rapidly, increases exponentially with time. During CRH experiments there is hardly enough time to obtain appreciable interference from grain boundary migration.

Initial sintering theory is based on spherical geometry for particles in contact with each other. Actually, spherical particles will be the equilibrium shape for particles of glass, but will not be the equilibrium shape for crystalline materials. In the case of crystalline materials, the equilibrium crystal morphology is the shape of powder particles that will best fit the initial shrinkage equation. This means that the particles having their equilibrium shape will not be changing shape during the sintering experiments.

Summary

It has been shown that equations for shrinkage of powder compacts heated at a constant rate show exponential relationships that describe rather accurately the experimental shrinkage of many types of materials. For the understanding and evaluation of the sintering characteristics of real powders, an effective activation energy and reactivity coefficient have been described which can correlate the initial shrinkage characteristics of powders and make available in nearly any laboratory a better evaluation of the reactivity of ceramic materials.

Acknowledgment

The authors gratefully acknowledge the financial support of the Atomic Energy Commission under Contract AEC–AT(11-1)–1122 and the U.S. Bureau of Mines under Contract SWD-5 at the University of Utah.

202ULTRAFINE-GRAIN CERAMICS

References

1. Frenkel, J., "Viscous Flow of Crystalline Bodies under the Action of Surface Tension," *J. Phys.* (USSR), 9 (1945), 385–91.
2. Johnson, D. L., "Solid-State Sintering," this volume.
3. Cutler, I. B., "Sintered Alumina and Magnesia," in *Refractory Oxides,* A. M. Alper, ed., Academic Press, New York, in press.
4. Johnson, D. L., and Cutler, I. B., "Diffusion Sintering: II. Initial Sintering Kinetics of Alumina," *J. Am. Cer. Soc.,* 46 (1963), 545–50.
5. Cutler, I. B., and Henrichsen, R. E., "Effect of Particle Shape on the Kinetics of Sintering of Glass," *J. Am. Cer. Soc.,* 51 (1968), 604.
6. Cutler, I. B., "Sintering of Glass Powders During Constant Rates of Heating." *J. Am. Cer. Soc.,* 52 (1969), 14–17.
7. Anderson, H. U., "Initial Sintering of Rutile," *J. Am. Cer. Soc.,* 50, (1967), 235–38.
8. Jorgensen, P. J., "Diffusion Controlled Sintering in Oxides," in *Sintering and Related Phenomena,* G. C. Kuczynski, ed., Gordon and Breach, New York (1967).

11. Hot Forming of Ceramics

ROY W. RICE*
The Boeing Company
Seattle, Washington

ABSTRACT

This chapter discusses hot-consolidation of ceramic powders with and without additives, and hot working of ceramic bodies. Hot consolidation (e.g., by hot pressing) of powders without sintering additives is shown to generally produce the finest grain sizes. The resultant bodies generally have the highest strengths to several hundred degrees Kelvin. Additives such as LiF and NaF to oxides (e.g., MgO, CaO, and Al_2O_3) and of metals (molten at pressing) to carbides are shown to often aid consolidation and formability (e.g., allowing hot rolling). However, the use of such additives also often leads to some penalties such as greater grain growth or lower elevated temperature strengths. The major problem of hot consolidation of oxides, with or without additives, is shown to be removal of the last traces of impurities such as H_2O and CO_2, which are usually present in reacted form (e.g., hydroxides and carbonates), while other anion impurities may cause problems in carbides and borides.

Hot working (i.e., causing plastic deformation within grains) is discussed both for starting with solid bodies, and starting with powders where consolidation and working are done in the same operation. Neither method has yet produced bodies of as fine a grain size as can be achieved by hot pressing. However, elevated temperature strengths of some hot worked bodies are shown to be much greater than those achieved by hot pressing. These higher strengths result from hot working bodies with substantially reduced anion impurity levels, as well as obtaining both texturing and slower grain growth in the hot worked bodies. Other advantages (e.g., improved optical properties) may also be obtained by hot working.

Possible solutions to the problems of impurities left after consolidation and of reducing grain sizes of hot worked bodies are discussed. Obtaining powders with more controlled composition and finer particle and agglomerate sizes will help both problems. Some comments on the means of obtaining such powders are briefly outlined.

---

*Presently: Naval Research Laboratory, Washington, D.C.

Introduction

Hot forming of ceramics encompasses a variety of materials, methods, operations, and goals. Materials can be categorized as all or mostly glass, crystalline with some glass, or all-crystalline. The latter is clearly most pertinent to this volume and to the author's experience, and hence will be emphasized. However, some informative examples from the former two categories will also be presented. The crystalline materials discussed most will be the oxides, primarily the simple oxides such as MgO, CaO, and Al_2O_3, because they have been studied most. However, some results will also be given for carbides and borides.

Powder or solid bodies may be used without additives, or with glass or liquid-phase additions. The choice of these methods depends on whether the forming operation is hot pressing, hot rolling, normal forging, press forging, or hot extrusion. Both the method and operation depend on whether the goal is primarily consolidation, or primarily working. Both can be done at once, but then working is the major factor in determining operational parameters. Working will normally be hot working, that is, plastic deformation by slip or twinning processes at such a temperature that most of the lattice strain from the deformation is removed by recrystallization at the completion of the process. This requires more extreme parameters of temperature, pressure, and time.

The above comments listed only hot consolidation or working as goals because these are the primary goals. Shaping is also an important goal, but only in conjunction with consolidation or working. While substantial reduction in costs should be achieved, hot forming operations will generally be more expensive than more conventional forming operations of cold consolidation (pressing, slip casting, etc.) and sintering. Thus, if shape is the primary requirement and properties are secondary, hot forming will usually not be justified. However, where the superior properties achievable by hot forming are needed, then shaping in the hot forming operation can, of course, be an important benefit.

All of these methods and operations to achieve consolidation or hot working will be reviewed. A great deal of progress has been made, and more should be feasible. However, the quality of starting materials, especially powders, will be shown to be a major problem in further development, thus making this volume timely. (Some of the author's thoughts on powder preparation and processing will be briefly outlined in an appendix to this chapter.) Because of the scope of the subject and other pertinent reviews [1-6], this chapter will emphasize recent developments and aspects not emphasized previously. In particular, some of the established or emerging problems of the various techniques will be emphasized, especially in the more established or conventional hot forming methods.

· Hot Pressing

· Technique

The general technique of hot pressing (or pressure sintering) is well known. In simplests terms it merely requires a method of heating and pressing a sample in a die. For most applications, the entire die is normally heated, either by a separate furnace or by direct heating of the die. The latter is usually done with a conductive die, where inductive heating is most common, but resistance (Joule) heating has been reported [7]. In some applications, especially those at lower temperatures, internal heating can be used by insulating the outer die from inner resistance windings [8].

Graphite is by far the most common die material, but others have been used or considered, as shown in Table I. Graphite is the least expensive, most readily available material (often coming in a wide range of rod stock sizes), and the easiest to machine and handle (because of its low density). The major reasons for not using graphite are its pressure limitations and its reduction of some materials, particularly oxides such as ferrites, titanates, NiO, and CeO_2. Commercial graphites are usually used to about 5,000 psi, though sometimes to about

TABLE I

Uniaxial Pressure-Sintering Die Materials*

Die Material	Maximum Use Temperature (°C)	Maximum Pressure (psi)	Remarks
Graphite	2500	10,000	Inert atmosphere usually required.
Al_2O_3	1200	30,000	Difficult to machine; careful align-
ZrO_2	1180	−†	ment and loading needed; limited
BeO	1000	15,000	thermal shock resistance; creep
MgO	−†	−†	limited.
SiC	1500	40,000	Difficult to machine; reactive with
TaC	1700	8,000	some materials; expensive.
WC, TiC	1400	10,000	
TiB_2	1200	15,000	Difficult to machine; expensive.
W	1500	3,500	Easily oxidized; creep limited.
Mo	1100	3,000	
Inconel X, Hastelloy,			Used primarily for halides; creep
Stainless Steel	1100	varies	limited.

*After Spriggs [4].
†Not known.

10,000 psi (usually only for thin specimens). A new higher strength, undoped graphite (POCO Graphite Inc., Garland, Texas) can be used to about 20,000 psi. Its cost is two to three times that of conventional graphites, but about half that of most metals for dies (at least to make parts about $1.5''$ in diameter). Higher thermal expansion and some creep of this high-strength graphite have created some problems in specimen ejection. Further development of the body and tapering and radiusing the ejection side and edge, respectively, can reduce this problem. Since ram failure is usually the main pressure limitation, even with short rams, this stronger graphite is sometimes used only for rams. (The high-strength rams must be undersize because of their greater expansion than the radial expansion of most die bodies from extruded graphites). Carbon or graphite bodies reinforced with carbon or graphite filaments (commercially available as Carb-i-tex, Graphite Specialties Div. of Carborundum) also show promise for higher pressures. Some problems of expansion under load and delamination have apparently occurred, but are being reduced. Such composite dies often run two to three times the cost of regular dies, but development of stock standard sizes of simple shapes, or usage in large numbers, might reduce prices substantially.

Metal dies, especially the molybdenum alloy dies, are the next most commonly used. Apparently the largest use of these is for the production of hot pressed polycrystalline optical materials such as MgF_2, ZnSe, and MgO. Their main advantage is higher pressure capability at lower temperatures (e.g., below about $1300°C$ for molybdenum-based materials), and general lack of effect on ceramics—especially the lack of reducing effects. Their cost is the main problem, and their weight can also be a factor. Welding of rams to the die can be a serious problem if not prevented by using a coating on either the die or ram, or an undersized ram. Spray graphite coatings are most common, but washes made from ceramic powders compatible with the material being pressed have also been used.

Oxide and other ceramic die materials have been used only for relatively small laboratory specimens; except for Al_2O_3, which is used for commercial hot pressing of titanate transducers. The main advantage of ceramic dies is their compatibility with a particular material or its required atmosphere. Thus, an oxide die such as Al_2O_3 can be used in an oxidizing atmosphere to hot press dense NiO bodies [9, 10]. However, compatibility with materials can also be a limitation of such dies. For example, the author has found that CaO hot pressed in Al_2O_3 dies bonded to the die, as did some MgO bodies from certain reagent-grade powders. The latter was attributed to the amount or form of CaO or SiO_2 impurities. Some of these problems can probably be solved by using suitable washes on the dies or by limiting temperatures. Ceramic dies can often be used at higher pressures, but this usually requires dies that have themselves been hot pressed, since sintered commercial bodies such as Al_2O_3 have generally not been found satisfactory unless provided with external support (e.g., a surrounding molybdenum sleeve). Use of ceramic dies can also be limited due to specimen ejection problems because of the higher thermal expansion of these dies. Costs can also be high, though wider use, or use in large numbers could make substantial reductions possible. Thus, use of ceramic dies is likely to be mostly for special applications for which they have particular advantages.

As noted above, washes or sprayed films (e.g., graphite) may be used on dies and rams to aid in the release of parts. Pyrolytic graphite (PG) disks or tape ("Grafoil": High Temperature Materials Div., Union Carbide Corp.) and graphite cloth have also been used, especially PG (in both graphite and metal dies), since it gives a smooth finish. In dies that are more expensive because of shape or materials used, replaceable liners are sometimes utilized. These can also aid release of the specimen, since the liner can be slipped out if the specimen cannot be ejected. Moss and Stollar have provided some design guides for hot pressing dies, especially those with liners [11].

The major parameters of hot pressing are of course temperature, pressure, and

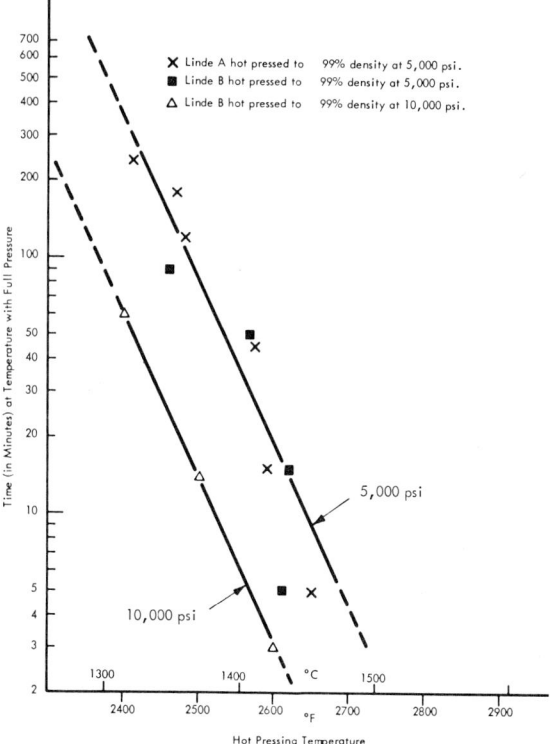

Figure 1. Sample data showing some hot pressing variables for Al_2O_3 (after Rice [13]).

time. One or two of these can generally be lowered by raising one or both of the others. Some more detailed studies of these relationships are beginning to be made (e.g., see Figure 1), but much more needs to be done. However, better control and characterization of powders is generally needed to eliminate variability. The method of applying pressure, the rate of heating, and their relationship have not been fully studied, but it appears that there is considerable latitude. Controlled hot pressing rates for $MgAl_2O_4$ have been proposed by Kriegel et al. [14], and studied by Rummler [15].

Heating rates are generally a compromise between fast rates to minimize particle growth and pre-sintering, and slower rates to reach thermal equilibrium and adequate outgassing, especially with fine powders which are needed to obtain fine grain sizes. Larger specimens may often require longer heating times because of poor thermal conductivity and outgassing of powders.

Hot Pressing Results

Materials Pressed with Densification Aids

The author [12, 16] (initially assisted by V. Edlin) observed that additions of LiF (typically 1/2-2 w/o) were even more beneficial to the densification of MgO in hot pressing than Atlas [17] had found it to be in sintering. Carnall and colleagues of Kodak were independently developing this [18], and Gray [19] of Alfred was also apparently using a similar approach. Since the author's original presentation [16], other investigators have further developed and improved this system [19, 20]. Typical parameters are shown in Table II. This appears to be the easiest method of producing transparent MgO, and apparently the most transparent MgO.

While transparency can be achieved without vacuum hot pressing [12, 16], such pressing does appear to improve both the quality and consistency of the body [21, 22]. Firing of hot pressed bodies to 1300-1400°C adequately reduces the residue of the additive and improves strengths, so that the resultant bodies are as strong as those of comparable (e.g., same density and grain size) MgO made without additives at all temperatures [12, 20, 22], at least in laboratory-size specimens. However, grain growth in the MgO made with LiF is higher, at least initially, so maximum strengths (to 60,000 psi) are not as high due to the larger grain size. MgO with LiF behaves in a fairly plastic fashion at low strain rates (e.g., 0.01 min^{-1}) and shows considerable formability [12, 16]. (This, however, could also be a problem for fabrication of large parts since this results in transmission of more radial force to the die.)

Rice [12, 22] showed that NaF also aided the densification of MgO, as shown

TABLE II

Hot Pressing With Densification Aids

Material	Additive	Hot Pressing			Percent of Theoretical Density	References
		Temp. ($^\circ$C)	Pressure (psi)	Time (min)		
MgO	1-3 w/o LiF	1000	1,500-5,000	5-15	T[1]	8, 9, 13, 14, 15
MgO	1 w/o LiF	850	10,000	–	T[1]	11
MgO	1-2 w/o NaF	1050	4,000	10	98-99	9, 15, 22
MgO	1-2 w/o NaF	1200-1300	3,000	240	T[1]	23
CaO	2 w/o LiF	1000	5,000	15	99	24
CaO	2 w/o NaF	1000	5,000	15	99	24
Al_2O_3	2 w/o LiF	1100	5,000	10	98-99	13, 22
UC	20-50 v/o U[2]	1135	2,000	5	–	26
TaC	20-50 v/o Co[2]	1500	2,000	5	99.5	26

[1]T indicates near theoretical density and some resultant transparency.
[2]Almost all additive is removed during the pressing operation—see text.

in Table II, and that residues from it can also be adequately reduced. Smethurst and Budworth [23] further perfected this, so transparency could also be achieved with NaF. They also observed that finer grain size bodies can be achieved with it (e.g., about 2 μ with NaF and 10-20 μ with LiF after firing to 1300-1400°C). Recent cursory checks by the author corroborate this.

Both LiF and NaF aid the densification of CaO [22, 24] (Table II), but the removal of residues is not nearly as easy as in MgO. Thus, while room-temperature strengths are the same for comparable bodies made with and without additives, elevated temperature strengths of CaO made with NaF or LiF are generally much lower due to residual grain boundary liquid phases [22, 25]. CaO also shows good formability with LiF or NaF [24], similar to that of MgO with LiF.

Aluminas (gamma or alpha) also densify at lower temperatures with LiF additions [13, 22], as shown in Table II. However, exaggerated grain growth even begins during hot pressing, which has probably been a factor in limiting strengths to maximums of 50,000-60,000 psi. Elevated temperature tests have not been made, but analysis indicates that additive residues may be adequately removed by firing to about 1300°C to prevent impairment of high-temperature strengths.

Stoops [26] has reported that carbides can be pressed to near-theoretical density by using molten metal additives that are squeezed out during the operation. (See Table II for sample results.) A reservoir is machined in the die, so the liquid can be squeezed out of the densifying body, down along the bottom ram and into this reservoir, apparently leaving only small amounts of metal (e.g., 0.1 percent of Co in WC), mostly at triple points. He reports that the material must be somewhat soluble in the liquid phase, and that powders with about 4 μ particles apparently gave best results.

Materials Hot Pressed Without Densification Aids

The data of Table III shows that near-theoretical density can be achieved by hot pressing a variety of materials with a variety of parameters. This data also shows that use of higher pressure generally allows finer grain sizes to be achieved through lowering of the temperature. Transparency (usually not quite as high as with LiF), has been achieved in some of the MgO bodies pressed by all of the combinations shown in Table III.

Data of Table III also show that room-temperature flexural strengths of the order of 100,000 psi can be achieved. Gazza [27] has reported diametral compression strengths of Al_2O_3 up to 45,000 psi, which is in the same range or higher considering the expected flexure/diametral strength ratio of 2 to 3. The author [33] has observed strengths of hot pressed Al_2O_3 of about 160,000 psi at liquid nitrogen temperatures, and as high as 63,000 psi at 2400°F (1315°C).

In most of this work, the best specimens have usually been produced by vacuum hot pressing.

Hot Pressing Problems

Oxides: A great deal of advancement has occurred in hot pressing, but some basic problems exist which should be examined.

The author has recently reviewed one of the major problems in hot pressing of oxides [22]: gaseous impurities such as carbonates and hydroxides. Sulfur also appears to be an additional, and possibly the major problem in Al_2O_3. Briefly, there are two extremes of this problem. The first, represented by CaO, is essentially one of eliminating gases during hot pressing. Very slow heating (e.g., 1000°C or less an hour) is necessary to prevent entrapment of gases, since this entrapment results in explosions that will break the specimen and die (but will usually not be too serious otherwise). The frequency and severity of these explosions tends to increase with finer powders and larger specimen size. Once fabricated, dense CaO bodies generally show little or no outgassing on subsequent firing.

The other extreme is Al_2O_3, which shows no problems of consolidation, but can manifest outgassing problems during subsequent firing. Both alpha and gamma aluminas hot pressed with LiF (see Table II) show weight losses, and development of up to several percent porosity due to outgassing of H_2O and CO_2, while heating to temperatures of only about 1100°C. On the other hand,

TABLE III

Hot Pressing Without Densification Aids

Material	Approximate Starting Particle Size (μ)	Temperature (°C)	Pressure (1,000 psi)	Time (min)	Percent of Theoretical Density	Approximate Typical Grain Size (μ)	References	Maximum Room Temp. Strengths (at Grain Size in Microns)(2)	Strength References
MgO	0.05	R.T.	1,000	–	T(1)	0.05	28	–	–
MgO	0.05	800–1000	250	10		0.05	14, 5	25,000	4
MgO	0.05	850	57	Not Reported	T(1)	–	18	–	–
MgO	0.05	1100–1400	10	15	T(1)	0.5–8.0	4, 5, 12, 29, 30	84,000(2)	31, 32
Al_2O_3 (ϵ)	0.03	R.T.	1,500–6,000	–	T(1)	0.03	28	–	–
Al_2O_3 (γ)	0.03	1200–1400	5–10	30–90	99+	0.5–4.0	4, 5, 13, 27	85,000(1)	33
Al_2O_3 (α)	0.5	1200–1400	5–10	30–90	99+	0.4	4, 5, 13, 30 34	100,000– 120,000(l)	4, 5, 33, 35
ZrO_2	0.01	1300	20–25	30–250	99+	0.3–0.5	–	–	–
BeO	3	1700–1800	4	60–240	99+	30–50	30, 37, 38	35,000(35)	30, 37, 38
HfC	3	2500	4	15	98–99	30	41	–	–
TaC	0.01–0.1	1600–2000	6.5–10	40–90	99	3–10	42	–	–
TaC	5	2200	6.5	60	98	30	42	42,000(32)	42
TiC	2–5	1800–1900	6	15–50	99+	4–12	43	96,000(6)	43
ZrB_2	3–5	2150	3–5	120	98–100	20–50	44, 45, 46	30–50,000	44, 45
ZrB_2+Si or C	3–5	1950–2050	3–5	120	98–100	3–5	44, 45, 46	40–60,000	44, 45

(1)T: near theoretical density so some transparency was achieved.

(2)Flexure strengths. Although test parameters will vary strengths, these figures give a general idea of the best strengths.

Al_2O_3 vacuum hot pressed without additives (see Table III) sometimes shows gross bloating or enormous grain growth above about $1650°C$. This occurs mostly in bodies pressed from gamma Al_2O_3 (converted to alpha during hot pressing) at lower temperatures with higher pressures. The grain growth will achieve grain sizes of several millimeters, but normally only in the central area of the section being fired. The ratio of center to surface grains may be 1000 or more. This problem has tentatively been attributed to sulfur (possibly as Al_2S_3) found in these bodies. Additions of MgO to the Al_2O_3 may control this exaggerated grain growth.

MgO shows some of both of the above problems whether pressed with or without LiF. Explosions are less frequent with MgO than CaO, even when faster heating rates are used. However, MgO will also sometimes have trapped gas bubbles (usually lenticular due to the pressure), whereas CaO usually does not. MgO that is successfully hot pressed will frequently cloud or bloat on subsequent exposure to temperatures of the order of $1000°C$, especially with faster heating rates. The last traces of hydroxide and carbonate impurities (typically a few hundred parts per million to begin with) are very tenacious. Some may still be present in bodies fired to $2000°C$. These impurities, which are quite variable, apparently can reduce as-hot pressed strengths by as much as 50 percent. Slow annealing can improve flexure strengths substantially, but at the penalty of some grain growth (and hence loss of potential strength). Further, such annealing may only be fully effective for bend strengths since the depth of such anneal-purification is limited, leaving only a strong surface, which carries the major load in bending, but would be of much less help in other stressing such as pure tension.

It is emphasized that the above problems are not restricted to the three oxides used as examples, but will occur in varying degrees with other oxides, generally more severe in finer powders used to get finer grain sizes. For example, anion impurities have been found or indicated to be problems in NiO [22], BeO [38] (e.g., sulfates and fluorides), and $MgAl_2O_4$ [22, 39].

Variability is also an important problem with fine oxide powders. Variations occur from lot to lot, within lots, and with storage, which limit repeatability in achieving the finest grain size with full density and, in fact, the reliability of achieving or maintaining transparency (with subsequent heating). A great deal of this variability appears to be due to variations in gaseous impurities.

Although finer grain sizes can be achieved by using higher pressures with lower temperatures, higher strengths are not obtained. Pressures beyond 5,000–10,000 psi have not produced better specimens. As the author has previously discussed [22], this can be directly attributed to the retention of greater quantities of gaseous impurities at progressively lower pressing temperatures with progressively higher pressing pressures. These impurities apparently occur at the grain boundaries, causing more weakening than the expected strengthening from finer grain size (e.g., compare MgO strengths in Table III). While no mechanical data is available for the MgO and Al_2O_3 bodies that Montgomery *et al.* [28] pressed at room temperature, they note that these crumbled and reverted to powder on annealing. These problems also increase with the size of the body to be hot pressed.

Vacuum hot pressing helps somewhat, but clearly does not eliminate the problem of gaseous impurities. In fact, most of the above problems were observed with vacuum hot pressed bodies. Slower heating cycles also help, but again do not solve the problem since this increases the cost of pressing, and probably increases resultant grain sizes.

Another problem is the voluminous nature of the fine powders needed to obtain fine grain sizes. For example, commonly used MgO powders undergo ten- to twentyfold reductions in volume during densification, and four- to fivefold from cold pressed to hot pressed bodies. Further, multiple cold pressing of one layer of powder on another has not been found acceptable in some cases. In MgO, for example, trace laminations between layers are apparently preserved by accumulation of gas released from the powder during hot pressing.

Carbides, Borides, and Nitrides: Three general improvements that would be desirable in these materials are powders of finer particle size, better controlled stoichiometry, and higher purity. For example, present ZrB_2 powders contain 6-10 v/o ZrO_2 impurities (about 1 percent O_2). (The SiC addition of Table III is used to react with impurities and control grain growth [46]). Finer particle size will make higher purity more difficult to achieve, since surface adsorption of gases such as O_2 and N_2 will be increased.

Future Development

Hot pressing is clearly the most widely used hot forming method discussed in this paper. Its use has expanded a great deal, not only in laboratory applications but in industrial use. Hot pressing of ceramic armour, cutting tools, and optical components each constitutes an annual sales level of from one to several million dollars, and the combination of several other less extensive industrial applications probably exceeds $1,000,000 annually. Thus, while small compared to major ceramic products, it clearly shows advancement beyond a laboratory process.

To expand these applications further, costs must be reduced, and hot pressed quality and uniformity must be improved. Some steps have been taken to reduce costs. Zehms and McClelland [47] have described a semi-continuous hot press where mating dies loaded with powder are loaded in one end of the press in tandem with other dies in various stages of pressing. By cyclic loading, and pressure transmitted through the chain of dies and specimens, the dies are cycled through the process. Thus, savings are made in the loading, heating, and cooling times, since loading a new die occurs simultaneously with moving another into the hottest zone, and heating and cooling are occurring during the pressing of another specimen. Gruintzes and Oudemans [48] have described a somewhat similar system for producing rods. One pressing ram is cycled so that small amounts of powder can be added and then rapidly pressed; the oxide rod is then advanced as the ram is retracted for another small loading of powder. A similar approach has also apparently been used in the production of some items at the Carborundum Company.

These approaches help, but gaseous impurities can also be problems if the highest quality hot pressed part is sought. The author has previously suggested that a basically different approach to heating may be helpful [22]. Presently specimens are heated from the outside-in, usually by so heating the die. This takes longer to reach thermal equilibrium, and may cause densification to progress from the surface inward, thus trapping unexpelled gases from the cooler interior. If powders were heated directly, they could be heated uniformly or progressively from the inside-out, depending on the use of other conventional

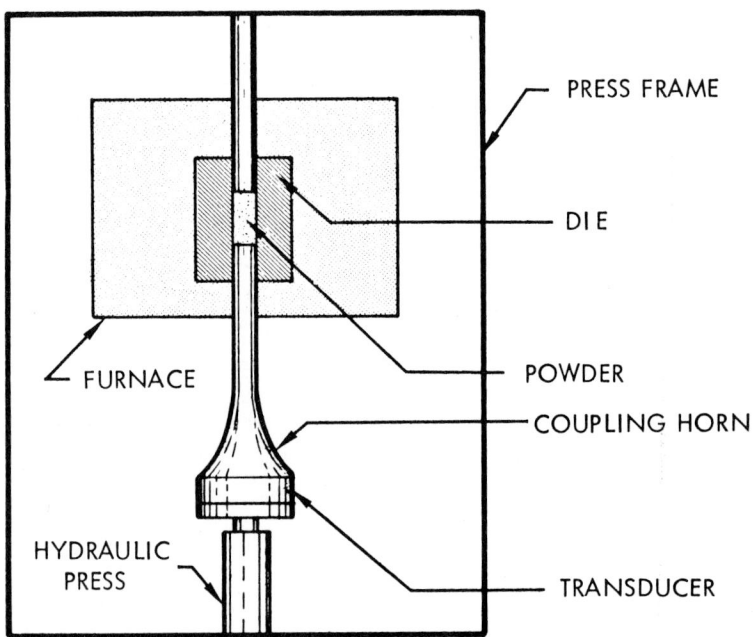

Figure 2. Schematic or ultrasonic hot pressing after Rice [22]. A furnace or other conventional heating would provide part of the heating (e.g., to within about $200°C$ of pressing temperature), and ultrasonic dissipation the remaining heating. Balance of the two could establish a controlled temperature gradient from the center of the specimen outward.

external heating. Use of ultrasonic energy, as sketched in Figure 2, has been suggested as one approach to accomplish this [22]. Dissipation of the ultrasonic energy in the powder would not only heat the powder directly, but may aid either compaction or removal of adsorbed species, or both. With conductive materials (e.g., carbides, borides, and nitrides), resistive heating using the rams as electrodes and an insulating die or a die with an insulating sleeve should also accomplish this. Such resistive heating may also aid densification by providing greater heating at the contact points of the particles.

Still another problem of hot pressing has been size. All of the technical and cost problems become compounded as one goes to larger sizes. Nevertheless, large sizes are being produced. Carbide hot pressings a foot or more in two directions and contoured in two directions have been made. Previously reported larger hot pressed sizes are shown in Table IV.

TABLE IV

Maximum Sizes of Hot Pressed Shapes*

Material	Diameter (in.)	Length (in.)
BN	14	14.0
BeO	10	10.0
	12	2.0
Al_2O_3	9	6.0
MgO	7	0.5
TiB_2	6	18.0
	8	8.0

*After Fulrath [1].

An area of application of hot pressing that may see further development is joining parts. Al_2O_3 [2], MgO [12, 29], CaO [24], NiO [49], CeO_2 [50], and TaC [51] are known to have been diffusion-welded by hot pressing parts, usually with a layer of powder in between. Weld strengths as high as 54,000 psi have been reported in such welds of Al_2O_3 [2], and the author has observed strenghts within 10-20 percent of this.

Hot Isostatic Pressing

Technique

Hot isostatic pressing is basically the same as normal isostatic pressing, except that the specimen is heated. This of course requires some basic changes, pri-

216   ULTRAFINE-GRAIN CERAMICS

marily in the pressure media. Gaseous media have been used almost exclusively, but granular or "sand" media have also been used [52].

Helium has been used almost exclusively for the gas media. With such gaseous media, the material to be pressed must be placed in a sealed metal "bag" or can. This can must be soft enough at pressing temperatures to transmit the pressure to the enclosed specimen, yet be non-reactive with the specimen. Canning is, in fact, one of the important skills to be mastered in this technique. The can and material are usually heated in a furnace inside of the pressure vessel, with the vessel walls kept "cold" (usually limited to temperatures of 100–200°C). This is accomplished by insulation between the furnace and the

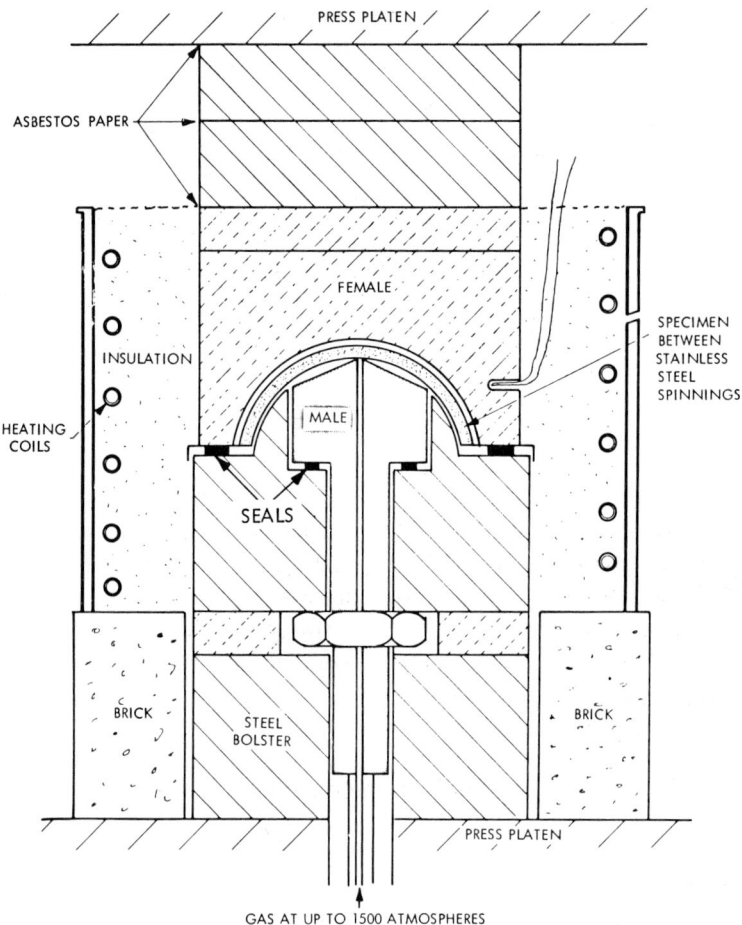

Figure 3. An example of specialized hot isostatic pressing. Compact simple facility for pressing irdomes (after Huffadine *et al.* [54]).

walls, and by water-cooling in the pressure vessel walls. A major problem can be extreme convection heat transfer in the gaseous media. In initial testing at Boeing, temperature variations of over 200°C from expected temperatures of 1200°C were experienced. Long and Snowden [53] report similar experiences, and showed that suitable baffling prevents this. The same approach is being followed in the redesign of the furnace in the Boeing facility.

For specialized configurations, the vessel can be designed to have little extra volume, which minimizes pumping and heating times. At lower temperatures, the entire vessel can be heated. These approaches are illustrated by Huffadine, et al. [54] in pressing of MgF_2 and ZnS irdomes, as sketched in Figure 3. They report using pressures up to 20,000 psi at temperatures to 850°C.

The second basic approach to hot isostatic pressing has been to use a granular or "sand" media for transmittal of the pressure. In this method the specimen is placed in a suitable sand which is in turn placed in a bag. The furnace is enclosed in the sand, which also acts as thermal insulation. This entire assembly is then placed in a conventional isostatic press using oil or water as the fluid. Various types of bagging for the sands have been successfully used, including

TABLE V

Hot Isostatic Presses*

Organization		Press Parameters			
Name	Location	Maximum[1] Temperature (°C)	Maximum[1] Pressure (in 1,000 psi)	Size (in.)[1] Diameter	Length
A. Gas Media:					
(1) Battelle Memorial Institute	Columbus, Ohio	750–2750[2]	10–50[2]	2.5–18[2]	3–58[2]
(2) NASA–Lewis Laboratory	Cleveland, Ohio	1650	50	3	12
(3) The Boeing Co.	Seattle, Washington	1300	30	3	10
(4) Atomic Weapons Research Establishment	Aldermaston, England	1250[3]	15	6	48
B. "Sand" Media:					
Atomic Energy Commission	Oak Ridge, Tenn.	3000[4]	30	42	42

*Known to the author.

[1]In many cases higher temperatures might be achieved using smaller diameter furnaces at the expense of specimen size, or at lower pressure.

[2]The range of about 8 presses, with the highest temperatures usually in smaller, or lower pressure units.

[3]See [53].

[4]See [52].

polyvinyl chloride, copper (O_2-free), lead, and stainless steel. The selection depends mainly on the temperature that the bag will reach. Various sands such as UO_2, BN, and MgO have been used, apparently mostly in 100–200 mesh sizes. A list of the hot isostatic press facilities known to the author, and their approximate parameters are shown in Table V.

Results

The results of hot isostatic pressing of ceramics are limited, both because the technique has not been extensively explored and because some of the applications have been proprietary, and many have probably been classified.

Huffadine *et al.* [54] report that domes of ZnS and MgF_2 up to 8″ in diameter have been pressed, using the apparatus shown in Figure 3. Cracking was sometimes a problem, but crack-free parts were obtained using boron nitride or talc as a mold release. Some domes were also transparent, but only if pressure was maintained during the first stages of cooling.

Leipold and Nielsen [55] have hot isostatically pressed MgO at 800°C with 15,000 psi for 30 minutes. The MgO had been isostatically pressed at room temperature at 90,000 psi (in rubber bags), then fitted in steel cans and vacuum outgassed at 250°C for 4 hours before loading in the hot isostatic press. About 90 percent of theoretical density was achieved, with resultant grain sizes of 0.09–0.3 μ, but cracking was a problem. Despite these low densities, the specimens were translucent, apparently because pores were below the size for efficient scattering. Reheating of specimens caused weight losses up to and over 3 percent. This was attributed mostly to decomposition of brucite in up to 10 w/o quantities. About 0.2 percent loss of carbon as CO_2 was observed.

Levey [52] reports several cursory trials where the sand media hot isostatic produced 95 percent of theoretical density BN, which appeared to be isotropic in contrast to partially oriented hot pressed BN. (84 percent density was achieved at 1500°C with 15,000 psi for 8 minutes). He also reports over 99 percent of theoretical density was achieved in UO_2 and W, and that WC with W additions showed good densification at 1800–2100°C for 30 minutes at 15,000 psi. The latter was found to have an average Knoop hardness of 2613.

Hodge [56] reports achieving 96–99.8 percent of theoretical density of Al_2O_3 at temperatures of 1150–1370°C with pressures of 10,000–20,000 psi for 30–180 minutes. Cracking was sometimes a problem, but could be minimized by using a carbon-free steel can and maintaining pressure during cooling. MgO and BeO were reported to be fully densified, and UO_2 up to 99.9 percent of theoretical density using similar parameters. WC and TaC were reported to fully densify between 1600 and 1760°C with pressures of 10,000–15,000 psi. High-density graphite structures were reported pressing at 1650–2300°C with 10,000–30,000 psi. The above materials were usually -325 mesh or coarser, with best densities achieved by prior cold isostatic pressing to 50 percent or more of theoretical densities.

Future Development

The opportunities for hot isostatic pressing for specialized jobs are good. The isotropic nature of the pressure, and often its higher levels, offer basic advantages. The capability of making simple shapes also hold much promise, and there are some opportunities in joining dissimilar materials. However, much more development is needed. Excess shrinkage means greater opportunity for can leakage or more non-uniform can deformation, with resultant cracking from constraint of the ceramic on cooling. Because of the very large shrinkages of fine powders needed for fine grain sizes, improved pre-consolidation techniques will be needed. In anticipation of this problem, the author has tried pre-hot pressing, then re-hot pressing with some degree of success [12]. Less success was obtained with an intermediate firing, but higher pressures of hot isostatic pressing may compensate for this. It may thus be possible to cold or warm press bodies, then fire them to achieve 70–90 percent of theoretical density before hot isostatic pressing. The re-hot pressing studies showed no significant increase in grain size, and this method should substantially reduce gaseous impurities.

However, gaseous impurities will probably be one of the most important problems in hot isostatic pressing bodies from fine powders to achieve fine grain sizes. The results of Huffadine *et al.* [54], showing the necessity of maintaining pressure during cooling (e.g., MgF_2), are probably due to trapped gases which weaken the body and limit its exposure to temperature. Leipold and Nielsen's [55] results clearly demonstrate this problem, and its much greater severity in hot isostatic pressing than in hot pressing. Since it is impractical to outgas the body during hot isostatic pressing, the body must be outgassed to the necessary level prior to hot isostatic pressing.

Hot Rolling and Swaging

Hot Rolling and Swaging Oxides

Technique and Initial Results

J. Hunt originally suggested that ceramics might be consolidated by encapsulating powders in suitable metal tubes and then consolidating the powder by drawing, swaging, or rolling operations at room and elevated temperatures. Evaluation of this idea with the author indicated that expected problems might be solved to make the process feasible, especially with bodies fabricated with LiF because of the formability of such bodies (MgO, CaO, and Al_2O_3). It was felt that hot rolling was the most desirable process since it did not require long-range stresses—as in drawing—to reduce the tubes, and could produce most of the shapes feasible with drawing or swaging, yet also provide wide pieces if desired. However, no simple method was seen of economically achieving the slow speeds and high temperatures that were expected to be necessary. Therefore, a year's study [57] was conducted on hot drawing and swaging. MgO + LiF was the main composition of study, but some work was also done on MgO without additives.

Hot drawing was tried first, since it was the easiest way to achieve accurate control of the temperatures and speeds expected. It was unsuccessful, since the steel tubes were not strong enough at elevated temperatures to be pulled or pushed through the die. Some success was achieved, however, by heating tubes to 700-1100°C in a furnace and rapidly feeding them (by hand) into a conventional swager. A few sections about 0.4″ in diameter and 0.5-1″ long, free of cracks and with some transparency, were achieved. However, repeatability was poor because of the difficult control of cooling and exact operation conditions. Some cursory work on cold rolling also showed promise.

Subsequently, a simple method of achieving both the slow speed and high temperatures for rolling was conceived and shown to be feasible. Since only consolidation and not hot working was sought, high drive forces were not needed, so a small, slow speed drive was satisfactory.

High temperatures were achieved by covering the rolls with insulation that had low compressibility, then heating the tubes in the roll and rolling with the insulation. A conventional mill with 6″-diameter rolls was used for a second year's study emphasizing rolling. Resistive heating of the tubes was found to be the best, with the method of Figure 4 being best of all.

Asbestos sheet insulation about 0.55″ thick was found satisfactory in preventing excessive heating of the rolls at the maximum tube temperatures of about 1100°C that could be achieved with the available power. However, heat loss from the tube near the roll caused a drop of 200-400°C in the vicinity of 1-2″ from the rolls with the selected resistive heating method, and 300-500°C with other resistive heating combinations that were tried.

Subsequently, transite pipe sections (8″ OD and 0.375″ thick) were found to work very well, producing the thermal profile shown in Figure 4B. There was some problem of delamination of the transite, but this was grossly reduced by baking it out at 200-300°C for several hours prior to use. Achieving this maximum temperature at the rolls resulted in increased densities from 96 to 99 or more percent of theoretical, and made these densities fairly reproducible.

Use of the above method and a special high-temperature rolling mill at the

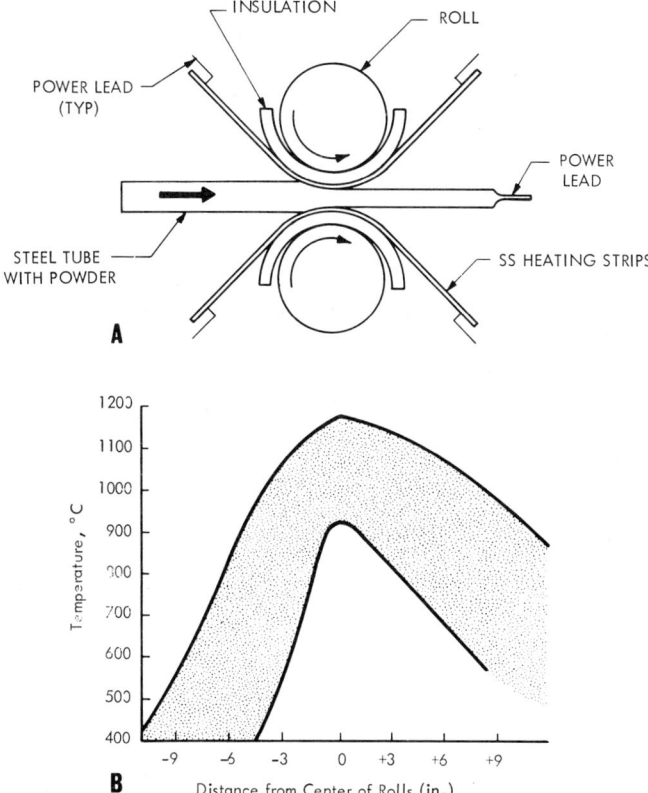

Figure 4. Hot rolling MgO + LiF. (A) Schematic of resistive heating system using a conventional rolling mill with 6″-diameter rolls, (B) Envelopes indicate temperature profiles achieved. Near-theoretical density was generally achieved with a maximum temperature of about 1000°C.

Bureau of Mines Research Station in Albany, Oregon (courtesy of H. Harris), showed that temperatures of 1000–1100°C and pressures of 3000–5000 psi were necessary, and that speeds should be slower than 1 in/min (the slowest that could be achieved without slipping of the tube). Higher initial packing densities, and avoidance of excessive preheating were found to aid the process. Figure 5 shows a sample of one of the better pieces that was obtained. Over twenty mechanical test bars were fractured at room temperature, showing strengths as high as 36,000 psi. These results, as well as other progress in the program, closely parallel the earlier development of hot pressing MgO + LiF. This is not surprising, since hot pressing parameters have been shown to be very similar to hot rolling parameters. Initial trials of MgO without additives suggest that it might also be hot rolled at about 1300°C, as often used for hot pressing.

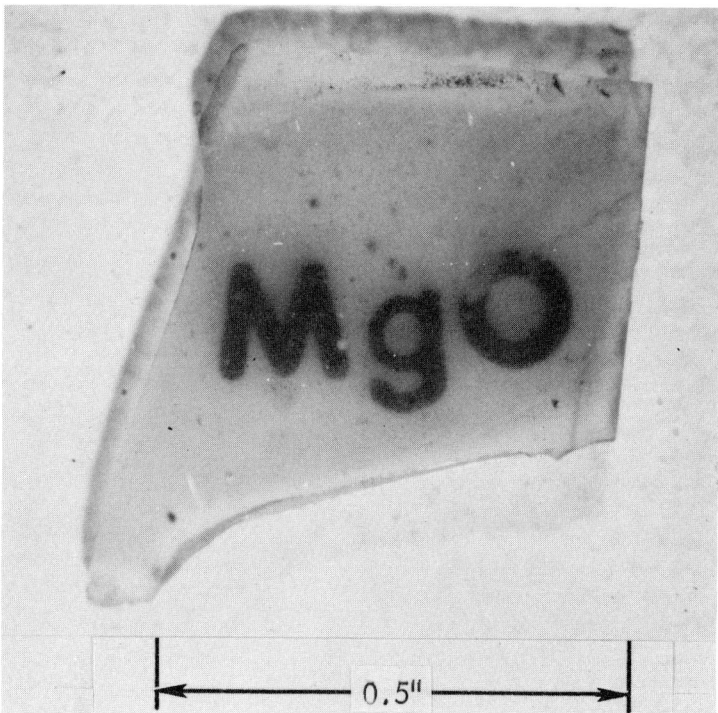

Figure 5. Sample of MgO + 2 w/o LiF hot rolled at approximately 1090°C, then fired to 1150°C in about 48 hours. Transparency is shown by letters MgO showing through specimen.

Problems and Future Development

The major remaining problems are gaseous impurities and cracking. Despite outgassing at 500°C for several days, gases were still trapped in rolled tubes. Putting venting holes in tubes just before hot rolling helped somewhat, but further improvements are necessary. Some cracking was caused by trapped gases, but differential stresses between the steel tube and the MgO were also a cause, especially when there were some irregularities in the MgO. Hot removal of tubes would eliminate this problem. The rate of consolidation also appears to have contributed to cracking. Larger diameter rolls would help, since the consolidation occurs over a longer length of the tube, and they allow slower speeds due to the greater friction from the longer length of contact. (Use of 10″ rolls is now being planned.)

In the above work, tubes were loaded by tamping, and higher starting densities generally gave better results, such as less laminar cracking. To be really economical, more efficient methods would be required. Green body extrusion shows some promise, and gives higher starting densities. Cold rolling of powders without tubes may be even more promising. Harris [59] reports that over 50 percent of theoretical density can be achieved (compared to about 30 percent in tamped tubes). Future development of consolidation by hot rolling might then entail cold rolling of powders, heating to remove binders, assembling a removable tube, then hot rolling in a turks' head (more than 2 opposing rolls). If necessary, sealed tubes could be used with continuous evacuation during final heating and rolling; however, suitable starting powders and control of powder handling might make this unnecessary. The hot rolling process is now showing good promise, but will require much more development.

Hot Rolling Oxide-Glass Bodies

Technique and Results

Harris *et al.* [60] constructed a special mill for hot rolling oxide-glass bodies. Stainless steel rolls (6.5″ diameter) for flat and shaped bodies were used with steady state roll temperatures up to 900°C and roll pressures of a few thousand psi. Cold pressed bodies were heated in a furnace, then transferred to the rolls. Best results were obtained with higher roll temperatures (700-900°C). Alumina bodies had to have at least 20 percent glass contents to roll well at the speeds used (typically 10 ft/min). Bodies having 30-50 percent feldspar additions were the most successful. With optimum specimen preheating temperatures (1400-1600°C), thickness reductions of at least 75 percent could be achieved. Some round shapes were also successfully rolled.

Limitations and Future Development

Apparently, the body requirements are that the compositions must not form interlocking grain structures, nor should the particle size be too small so the rate of solution in the glass phase is too high (e.g., difficulties were encountered using -325 mesh Al_2O_3 instead of -90 mesh). Thus, fine-grain bodies cannot be achieved, nor can high purity, because of the large amount of glass needed. However, further development at slower speeds and higher temperatures may reduce these limitations. Further, the existing process may have some applications because of the potential size and speed (i.e., lower cost) of production.

· Hot Forging

· Technique

Some forging has also been done by Harris *et al.* [60] with oxide-glass compositions using conventional metal forming equipment and techniques. Blanks of alumina-feldspar bodies similar to those they hot rolled were heated separately, then rapidly transferred by tongs to a forging press where they were forged in a fraction of a second.

Most forging work has been on pure refractory oxides. This has been done at what might be described as fast creep rates, i.e., much slower speeds than used for conventional metal forging, so the term "press forging" [4] has been suggested to distinguish this from conventional forging. In this process, conventional hot pressing equipment with graphite tooling has been used. Washes (e.g., BN), graphite cloth, pyrolytic graphite (solid or tape), or molybdenum foil (for Al_2O_3) have often been used on the ends of rams to prevent excessive reaction between the specimen and the rams. An oversized die is used so there is unconstrained lateral flow of the material. Some of this work has been extended to approach deep drawing, that is, to forge a dish- or cup-shaped body.

· Results

Oxide-Glass Bodies

Harris *et al.* [60] report that 7:3 and 8:2 alumina–feldspar bodies could be reduced as much as 80 percent in thickness (from 0.4″) after heating to 1300–1450°C. From 6:4 and 7:3 alumina–feldspar bodies they formed crucibles with diameters and heights of about 1.5″ (about 0.25″ wall thickness). This was done by heating the bodies to between 1470 and 1600°C and forming with tool steel dies heated to between 400 and 600°C.

Oxide Crystals

Rice [61] has forged a number of crystals, as shown in Table VI (see also Figure 6). Sapphire was the most difficult to forge since temperatures had to be within 100–200°C of the melting point to get deformation without cracking. Alumina-rich spinel was the next most difficult, with MgO and CaO being relatively easy. Cylindrical sapphire and ruby boules deformed to elliptical bodies

TABLE VI

Sample Crystal Forging Parameters

Material	Temperature	Percent of Melting Temp. (T_m)	Pressure (1,000 psi)	Percent Original Height	Strain (Ram Travel) Rate in 10^{-2} in/in/min	Typical Resultant Grain Sizes (μ)
MgO Crystals	1850°–2100°C	62–70	3–5	16	1–6	20–1,000[1]
CaO Crystals	1630°–1700°C	67–70	5	30	1–3	20–40
Spinel Crystals	1700°–1860°C	84–91	5	30	2–4	100
Ruby Crystals	1810°–1970°C	90–97	6	50	2	∞[2]
Sapphire Crystals	1700°–1960°C	85–96	6	45	2	∞[2]
TiC Crystals	1930°C	65	3–5	25	2	25

[1] Recrystallization was irregular, and it was difficult to distinguish between substructure and grain structure in some cases.
[2] Did not recrystallize, and so remained a single crystal.

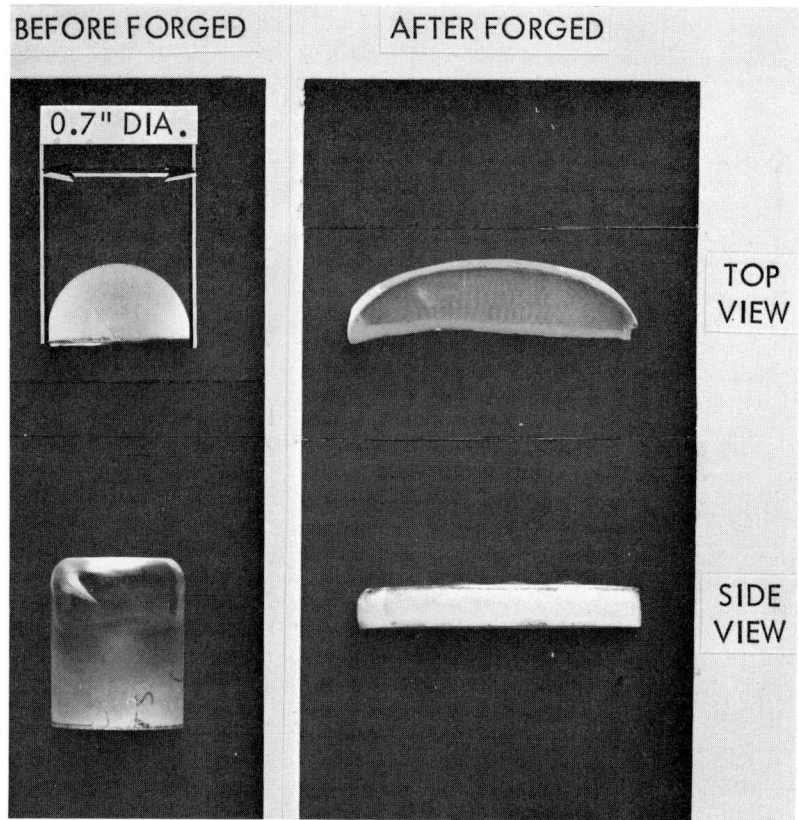

Figure 6. Sapphire forging. Before and after views of a split boule of sapphire forged at 1950°C. Note that lateral deformation occurred in only one direction.

indicating deformation only on the basal plane. Various alloy elements such as TiO_2 (in MgO), CoO (in $MgAl_2O_4$), and Cr_2O_3 (in Al_2O_3, MgO, and $MgAl_2O_4$) made no obvious differences in the deformation or recrystallization. Crystals were not successfully formed into shapes by lateral constraints, since required pressures rose substantially with much lateral constraint.

Sapphire and ruby did not recrystallize (i.e., remained highly deformed single crystals) while all other crystals underwent some recrystallization, with CaO recrystallizing the most. This was related to the number and type of slip intersections. For example Al_2O_3, with only basal slip, had no slip intersections, hence no nuclei for recrystallization. On the other hand, CaO, which has intersecting slip, should have easier interpenetration of such slip than in MgO due to the higher polarizability (i.e., deformability) of the Ca^{++} compared to the Mg ion [25]. Since Day and Stokes [62] have found proper interpenetrability of slip systems necessary to form stable recrystallization nuclei in MgO, this explains the greater recrystallization of CaO.

The recrystallized CaO bodies, though not fully random, showed reasonable agreement with the strengths of dense hot pressed bodies, but a much lower ductile-brittle transition temperature (about 1100°C).

Polycrystalline Oxides

A number of polycrystalline oxides have been press forged, usually at pressures of a few thousand psi, but some data using over 100,000 psi have been reported with temperatures in the range of 1000–1500°C [4,63]. MgO and Al_2O_3 have been investigated the most, but some trials have been made on other oxides (see Table VII). It has also been reported that La_2O_3, CeO_2, and $BaTiO_4$ have been press forged [3]. Some transparency has been achieved in forged MgO, Y_2O_3, and Al_2O_3 [3]. While this was neither new nor an improvement for MgO, it was new for Y_2O_3 [63], and an advance over other translucent polycrystalline Al_2O_3 [64–67] bodies (e.g., Lucalox, and hot pressed Al_2O_3 [30]).

Both MgO and Al_2O_3 [64] have been forged without prior hot pressing; in fact, starting with cold pressed bodies (which sinter during heating) allows easier forging. Al_2O_3 with 0.25% MgO additions can also be forged, but reaction with graphite (even with molybdenum spacers) is a problem, so dense starting billets

TABLE VII

Polycrystalline Oxide Forging

	Approximate or Typical Parameters Forging Conditions				Results		
Oxide	Temp. (°C)	Pressure (1,000 psi)	% Height Reduction	Strain Rate (10^{-4} min.)	Grain Size or Range (μ)	Maximum Strength in 1,000 psi	References
NiO			70	4–18	1	37–1000°C	
Y_2O_3	1000–	1–100	55	4–18	1	42–R.T.	3, 63
TiO_2	1500		70	4–18		26–R.T.	
CeO_2						26–R.T.	
Al_2O_3	1850–1900	5–20		4–18	5–1,000	70–R.T.	64
Al_2O_3 + MgO	1850–1900	5–20		4–200	5–10	70–R.T.	63–68
MgO	1100	100	50		2		3
					40	30–R.T.	
MgO	1350–	5–10	5–20	1–5			68, 69
MgO + NiO	1400				20	40–R.T.	
MgO	2150	4	57	200	75	24–33–R.T.	61

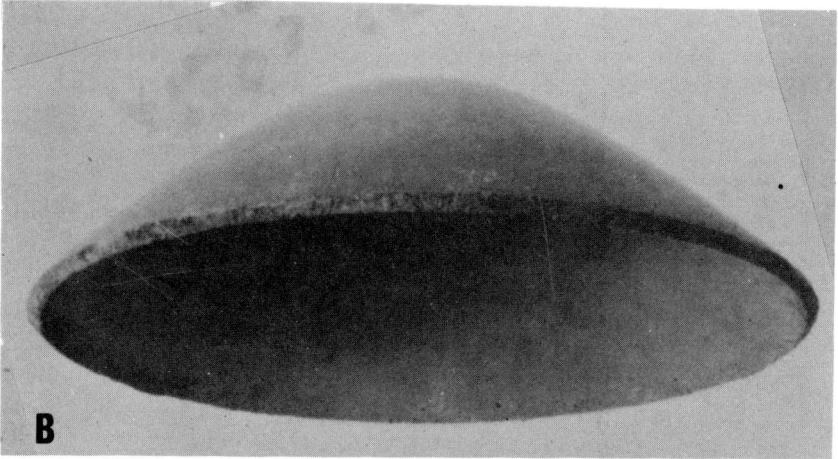

Figure 7. Press forged (deep drawn) Al_2O_3 shapes. (A) Cone, (B) Dome. Both are about $3''$ in diameter. (Photos courtesy D. Sellers, Avco Corporation.)

and shorter times must be used. (Transparency has not been achieved in Al_2O_3 with MgO.)

Some forging of Al_2O_3 shapes not requiring lateral constraints (e.g., approaching deep drawing) has been accomplished. Blunt cones and shallow domes have been successfully forged (e.g., Figure 7), and forging of small (about $2''$ diameter) thin walled hemispheres have shown some success [67].

Some evidence of grain elongation perpendicular to the forging direction has been found in MgO and Y_2O_3, but this has been slight. On the other hand, forged Al_2O_3 with MgO additions had grains that were substantially longer perpendicular to the forging direction, but had a normal grain structure parallel with the forging direction, as shown in Figure 8. The microstructure of forged

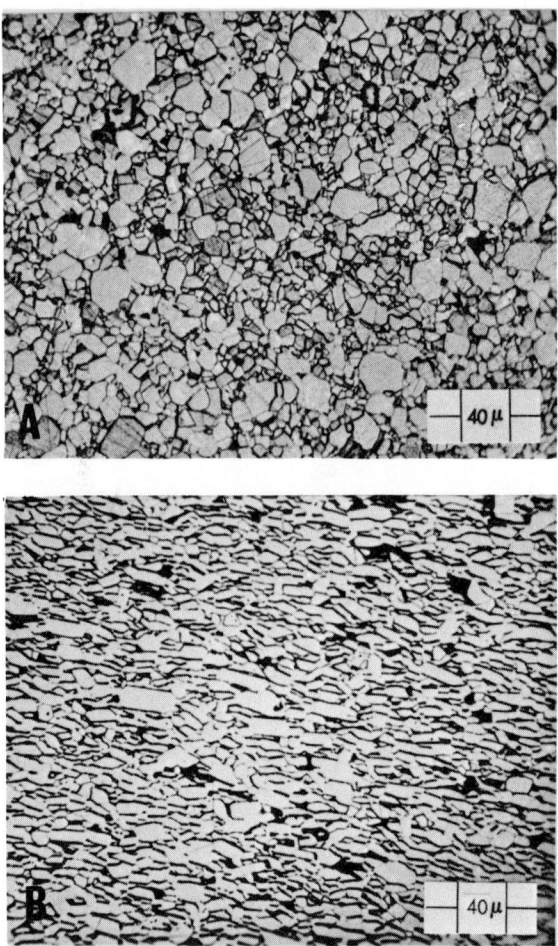

Figure 8. Microstructure of forged Al_2O_3. Etched microstructure of Al_2O_3 forged at 1900°C with 6,500 psi for 10 minutes. (A) parallel, and (B) perpendicular to forging direction. (Photos courtesy D. Sellers, Avco Corporation.)

Al_2O_3 often varied from nearly single-crystal, to equiaxed, to tabular grains. The tabular grains, like those of Figure 8B, have been reported to be the true forged microstructure, and the others represented various degrees of recrystallization. Thus, MgO additions apparently suppressed recrystallization. Recrystallization in forged Al_2O_3 is not at variance with the lack of it in forged sapphire, since the grains should provide mutual constraints to activate other slip or twinning that would intersect with themselves, one another, or with basal slip to nucleate recrystallization. The forged Al_2O_3 bodies have a basal texture (i.e., with the c axes approximately parallel to the forging direction). Sample mechanical properties of forged Al_2O_3 are shown in Table VII. Strengths of forged Al_2O_3 bodies have been as high or greater than those hot pressed bodies of the same grain size, and relatively insensitive to grain size [65] and texture [67]. Further analysis of these grain size and texture effects should lead to increased understanding of fracture mechanisms.

MgO has been forged with several additions (e.g., TiO_2, MnO, and NiO) [69]. One mole percent or less additions of TiO_2 or NiO aided forging, and small NiO additions gave finer grain sizes. This was attributed to work softening and enhanced recrystallization from reduced NiO. Preliminary tests on polycrystalline MgO forged by the author indicate that it still retains some gaseous impurities.

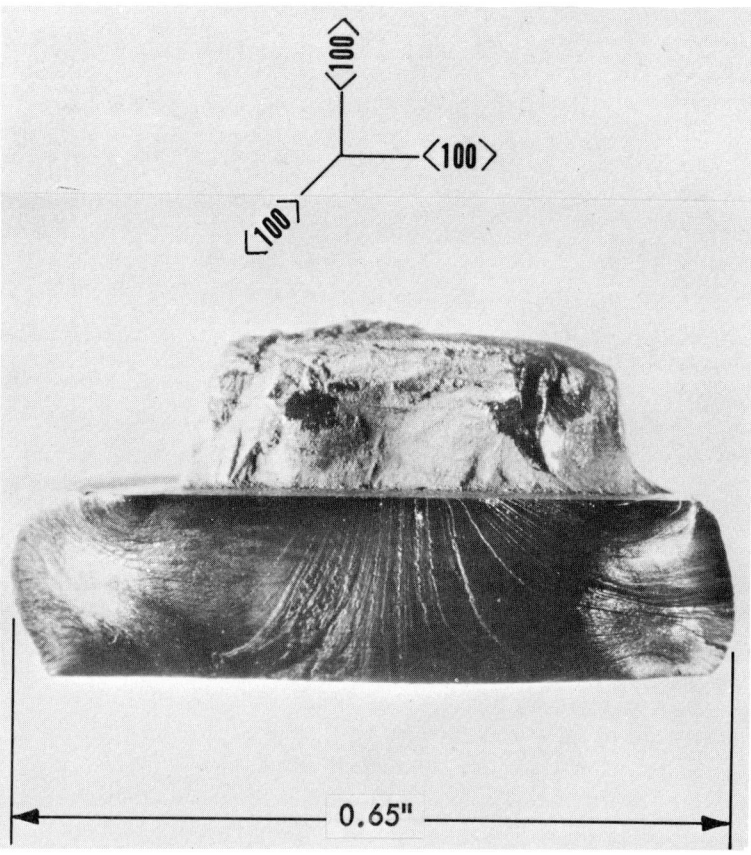

Figure 9. TiC crystal forging. Specimen was a cylindrical crystal cleaved in half along the axis. Its thin top was placed in a depression in the top ram to keep it from tipping over, thus shows the original diameter.

Carbides and Borides

Little work has been done on these systems. However, Rice [61] has forged TiC crystals with graphite tooling (see Table VI and Figure 9). Fair recrystallization was observed, and resultant grain growth was essentially zero to temperatures of at least 1425°C, similar to the oxides. Sellers [70] notes that preliminary forging trials with fairly dense ZrB_2 bodies (with grain sizes of about 10 μ) shows signs of possible formability at about 2150°C at pressures of a few hundred psi with graphite tooling.

Limitations, Advantages, and Future Development

A general limitation of forging is the aspect (length/diameter) ratio. The author [61] has found that billets (usually 0.5–1.5″ diameter) often could not be successfully forged with aspect ratios over 2. Single crystals tended to buckle, especially in certain directions (e.g., ⟨110⟩ in MgO). Polycrystalline specimens with aspect ratios over 1 fractured and crumbled, though more recently Heuer [71] has successfully forged Al_2O_3 with aspect ratios of 2 by using careful alignment. However, this adds to the cost, and can only be extended to a limited degree, so that unless techniques for movable lateral support are developed, aspect ratios of forgeable bodies will be limited.

Another related problem is grain boundary sliding. In forging polycrystalline materials, this may accentuate fracturing and crumbling. Sliding may be enhanced by impurities, especially gaseous impurities. However, reforging of previously forged MgO crystals showed extensive boundary sliding, indicating that this is an intrinsic problem [61].

Grain size has been a basic limitation. Because of the high temperatures required, considerable grain growth occurs during the process, even if powder rather than solid billets are used. No material has yet been forged to as fine a grain size as can be achieved in that material by hot pressing (e.g., compare grain sizes of Tables III, VI, VII, for comparable pressure ranges). However, forged bodies have been reported to have much slower grain growth, at least for periods of the order of an hour, to temperatures of 1600°C or more [61]. Thus, for applications at or after exposure to elevated temperatures, finer grain sizes may be sustained in some forged bodies than in their hot pressed counterparts. Further, less decrease in strength has been observed in forged bodies of increasing grain size [62, 63]. Therefore, at higher temperatures where grain growth results in larger grain sizes, forged bodies may be stronger than hot pressed bodies.

Since many ceramic powders have gaseous impurities, the higher temperatures of forging can lead to greater purification, especially in forging cold pressed billets (which sinter during heating). This capability of forging flat bodies (not shapes) from powder billets is also an advantage since intermediate fabrication is eliminated.

Hot Extrusion

Technique

Ceramic extrusions have been conducted using conventional metal extrusion presses. This has been accomplished by placing the ceramic in a suitable metal can and heating the can and ceramic together, then co-extruding them. The can

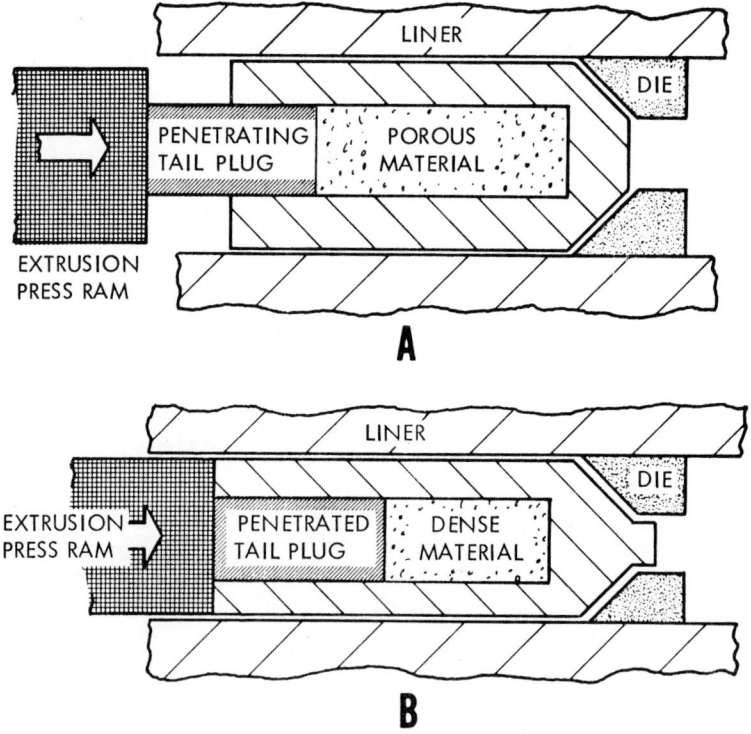

Figure 10. Densification of material for extrusion with a penetrator. (A) The tail plug extends out of the can so the extrusion press ram contacts it first, pushing it into the can. It should be designed so the porous material just reaches full density as the plug is flush with the end of the can so the ram starts uniform extrusion by pressing over the whole can-plug surface. Cup-shaped penetrators can be used to densify porous material around a solid core.

(removed later by machining or by acid) allows the use of established metal extrusion lubricants (mostly glasses), protects the ceramic from contamination (from the die and lubricants), and limits surface cooling of the ceramic by the relatively cold (e.g., 400°C) tooling. Thick walled cans (e.g., 3″ OD, 1.5″ ID—see Figure 10) are usually used to substantially reduce surface cooling and shear. Shear on the ceramic is reduced since shear is highest on the material in contact with the die, which is now the can rather than the ceramic. The can and ceramic must be both chemically and mechanically compatible.

Mechanical compatibility is defined as a suitably close matching of stiffness where each stiffness, k, of each component is defined by the equation:

$$P = k \log_e R$$

where P is the unit ram pressure on each component, and R is the ratio of the cross-sectional area before and after extrusion of the component. Molybdenum- and tungsten-base materials are generally suitable can materials for the refractory ceramics.

Both solid (e.g., pre-sintered or hot pressed) and powder billets have been

TABLE VIII
Hot Extrusion of Refractory Materials[1]

Material	Approximate Heating Temperature (°C[1])	Approximate Fraction of Absolute Melting Point
MgO	2050–2200	0.8[2]
ZrO$_2$	2000–2100	0.8[2]
CaO	1700–2000	0.7–0.8[2]
CaO + MgO	2100–2300	0.8–0.9[2]
MgO + 1 to 5 w/o ZrO$_2$ or CaO	2300–2600	0.8–0.9[2]
UO$_2$	>1900	0.7–0.8[3]
SiC	>2200	>0.8[4]
UC	>1800*	>0.8[4]
TiC	2200–	0.7–0.8[5]
NbC	2400–	0.7[5]

[1]Based on transferring the can and billet from the furnace to the press and extruding in about 30 seconds or less.

[2]After Rice et al. [61].

[3]After Hunt and Lowenstein [75]. (See also [61].)

[4]After Hunt (see [61]).

[5]After Doloff and Probst [76].

*Since the preparation of this chapter, successful extrusion of uranium carbide compositions has been reported [82], with the minimum extrusion temperature being reported as about 2200°C. These investigators report a radial (100) texture—that is (100) planes perpendicular to the extrusion axis—based on UC$_2$ percipitation.

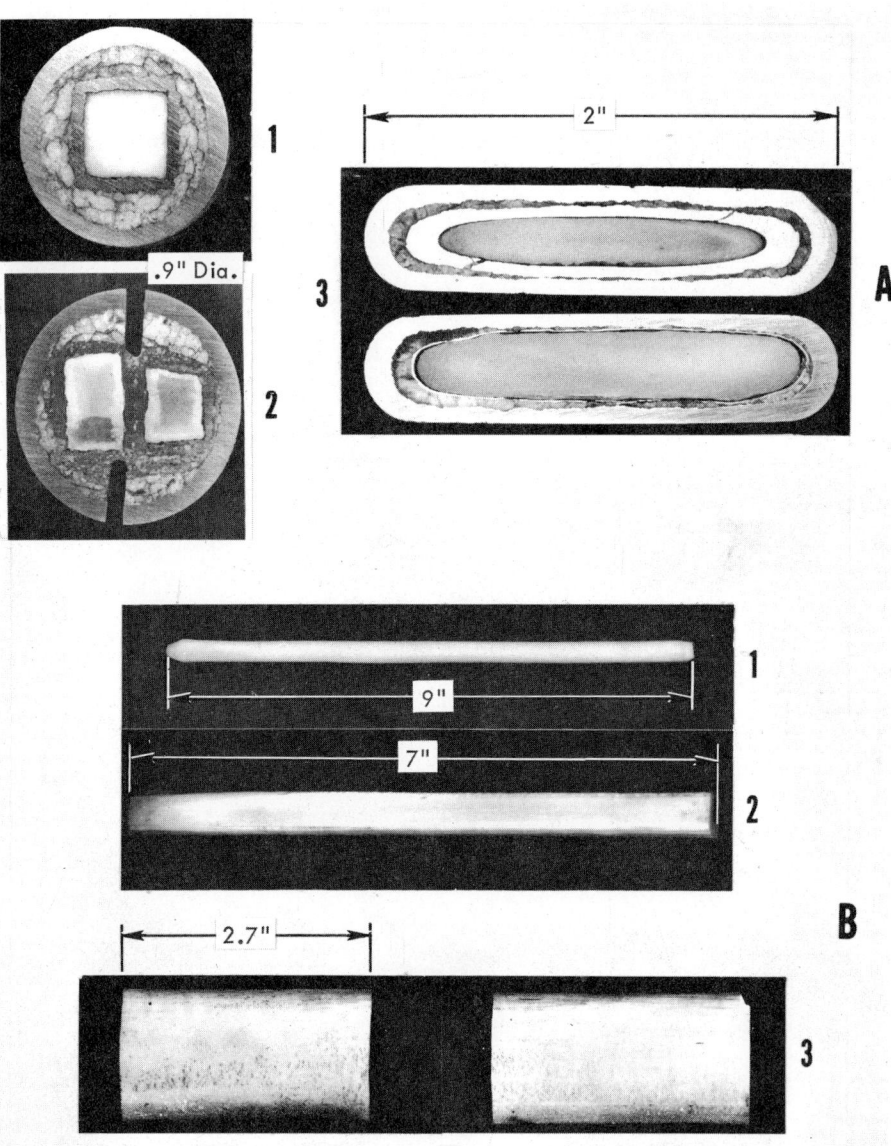

Figure 11. Sample extruded MgO parts. (A) Cross-sections of various extruded shapes: (1) square, (2) two offset rectangular bars (note slotting of can for removal), (3) slab. (B) Crack-free extrusions of sections removed from the cans: (1) round rod, (2) square rod, (3) slab sections.

used in extrusion with thick walled cans. The former makes it particularly easy to extrude several bodies either in tandem or parallel to one another in the same can. The powder billets bypass the fabrication step and eliminate possible costs of matching billets and can sizes. The powder will sinter only partially during heating in the can, but can be densified by extrusion or just prior to extrusion. The latter is often preferable, and can be done by using a penetrator, as shown in Figure 10. The problem with this method is that up to 50 percent of the can volume may be wasted. Since cans cost on the order of $500 or more, this loss may be more costly than fabricating dense billets.

Some heating temperature data are shown in Table VIII. Extrusion with area reduction ratios up to 16 to 1 has been successful, and various shapes have been extruded (e.g., see Figure 11). Typical pressures are 60 to 90 tons per square inch. Ram speeds of a few inches a second are usually used to limit cooling.

Results

MgO [61]

The oxides, particularly MgO, have been studied most. These show that substantial grain size reduction occurs during extrusion. However, this reduction is from the size the grains reached during heating. Because of the high heating temperatures, the resultant minimum extruded grain size is 10–20 μ, which is nearly an order of magnitude higher than in hot pressed bodies but lower than many sintered bodies. Generally, the finest starting grain size (e.g., from powder billets) gives the finest extruded grain sizes. However, there is a great deal of variation; e.g., some single crystals have been extruded to produce bodies with grain sizes of 20–30 μ. These finer grain sizes from crystals were more common in one type of MgO crystal, indicating a purity effect. However, addition of ½ w/o ZrO_2 or 1–2 w/o CaO to hot pressed MgO did not reduce grain sizes very much, though higher additions did lower grain sizes to about 10 μ.

Forging of MgO crystals prior to extrusion made only limited reduction in the resultant of extruded grain size. In almost all cases, the extruded grain structure is almost completely recrystallized by the time the extrusion has cooled. Wrought structures, that is, elongated grains, are seen only in extrusions made at the lowest temperatures where cracking was very extensive. Though the resultant grains were larger than in hot pressed bodies, their rate of growth was much slower—at least to temperatures of 1600°C or more—so that after exposure to elevated temperatures, hot extruded oxides can have finer grain sizes than hot pressed bodies.

The densities achieved in hot extrusion also vary somewhat. Generally, the densest starting bodies produced the densest extruded bodies. Transparency was observed mostly in extruded crystals, but even some of these were clouded after hot extrusion.

Cracking was a major problem, but crack-free MgO parts from a few to several inches long have been obtained (e.g., see Figure 11). Larger crack-free parts required slower cooling than could be achieved by extrusion into an insulating catch tube. Extrusion into a heated furnace helped, but was technically difficult. Use of insulating cans, that is, cans consisting of alternate metal and ceramic layers, was equally or more successful. This also limited heat losses during transfer and extrusion (probably lowering required heating temperatures about 100°C), as well as after extrusion. To maintain uniformity of the extrusion, it was found best to have the ceramic insulation relatively dense before extrusion (either by using dense material or by using a penetrator). Comparison of results of MgO with other oxides indicates that those that can be extruded could be obtained free of cracks, at least in limited sizes.

Extruded MgO has a ⟨100⟩ fiber texture along the extrusion axis. The (200) intensity of a cross-section of extruded MgO is 10 to 20 times a random body. Extrusion of MgO as a slab also developed ⟨100⟩ textures perpendicular to the slab sides, giving it a "cube" or "pseudo-crystal" texture. Annealing of MgO left the same or greater texture.

Extruded bodies of MgO, with a fiber texture, that were free of cracks generally averaged 50–100 percent stronger parallel to the extrusion axis than unextruded bodies of the same grain size at room temperature. Extruded bodies were also stronger in one direction perpendicular to the extrusion axis, and as strong as hot pressed bodies in the other perpendicular direction. However, the larger grain sizes of the extruded MgO limited their maximum strengths to 60,000 psi compared with 84,000 psi for hot pressed bodies. Some extruded bodies were also weakened by residual gaseous impurities. Extruded MgO alloys with second-phase additions (e.g., ZrO_2, CaO, or Al_2O_3) were sensitive to heat treatment, but were at best about half as strong as unalloyed, extruded MgO.

At elevated temperatures, extruded MgO crystals had strengths as high as 64,000 psi (at 1315°C), and a ductile–brittle transition temperature of about 1400°C, which is about 300°C lower than for unextruded MgO. Most extruded hot pressed bodies were much weaker (10 to 20 times weaker). This was attributed to gaseous impurities such as carbonates. Only a few hot pressed billets made from high-purity MgO (from Leipold [29]) or reagent-grade MgO made with LiF approached the behavior of extruded crystals. This occurred only with dense bodies that had been slowly fired to 1600°C for long periods to remove most of the gaseous impurities prior to heating for extrusion. (Similar firing of other hot pressed bodies to over 1600°C prior to heating for extrusion still left such impurities in them.)

The reason for strengthening in MgO was attributed to the texturing which reduced average grain misorientations and hence the propensity for crack nucleation due to grain boundary sliding or dislocation pile-ups. The extruded material also provided much information on fracture that is too extensive to review here.

Other Oxides

Pure Al_2O_3 and $MgAl_2O_4$ were not successfully extruded because of the combination of stiffness, cooling during loading, and the high extrusion strain rates necessary to prevent further cooling [61]. Cheney [72] has also been unsuccessful in extruding pure Al_2O_3, and it has not been elongated in extruded metal-oxide composites as have other oxides [73]. Commercial aluminas with 0.3-1.0 percent impurities were extruded. Apparently the impurities allowed enough grain boundary sliding and crack healing to give coherent bodies. No evidence of hot working was found in these bodies.

Extrusion trials of NiO and CeO_2 were unsuccessful, due to decomposition of the oxides. Trials with titanates have also shown some problems of reduction [61, 74]. They have been extruded, however, and these problems might be overcome with further development. UO_2 [61, 75] has been extruded for fuel rods clad in stainless steel. This was done by heating the UO_2 and steel cans separately to different temperatures, and then assembling them moments before extrusion. The UO_2 was dense, but cracked, and had a wrought structure.

CaO and ZrO_2 have both been successfully extruded (see Table VIII). Though cracking occurred, this should be eliminated with further development. CaO behaved similarly to MgO, developing a similar, but about half as intense (i.e., 5-10 times random) $\langle 100 \rangle$ texture. ZrO_2 extruded at the lowest temperatures showed some signs of this texture (about 1.3 times random), but not when extruded at higher temperatures, indicating that the texture is destroyed by recrystallization.

Some small crack-free specimens of extruded CaO and ZrO_2 were obtained for room-temperature testing. CaO showed strengths about 50 percent stronger than comparable hot pressed bodies. However, again, the larger grain size of the extruded bodies limited maximum strengths to about 20,000 psi, compared to about 30,000 psi for smaller grain hot pressed bodies. The strengthening was attributed to the texture, as in MgO.

The extruded ZrO_2 (Zircoa C—partially stabilized with 2.9% CaO) showed strengths to about 75,000 psi, higher than any other known strengths for ZrO_2. Part of the strengthening of the extruded ZrO_2 is due to reduction of grain size and elimination of porosity from the starting commercial body. Some additional strengthening may have occurred, possibly due, for example, to some mechanism involving dislocation debris, fine precipitation, or order–disorder phenomena.

Both the extruded CaO and ZrO_2 showed little or no grain growth on annealing to temperatures of about 1600°C. Considerable information on the fracture was also obtained.

Carbides

Extrusion of other materials has been very limited. Some borides have been tried, but the single trial was aborted due to improper loading. Hunt [61] made cursory trials of SiC and UC powders, as shown in Table VIII, after heating to 2200° and 1800°C, respectively. The UC was dense and cracked, indicating that it could be extruded after heating to about 1800°C or higher. The SiC was cracked and not fully dense, indicating that it must be heated above 2200°C.

Probst *et al.* [72] have made a few extrusions of TiC and NbC at 8 to 1 reductions in thick walled tungsten cans. These bodies had to be heated respectively to 2200°C and 2400°C before extrusion to obtain coherent bodies. Cracking was also a problem, but some apparently crack-free pieces 0.25″ in diameter and a few to several inches long were obtained (up to 8″ in NbC). Use of a Ta foil insert in the can appeared to reduce cracking of TiC. Remaining cracking was probably due to stiffness mismatch and cooling. The starting powders had about 1 μ particle size, and the extruded bodies had grain diameters of 5–10 μ. At lower extrusion temperatures many grains were elongated (aspect ratios of 5–10) in a typical metal wrought structure, and the fiber texture was mostly ⟨111⟩, with some ⟨100⟩ texture. At higher temperatures there was less wrought structure (bodies extruded after heating above 2500°C and 2650°C, respectively, for TiC and NbC were essentially all recrystallized), and the ⟨100⟩ texture increased and the ⟨111⟩ texture decreased. It was thus concluded that ⟨111⟩ is the extrusion texture and ⟨100⟩ the recrystallized texture.

Sample strength testing showed strengths at least as good or better than unextruded material despite some microcracks in some extruded specimens.

Limitations and Future Development

Substantial improvement in extrusion of ceramics can be expected from refining the existing techniques. The insulating-can approach appears to be one of the more promising avenues for refinement. However, results indicate that the billets must be sealed from the insulation (unless it is fully dense) due to gaseous impurities (e.g., H_2O and CO_2) given off by it during heating [61]. Development of higher temperature tooling would also be quite helpful.

Another very promising, and possibly by far the most significant new approach is the use of hydrostatic and fluid-to-fluid extrusion [61,71–81]. These are illustrated in Figure 12, along with conventional extrusion. Substantial advantages have been found in extruding metals with hydrostatic rather than direct ram pressure, as shown in Figure 12. Further benefits, mainly in controlling cracking, have been found in extruding into a second fluid at lower temperatures [81]. This metal work was limited to low temperatures by use of hydraulic oils, however.

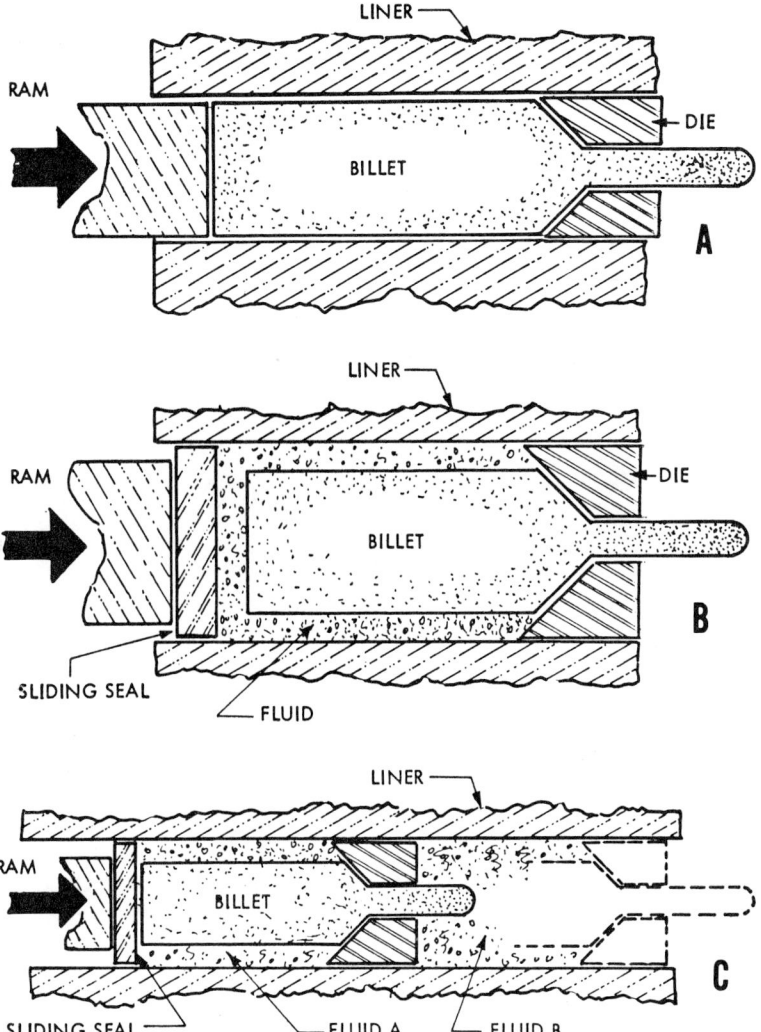

Figure 12. Extrusion techniques. (A) Conventional, (B) Hydrostatic, (C) Fluid-to-fluid. If the second fluid (B) is a soft solid, then its pressure is controlled by extrusion through a second die as dashed in (C).

Recently, use of molten salts as fluids, or soft solids (near, but below their melting points), as "quasi-fluids" has been shown to be feasible [61]. Fluid-to-fluid extrusion using soft solids (e.g., NaCl at 700–800°C as the pressing media and NaCl at about 500°C as the receiving media) also showed promise in some trials. Most of this work was on metals, but some work has been done on $CaCO_3$ and CaO. The latter was extruded using CaF_2 at about 1200°C as the quasi-fluid. Some cracking occurred, but bodies were extruded and appeared to be cold worked. The results suggested that such fluid extrusion of CaO could be successfully done at 300°C or more below the temperatures needed for conventional extrusion. Further development of this technique may thus reduce ceramic extrusion temperatures a few to several hundred degrees. This would make many more ceramics extrudable, give finer grain sizes, and probably reduce extrusion costs. Lowering temperatures could increase the versatility of extruding ceramics similar to that of glasses, as shown in Figure 13.

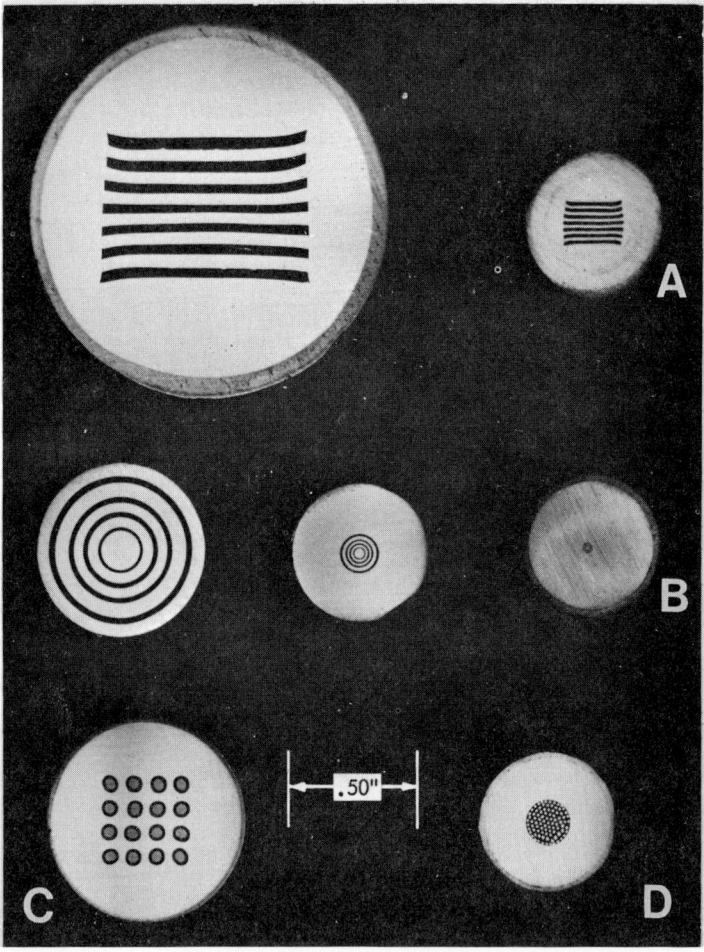

Figure 13. Extrusion of glass in steel. Extrusion of several configurations and re-extrusion of some (e.g., in A and B) show the possible versatility of extruding other ceramics, especially if extrusion temperatures can be lowered by hydrostatic extrusion techniques.

The major limitations of extrusion are cost and size. Improvements are quite feasible with further development, but there will always be some restrictions. However, costs on small components need not be greater, since many can be extruded in a single extrusion (e.g., several rods can be extruded in parallel in a can, with resultant lengths of 10–20 feet).

While strengthening has been investigated, and found feasible, many other applications should be feasible. Development of favorable textures by extrusion can allow some of the ferroelectric, ferromagnetic, optical, and friction and wear characteristics of favorably oriented single crystals to be approached in polycrystalline bodies. This allows retaining the size, strength, domain limitations, and some cost advantages of the polycrystalline body.

Discussion and Conclusions

A variety of hot forming processes have been shown to be feasible on a variety of materials. Processes for densification alone are more developed than those involving hot working. All have their advantages and limitations. Besides further development of each process, interplay and competition between processes will probably occur. Thus, press forging of flat disks may offer advantages over hot pressing and thus compete with hot pressing in such cases. Resistive heating of conductive specimens may be more easily accomplished in forging, where they do not contact the die walls and guides for the rams can be more easily electrically insulated. On the other hand, hot isostatic presses might be used for forging using Hunt's quasi-fluid approach, as sketched in Figure 14. The future importance of joining bodies by hot pressing rests partially on the costs and problems of the hot isostatically pressing of more complex parts. If progress continues on development of hot rolling, this could replace some hot pressing applications.

One of the major problems is the quality of materials available. Gaseous anion impurities such as OH^-, CO_2^{--}, and S^{--} are major problems with oxides across the spectrum of forming processes. This is most critical with hot isostatic pressing where little outgassing during densification can occur. While hot (press) forging may reduce this problem somewhat, it apparently does not eliminate it. With carbides and borides, finer and purer powders would be desirable, but this will probably increase the problem of impurities such as oxygen and nitrogen in these materials. One of the basic problems is that powders commonly used for hot forming were not developed for this purpose, and often were not even developed for any ceramic fabrication. No completely integrated study of powder preparation, characterization, and hot consolidation (pressing, etc.) has yet been made. However, interest is growing, and some good preliminary work has been done. Much attention has been paid to cation impurities, while the anion impurities

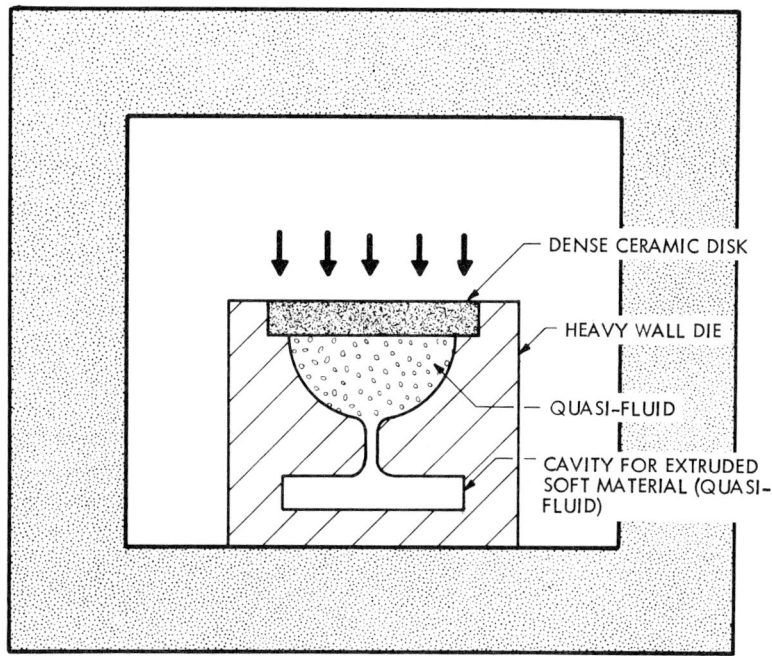

Figure 14. Hot isostatic forging. Schematic representation of how the "quasi-fluid" approach to hot extrusion could be adapted to forging in a hot isostatic press.

have been somewhat neglected. The author presents some thoughts on reducing these, as well as cation impurities, in an appendix to this chapter.

A problem with hot working is that very fine grain sizes have not yet generally been achieved because of the high temperatures required. For elevated temperatures, the slower grain growth of hot worked bodies compensates somewhat for this. Some further reduction in grain sizes can probably be achieved by the study of additives and refinement of techniques. One of the most promising developments would be reducing the promising trials of hot hydrostatic extrusion to practice. This, and canning techniques from hot extrusion, may find application in other fields. Both might be used for forging, and canning might be used for hot rolling for working (not just consolidation as presently explored).

Some of these techniques will continue to be developed simply because of the scientific results that the processes and the resultant bodies provide. The latter is shown by the effects or lack of effects of texturing on strength and fracture of different materials. However, the acid test of these techniques will be producing bodies to meet particular applications. Not all will necessarily survive this test, but clearly some will do so for a permanent advance in ceramic fabrication technology.

Acknowledgment

The author wishes to acknowledge the aid of Messrs. C. D. Burns and M. W. Benecke in editing the manuscript, and of various investigators for providing photos and information on current developments. Permission of James Gangler to use information from NASA-supported programs, and of Charles Bersch to use information from Naval Air Systems Command-supported programs is also appreciated.

References

1. Fulrath, R. M., "A Critical Compilation of Ceramic Forming Methods. III. Hot Forming Processes," *Am. Ceram. Soc. Bull.*, 43 (1964), 880–85.
2. Vasilos, T. and Spriggs, R. M., "Pressure Sintering of Ceramics," in *Progress in Ceramic Science*, Vol. 4, J. E. Burke, ed., Pergamon Press (1966), 95–132. See also: Spriggs, R. M. and Vasilos, T., "Hot Pressing (Pressure Sintering) of Nonfissionable Ceramics," in *Nuclear Applications of Nonfissionable Ceramics*, A. Boltax and J. H. Handwerk, eds., Am. Nuclear Soc. (1966), 381–417.
3. Spriggs, R. M. and Atteraas, L., "Densification of Single Phase Systems Under Pressures," in *Ceramic Microstructures – Their Analysis, Significance, and Production*, J. A. Pask and R. M. Fulrath, eds., to be published by John Wylie and Sons, Inc.
4. Spriggs, R. M., "Microstructure and Its Influence on Strength" in *Strengthening Mechanisms – Metals and Ceramics*, J. J. Burke, N. L. Reed, and V. Weiss, eds., Syracuse University Press (1967), 181–218.
5. Vasilos, T. and Spriggs, R. M., "Microstrucutre in Oxides," in *Sintering and Related Phenomena*, G. C. Kuczynski, N. A. Hooton, and C. G. Gibbon, eds., Gordon and Breach (1967), 301–27.
6. Rice, R. W., "Hot Working Oxides," submitted for *Refractory Oxides*, A. Alper, ed., to be published by Academic Press.
7. Roeder, E. and Scholz, S., "A Simple Hot Press for Laboratory Investigations," in *Special Ceramics*, P. Popper, ed., Academic Press (1965), 341.
8. Zneimer, J., Lepore, D. A. and Lehman, K. A., "The Preparation of Ferrites by Flame Spraying and Hot Pressing," Presented at the 66th Annual Meeting of the Am. Ceram. Soc. in Chicago, Ill. (April 1964). Abstract: *Bull. Am. Ceram. Soc.,* 47 (1968), 277.
9. Spriggs, R. M., Brissette, L. A. and Vasilos, T., "Pressing Sintered Nickel Oxides," *Bull. Am. Ceram. Soc.*, 43 (1964), 572–77.
10. Spriggs, R. M., Brissette, L. A., Rossetti, M. and Vasilos, T., "Hot Pressing Ceramics in Alumina Dies," *Bull. Am. Ceram. Soc.,* 42 (1963), 477–79.
11. Moss, H. I. and Stollar, W. P., "Die Design for Pressure Sintering," *Bull. Am. Ceram. Soc.*, 45 (1966), 792–93.
12. Rice, R. W., "Fabrication of Dense MgO," to be submitted for publication.
13. Rice, R. W., "Hot Pressing Al_2O_3," to be submitted for publication.
14. Kriegel, W. W., Palmour, H. III and Choi, D. M., "The Preparation and Mechanical Properties of Spinel," in *Special Ceramics*, P. Popper, ed., Academic Press (1965), 167–86.
15. Rummler, D. R., "Vacuum Hot-Pressing of Magnesium Aluminate." Presented at the Basic Science Div. Meeting Am. Ceram. Soc. at Univ. Park, Penn. (October 1966).

16. Rice, R. W., "Production of Transparent MgO at Moderate Temperatures and Pressures," Presented at the 64th Annual Meeting Am. Ceram. Soc. in New York, N.Y. (May 1962). Abstract: *Bull. Am. Ceram. Soc.*, 41 (1962), 271.

17. Atlas, L. M., "Effect of Some Lithium Compounds on Sintering of MgO," *J. Am. Ceram. Soc.*, 40 (1957), 196–99.

18. Carnall, E. Jr., "The Densification of MgO in the Presence of a Liquid Phase," *Materials Research Bull.*, 2 (1967), 1075–86.

19. Gray, T. J., of Alfred University, unpublished discussion at the Conference on The Role of Grain Boundaries and Surfaces in Ceramics at the University of North Carolina State University at Raleigh (November 1964).

20. Copley, S. M. and Pask, J. A., "Deformation of Polycrystalline MgO at Elevated Temperatures," *J. Am. Ceram. Soc.*, 48 (1965), 365–68.

21. Miles, G. D., Sambell, R. A. J., Rutherford, J. and Stephenson, G. W., "Fabrication of Fully Dense Transparent Polycrystalline Magnesia," *Trans. Brit. Ceram. Soc.*, 66 (1967), 319–35.

22. Rice, R. W., "The Effect of Gaseous Impurities on the Hot Pressing and Behavior of MgO, CaO, and Al_2O_3," to be published in the *Proceedings of the Brit. Ceram. Soc., No. 12, Fabrication Science.*

23. Smethurst, E. and Budworth, D. W., private communications (1967), and Progress Reports for Ministry of Tech. Agreement No. PD/31/028/AT, "Preparation of Transparent Ceramics by Hot Pressing" (1968).

24. Rice, R. W., "CaO: I–Fabrication," to be published in *J. Am. Ceram. Soc.* (August 1969).

25. Rice, R. W., "CaO: II–Properties," to be published in *J. Am. Ceram. Soc.* (August 1969).

26. Stoops, R. F., "Liquid-Phase-Extrusion Hot Pressing," presented at the 68th Annual Meeting, Am. Ceram. Soc., Washington, D. C. (May 1966), submitted for publication.

27. Gazza, G. E., Barfield, J. R. and Preas, D. L., "Reactive Hot-Pressing of Alumina with Additives," *Bull. Am. Ceram. Soc.*, 48 (1969), 606–10.

28. Montgomery, P. W., Stromberg, H. and Jura, G., "Sintering of Refractory Materials at Room Temperature by High Pressures," *Solid Surfaces and the Gas-Solid Interface—Advances in Chemistry No. 33*, Am. Chem. Soc. (1961), 18–22.

29. Leipold, M. H. and Nielsen, T. H., "Hot-Pressed High-purity Polycrystalline MgO," *Bull. Am. Ceram. Soc.*, 45 (1966), 281–85.

30. Gardner, W. J., McCleland, J. D. and Richardson, J. H., "Translucent Oxides," Aero-Space Corp., Report for Contract AF 04(695)–269 (1963).

31. Rice, R. W., "Strength and Fracture of Dense MgO," in *Ceramic Microstructures, Their Analysis, Significance, and Production,* J. A. Pask and R. M. Fulrath, eds., John Wylie and Sons, Inc. (1968), 579–93.

32. Rice, R. W., "Strength and Fracture of Hot Pressed MgO," to be published.

33. Rice, R. W., "Strength and Fracture of Hot Pressed Al_2O_3," to be published.

34. Rossi, R. C. and Fulrath, R. M., "Final Stage Densification in Vacuum Hot-Pressing of Alumina," *J. Am. Ceram. Soc.*, 48 (1965), 558.

35. Simpson, F. H., The Boeing Company, Private communication (1966).

36. Rhodes, W. H., Haag, R. M. and Burnett, P. L., "Ultra-High Purity Fine Particulate Stabilized ZrO_2," Progress Reports for Contract AVSSD–0135–68–CR (1968).

37. Johnson, R. E., "Hot-Pressing High-Density Small Grain Size Beryllia," *Bull. Am. Ceram. Soc.*, 43 (1964), 886–88.

38. Bentle, G. G. and Kniefel, R. M., "Brittle and Plastic Behavior of Hot-Pressed BeO," *J. Am. Ceram. Soc.*, 48 (1965), 570–77.

39. Adams, R. B. and Stuart, W. I., "Effect of Adsorbed Sulfate and Fluoride on the Density of Hot-Pressed Beryllium Oxide," *J. Am. Ceram. Soc.*, 50 (1967), 685.
40. Rice, R. W., unpublished data.
41. Sanders, W. A. and Probst, H. B., "Mechanical Properties of Hafnium Carbide to 2635°C," presented at the 70th Annual Meeting of the Am. Ceram. Soc. in Chicago, Ill. (April 1968). Abstract: *Bull. Am. Ceram. Soc.*, 47 (1968), 354.
42. Leipold, M. H. and Becher, P. F., "Fabrication and Characterization of Hot-Pressed Tantalum Carbide", NASA Tech. Report 31–1204 (1967). See also: Leipold, M. H. and Nielsen, T. H., "The Mechanical Behavior of Tantalum Carbide and Magnesium Oxide," NASA Tech. Rept. 32–1201 (1967).
43. Harrison, W. B., "Mechanical and Electrical Properties of TiC," Second Interim Tech. Report for Contract No. DA–11–022–ORD–3441 (1965).
44. Clougherty, E. V. and Kalish, D., "Verification of Property Trade-Off Consequences in the Design and Development of Refractory Diboride Materials," presented at the Composites Working Group Meeting (1968).
45. Clougherty, E. V., Kalish, D. and Peters, E. T., "Research and Development of Refractory Oxidation Resistant Diborides," Tech. Report AFML–TR–68–190 (1968).
46. Rhodes, W. H., AVCO Corporation, Private communication (1968).
47. Zehms, E. H., and McClelland, J. D., "Semi-Continuous Hot Pressing," *Bull. Am. Ceram. Soc.*, 42 (1963), 10–12.
48. Gruintjes, G. S., and Oudemans, G. J., "Continuous Hot Pressing," in *Special Ceramics 1964*, P. Popper, ed., Academic Press (1965), 289-98. See also: Oudemans, G. J., "A Continuous Hot Pressing Method," to be published in the *Proceedings of the Brit. Ceram. Soc., No. 12, Fabrication Science.*
49. Nielsen, T. H. and Leipold, M. H., "Thermal Expansion of NiO," *J. Am. Ceram. Soc.*, 48 (1965), 164.
50. Leipold, M. H. and Nielsen, T. H., "Ceramics," in Jet Propulsion Laboratory Space Programs, Summary 37-37, Vol. IV (1964), 47–52.
51. Leipold, M. H., Jet Propulsion Laboratory (now at the University of Kentucky) (1966).
52. Levey, R. P., "Isostatic Hot Pressing," AEC R & D Report Y–1487, Contract W–7405, eng. 26 (1965).
53. Long, W. M. and Snowden, P., "Some Practical Aspects of Isostatic Hot-Pressing," Pulvermetallurgie in Stuttgart, Vol. 1 (May 1968).
54. Huffadine, J. B., Whitehead, A. J. and Latimer, M. J., "The Application of Hot Isostatic Gas Pressing to the Fabrication of Irdomes," to be published in the *Proceedings of the Brit. Ceram. Soc., No. 12, Fabrication Science.*
55. Leipold, M. H. and Nielsen, T. H., "Fabrication and Characterization of Isostatically Hot-Pressed MgO," *J. Am. Ceram. Soc.*, 51 (1968), 94–97.
56. Hodge, E. S., "Elevated-Temperature Compaction of Metals and Ceramics by Gas Pressures," *Pwd. Metallurgy*, 7 (1964), 168–201.
57. Rice, R. W., and Hunt, J. G., "Semi-Continuous Pressure Sintering," report for Contract NOw 65–0382-c (1966).
58. Rice, R. W., Benecke, M. W., Sliney, J. L. and Friedman, G. I., "Semi-Continuous Pressure Sintering," report for Contract N000 19–67–C–0314 (1968).
59. Harris, H., Bureau of Mines Research Station, Albany, Ore., Private communication (1968).
60. Harris, H. M., Kelley, J. E., Sunset, P. H. and Kelly, H. J., "Hot Rolling of Oxide-Glass Compositions," Bureau of Mines Report of Investigation 6967 (1967).
61. Rice, R. W., Hunt, J. G., Friedman, G. I. and Sliney, J. L., "Identifying Optimum Parameters of Hot Extrusions," Final Report for Contract NAS7–726 (1968).
62. Day, R. B. and Stokes, R. J., "Grain Boundaries and The Mechanical Behavior of

Magnesium Oxide," in *Materials Science Research*, Vol. 3, W. W. Kreigel and H. Palmour, III, eds., Plenum Press (1966), 355–86.

63. Brissette, L. A., Burnett, P. L., Spriggs, R. M. and Vasilos, T., "Thermomechanically Deformed Y_2O_3," *J. Am. Ceram. Soc.*, 49 (1966), 165–66. See also: Vasilos, T., Brissette, L. A. and Burnett, P. L., "Effects of Processing on the Microstructure and Mechanical Properties of Polycrystalline Ceramics Prepared by High Pressure," Report for Contract No. DA–31–124–ARO–(D)–168 (1965).

64. Rhodes, W. H., Sellers, D. J., Vasilos, T., Heuer, A. H., Duff, R. and Burnett, P., "Microstructure Studies of Polycrystalline Refractory Oxides," Summary Report for Contract NOw 65–0316–f (1966).

65. Heuer, A. H., Rhodes, W. H., Sellers, D. J. and Vasilos, T., "Microstructure Studies of Polycrystalline Refractory Oxides," Summary Report for Contract NOw 66–0506–d (1967).

66. Rhodes, W. H., Sellers, D. J., Heuer, A. H. and Vasilos, T., "Development and Evaluation of Transparent Aluminum Oxide," Final Report for Contract N178–8986 (1967).

67. Rhodes, W. H., Sellers, D. J., Cannon, R. M., Heuer, A. H., Mitchell, W. R. and Burnett, P., "Microstructure Studies of Polycrystalline Refractory Oxides," Summary Report for Contract N000 19–67–C–0336 (1968).

68. Spriggs, R. M., Atteraas, L. and Runk, R. B., "Thermochemical and Thermomechanical Adaptations of Pressure Sintering of Ceramics," to be published in the *Proceedings of the Brit. Ceram. Soc., No. 12, Fabrication Science*.

69. Atteraas, L., "Mechanisms of Densification, Deformation and Strengthening in Alloyed, Thermomechanically Worked MgO," Ph.D. Thesis, Lehigh University (1967).

70. Sellers, D. J., AVCO Corporation, Private communication (1968).

71. Heuer, A. H., Case Western Reserve University, Private communication (1968).

72. Cheney, R., "Hot Extrusion of MgO and Al_2O_3 Powders," *Studies of the Brittle Behavior of Ceramic Materials*, N. M. Parikk, ed., Tech. Doc. Report No. ASD–TR–61–628, Part III (June 1964), 387–96.

73. Bruckart, W. L., Craighead, C. M. and Jaffee, R. I., "Investigation of Molybdenum and Molybdenum-Base Alloys Made by Power-Metallurgy Techniques," WADC Tech. Report 54–398 (January 1955). See also: Jech, R. W., Weeton, J. W. and Signorelli, R. A., "Fibering of Oxides in Refractory-Metal Matrices", NASA Tech. Note NASA–TN–3923 (May 1967).

74. Kastenbein, E. L., Pitetti, R. C., Rankin, D. T., Wechansky, H., Rooda, J., Lupfer, D. A., Smoke, E. J. and Phillips, C. J., "Inorganic Dielectric Research," Reports for Contract DA–36–039–AMC–03403(E) (1963–68).

75. Hunt, J. and Lowenstein, P., "Hot Extrusion of UO_2 Fuel Elements," *Bull. Am. Ceram. Soc.*, 43 (1964), 562–65.

76. Dolloff, R. R. and Probst, H. B., "Hot Extrusion of Carbides," presented at the 70th Annual Meeting of the Am. Ceram. Soc. in Chicago, Ill. (April 1968). Abstract: *Bull. Am. Ceram. Soc.*, 47 (1968), 441.

77. Beresnev, B. I., Vereshchagin, L. F., Ryabinin, Yu N. and Livshits, L. D., "Large Plastic Deformation of Metals at High Pressures," Akad. Nauk Izdat., Moscow (1960).

78. Pugh, H. L. D. and Low, A. H., "The Hydrostatic Extrusion of Difficult Metals," *J. Inst. Met.* 93 (March 1965), 201–17.

79. Fiorentino, R. J., Sabroff, A. M. and Boulger, F. W., "Investigation of Hydrostatic Extrusion," Tech. Report AFML–TR–64–372 (January 1965).

80. Randall, R. N., Davies, D. M., Surgief, T. M. and Lowenstein, P., "Experimental Hydro-

static Extrusions Point to New Production Techniques," *Modern Metals,* 17 (1962), 68–73.

81. Gelles, S. H., "Hydrostatic Forming–A Survey," *J. Met.,* (1967), 35–46.

82. Magnier, P., Marchal, M. and Accary, A., "High-Temperature Compressive Creep and Hot Extrusion of U–C Alloys," *Proceedings of the Brit. Ceram. Soc., No. 7, Nuclear and Engineering Ceramics* (1967), 141–58.

Appendix

NOTES ON POWDER PREPARATION AND PROCESSING

Oxide Preparation

Oxide powders are normally prepared by calcining an appropriate precursor salt. To achieve high purity, these salts have sometimes been prepared by acid digestion of pure metals. However, regardless of the cation purity, past calcining practice has still left anion impurities such as OH^- and CO_2^{--} which have been found to be basic problems of fabrication. Improved calcining techniques, such as using rotary kilns of fluidized beds to limit particle growth while getting better removal of gases, may help. As previously discussed [1], the ultimate of this would probably be to drop individual particles or agglomerates down a tube through a hot zone for decomposition. Another approach to preparing powders is through chemical vapor deposition. Tonnage quantities of colloidal (e.g., 0.02 μ) TiO_2 and SiO_2 are produced in this fashion.

However, the last monomolecular layer of OH^- is estimated to be stable to temperatures of the order of $2000°C$ in MgO [2, 3]. This is corroborated by OH^- observed being given off by MgO to temperatures of the order of $2000°C$ [1, 2, 4]. It has also been estimated that 15 percent of the surface of Linde A Al_2O_3 powders is still covered with H_2O at $1250°C$ [5]. Similar tenacity may occur with the F, Cl, HCl, etc., involved in chemical vapor deposition.

Because of this concern, the author proposed and has made some preliminary tests on a basically different approach to making oxide and nitride powders. This approach is to vaporize the metal and oxidize it with oxygen or nitrogen to produce the desired oxide or nitride. The vaporization can be done by an induction plasma, or an arc plasma (from a plasma torch, or an arc–preferably a high-intensity arc [6]), or by simple boiling. The advantage of the plasma approaches is that they can be used to vaporize all materials. Thus mixtures could be produced. Using a simple arc, the author produced small quantities of MgO,

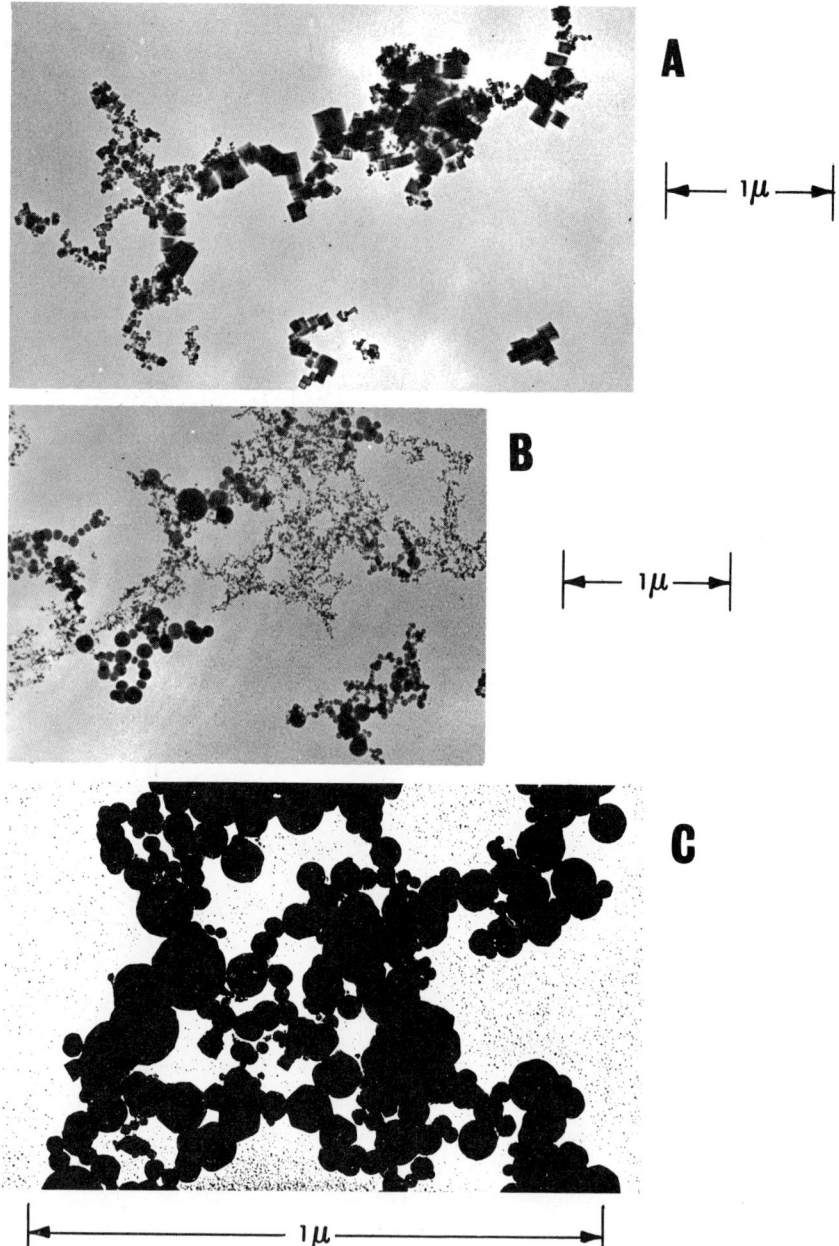

Figure 1. Sample powders made by vaporizing metals, then reacting the vapor with the appropriate gas to form oxides or nitrides. (A) MgO, (B) Al_2O_3, (C) TiN + AlN (from an alloy of Ti + Mn + Al).

CaO, and Al_2O_3—as well as MgO with ZrO_2, HfO_2, CaO, or Al_2O_3—by using appropriate metal (often pressed powder) electrodes for the arc (see Figure 1). The Al_2O_3 addition reacted, apparently in the vapor phase, to form $MgAl_2O_4$, confirming that alloying was feasible. Using titanium alloy (Ti + Mn and Al) electrodes in N_2 atmospheres, mixed nitrides with fine particle sizes (see Figure 1) were produced. T. Ross and M. Bencke working with the author have since shown that magnesium can be safely boiled and the vapor reacted with oxygen to form a stable flame. Plugging of nozzles and incomplete combustion have been problems, but appear to be solvable. For pure, undoped materials, this method may have the advantage of further purification by selective boiling. Resultant powders though made only in small quantities, showed much less tendency for agglomeration (e.g., when mounting on electron microscope grids). This may be due to less susceptibility to atmospheric contamination.

Processing Powder

Recently, the author has considered further processing of powders. Since powders can often be charged (e.g., as often occurs in a plastic bag), it may be feasible to charge them purposely. This could aid many processing steps. Thus, oxide particles such as those produced by metal vapor oxidation could be given a like charge (if they are not procuced with a charge) to inhibit agglomeration. Reactions could be assisted by charging reacting particles with opposite charges. For example, collodial carbon and metal particles could be given respectively opposite charges, so that when they were mixed in gas streams there would be primarily carbon-metal particle attractions. Reaction time could be increased by passing such a stream through a magnetic field so the positive (e.g., carbon) particles would spiral in one direction, and negative particles (e.g., Ti) in the other direction. Multiple overlaps of the helical paths would increase collision possibilities.

It may also be that charges on fine non-conductors such as oxide particles may contribute to the voluminous character of such fine powders. Properly neutralizing such powders before or during cold pressing might thus be a substantial aid in cold pressing.

References

1. Rice, R. W., "The Effects of Gaseous Impurities on the Hot Pressing and Behavior of MgO, CaO, and Al_2O_3," in *Proceedings of the Brit. Ceram. Soc., No. 12, Fabrication Science,* to be published.
2. Nielsen, T. H. and Leipold, M. H., "Surface Hydroxyl in MgO," *J. Am. Ceram. Soc.,* 49 (1967), 626–27.

3. Freund, F., "Retention of Hydroxyl Groups in Magnesium Oxide," *J. Am. Ceram. Soc.,* 50 (1967), 493–94.

4. Rice, R. W., "Characterization of Hot Pressed MgO," to be published.

5. Rossi, R. C. and Fulrath, R. M., "Final Stage Densification in Vacuum Hot-Pressing of Alumina," *J. Am. Ceram. Soc.,* 48 (1965), 558–64.

6. Sheer, C. and Korman, S., "The High-Intensity Arc in Process Chemistry," in *Arcs in Inert Atmospheres and Vacuum,* W. E. Kuhn, ed., John Wiley and Sons, Inc., New York (1956), 169–83.

12. Superplasticity in Ceramics

PETER E. D. MORGAN

The Franklin Institute Research Laboratories
Philadelphia, Pennsylvania

' ABSTRACT

Superplasticity in metals has received a considerable amount of attention in recent years. However, the possibility that ceramic materials also may show anomalous deformational properties coincident with a phase transformation has been neglected. Examples will be given where ready deformation of ceramic bodies may occur with (1) simple allotropic phase transformation, (2) eutectic exsolution, (3) exsolution from solid solution, (4) exsolution by decomposition of mixed crystal systems, and (5) simple decomposition producing one or more solid-phase products and a gas phase; these will be related to the better known processes in metallic systems. It will be shown that the unifying feature of these phenomena in both metals and ceramics is achievement and retention of extremely fine crystal size during deformation.

Underwood reported in 1962 [1] : "Numerous reports of mechanical weakness in metals and alloys undergoing allotropic transformations are available in the literature. This effect should also apply, in principle, to other crystalline structures, e.g., ceramic oxides." In fact, of course, the superplastic behavior of one well-known oxide, glass, had been known for a very long time and apparently caused very little comment or excitement. Explanations, or rather rationalizations, generally emphasized the fact that glass was a supercooled liquid and was amorphous. By superplasticity then, well exemplified by this material, we mean to include cases of exceptional elongation, in the hundreds to thousands percent range, without necking down. But superplasticity, as it was identified and defined by the early Russian investigators in the field, especially Bochvar [2], was intended to cover various anomalous deformational processes, thought at the time to be related to phase transformation but now known to be a result of extremely fine microstructures. Thus the finding by Sauveur in 1924 [3], of an anomalous torsional creep in iron at the γ–α transformation temperature, was the first example related to phase transformation. Sauveur placed an iron bar in torsion and established a temperature gradient along its length. A line was inscribed along the bar. At the point on the bar corresponding to the temperature of the γ–α transformation a large torsional creep occurred, twisting the line around the bar, while elsewhere the line was relatively straight.

Later, in 1934, Pearson [4] found an effect in a quenched eutectic composition similar to the glass phenomenon. He was able to pull extruded quenched eutectic compositions of Sn/Pb and Sn/Bi to up to 2,000 percent elongation. Subsequently, there was much work on iron-carbon alloys and, in Russia, extensive investigations commenced in the war years on effects in many eutectic and eutectoid alloys, notably Al/Zn, Al/Cu, Al/Si, Al/Ni, Al/Fe, Cu/Zn, and Cu/Ni. This work is well covered in Underwood's review [1]. The general opinion existed at this time and until much later that recrystallization was a necessary concommitant of superplasticity, whether by exsolution or by phase transformation.

Recent work [5, 6] has shown that there are two types of superplasticity:

1. Systems in which a phase transformation produces stress concentration around transforming grains where the difference in density of the two allotropic forms is significant. Stresses generated during the transformation act to facilitate flow directed by the primary applied stress and extension of up to 10 percent without necking is possible. However, by traversing the phase boundary many times with incremental extensions, cumulative extension may be 1,000 percent. Recently, Weiss [7] has seen this effect also in single crystals. We may term this type [5] "transformation plasticity."

2. The main type of superplasticity treated here is where the unifying feature is a very fine grain size of the order $10\,\mu$ or less in metals, and $1\,\mu$ or less in ceramics. This has been termed "micrograin plasticity" [5].

The important feature of this latter type of superplasticity that is of interest to ceramics research, both practically and theoretically, is that very large amounts of grain boundary sliding are observed in many cases in metal systems. This is interesting enough in metals where other modes of deformation can easily occur also; however, we might speculate that in ceramics, where other modes (i.e., dislocation motion) may be much more difficult, the phenomenon could be of even more importance if suitably optimized.

Early on, Pearson [4], and much later, Gifkins [8], in 1952, showed that grain boundary sliding was characteristic of superplastic deformation of the micrograin variety. More recently the work has been extended, especially by Alden and co-workers [9, 10]. The sliding effects were especially noted at low strain rates and correspondingly low stresses. Under these conditions, the grains remain equiaxed even to extensions of 2,000 percent. Even in ordinary creep, total strains of 80 percent by grain boundary sliding alone have been seen [11, 12]. Grain growth may occur, but is not necessary. In fact, growth, if it does occur, progressively lowers the ability of continiung superplastic deformation. However, it has been observed that grain growth may sometimes be stimulated by superplastic deformation [9] and this is probably related to a dislocational accommodation mechanism which will be mentioned later. Grain rotation is very significant, e.g., 25-30° after 45 percent extension has been seen with retention of equiaxed grains [13].

Mott [14] has theorized on slip at boundaries with plastic deformation of protrusions and jogs, and Gifkins [15] has extended the ideas to the possibility of diffusion-controlled boundary sliding where the movement of boundary ledges, at most a few atoms thick, is controlling. Local deformation of grains in the vicinity of boundaries; manifesting itself in an apparent widening of the boundary region, has been observed in both conventional creep [16, 17] and in superplastic deformation, but particularly in the latter, where peculiar rounded grains with apparent widening at corners is often seen [8].

Even in the conventional model of Nabarro–Herring creep, grain boundary sliding must occur, as shown by Lifshitz [18], Gifkins [19], Backofen [6], etc., and up to 50 percent of the total strain may be by this mode alone [19]. Recent work on the apparently conventional creep of MgO by Hensler and Cullen [20] has shown that the main mode of deformation, even here, involves boundary sliding, maintaining an equiaxed grain structure, at up to 58 percent total deformation. This must cast serious doubt, not only on analyses of compressive creep in ceramics by Nabarro–Herring diffusional creep where anisotropic grains must result, but also on such analyses in hot pressing where sliding is even more likely to be occurring.

Grain boundary sliding in a dense body, if it is not to lead to void formation, must be accompanied by accommodation processes, either of a diffusional, or of a plastic (i.e., dislocation) type. At very fine grain size the diffusional creep may be very large, obeying approximately an inverse (grain-size)^3 law, if, as is widely suspected, grain boundary diffusion is predominant at lower temperatures. Thus, it is possible to get diffusion-controlled rate equations of the right grain size dependence for Coble creep [21], for instance, but where sliding, rotation, and limited grain boundary movement allows the total strain to be much larger than would be predicted using conventional models and normal diffusion values; the failure to appreciate this effect has led almost certainly to erroneous evaluations of "stress-enhanced" diffusion values in hot pressing and creep.

Alden and others have seen the 1/G.S.^{3} relation holding for some cases of superplastic creep suggesting diffusional control, but Gifkins has extended the idea to diffusional control around ledges on the boundary also predicting a 1/G.S.^{3} relation [15]. Gifkins, in particular, has suggested that the feature of superplastic flow is that control by diffusion at triple points is replaced, below a critical grain size of 1-10 μ, by control by this ledge or protusion mechanism. If this is true, then the real nature of this kind of superplastic flow is governed by the grain size rather than, as had been suspected earlier, by the strain rate used. In fact, it is still ignored that superplasticity in glass occurs at very high strain rates. Thus Alden's data [22] on the Sn/Bi system predict that, if he could get to a grain size of 0.1 μ, superplastic flow would occur at a strain rate of 10^2 min^{-1}.

That fine grain size alone is responsible for the phenomenon was shown in this system and was also nicely demonstrated in the work of Floreen [23], where cold worked pure nickel was heated at 300°C per minute and simultaneously strained at up to 65 inches per minute. During the rapid heating, recrystallization giving the fine grain size suitable for superplastic deformation occurred, and the strain rate was so rapid that the material deformed before grain growth could markedly reduce the effect. Thus, the effect increased up to 820°C where elongations of 250 percent were achieved. Holding at temperature for only 10 seconds before straining caused the elongation before fracture to decrease to one-half of the value without a hold. It is clear that very special effects can be seen with very rapid heating and deformation schedules, another observation made occasionally in ceramics but never adequately explained.

Another phenomenon, rarely observed earlier in conventional creep, that of rumpling of boundaries, is often seen [8, 24] in superplastic deformation and is immediately suggestive of grain boundary sliding with some dislocation activity. Figure 1 shows some of the possible conformations of grain boundary sliding, slip

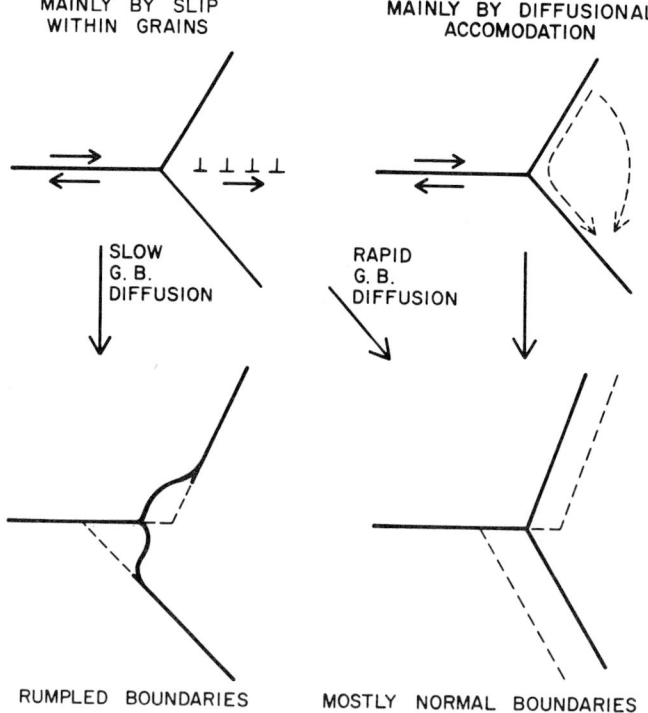

Figure 1. Grain boundary sliding with accommodation by diffusional or dislocation modes.

within grains, and diffusional phenomena, leading to differing microstructural effects. In cases where accommodation is by dislocation mechanisms, additional effects due to grain boundary migration into strained areas is especially likely to occur. Recent papers [24, 25] have reported the extreme complexity of the superplastic flow phenomenon as directly observed by time-lapse cinematography but, nonetheless, the boundary sliding part increases as the grain size diminishes, and it is now realized that the mode of arriving at a fine structure may be immaterial to the effects which follow. If we assume that creep, or superplastic flow, involves large grain boundary slippage, then we can be sure that in porous materials this effect will be even larger; for example, we can expect a large contribution from this source in hot pressing.

Very little work has been done as yet in ceramic systems to show that superplastic phenomena may be utilized in these materials. However, Table I lists a selection of identified types of superplastic deformation observed for metal systems, and shows the possible types of changes in ceramic systems which may show the analogous deformational effects. Listed are some of the ceramic systems already studied which appear to be manifesting unusual properties, and also conjectured systems. (Numbers refer to references.) The only clear example of superplasticity without the interfering effects of porosity, etc., in a ceramic system is that observed during the creep deformation of the ZrO_2, monoclinic $\longleftrightarrow$ tetragonal, case cited by Hart and Chaklader [27]. A clear relationship between creep deformation and the allotropic transformation was seen. However, in earlier work the possible operation of superplastic effects had been invoked in the hot pressing of ZrO_2 [31] and in many other systems [29]. Most of the ceramic cases, therefore, refer to work on hot pressing* the various systems (including the effects variously called reaction hot pressing, reactive hot pressing, and pressure calcintering). An exception is the example of $MgAl_2O_4/Al_2O_3$ cited by Newey [35]. Newey reports that compositions of spinel, with a large excess of alumina in solid solution (plasma-grown single crystals, therefore almost certainly quenched material), showed anomalously large compressive creep deformation; almost certainly related to subsequent exsolution of Al_2O_3 in the spinel phase. The effect increased with increasing amount of Al_2O_3, and testing was carried out in the region of the phase diagram showing the coexistence of the two phases $MgAl_2O_4$ solid solution + Al_2O_3.

Of the other examples already studied, in Table I, section 1, the work or compressive creep of MgO by Vasilos et al. [40, 41] is of interest because, in particular, they noted a large change in mechanism on going to grains less than $5\,\mu$ in size, which has since [41] been associated with the onset of larger amounts of boundary sliding. In the case of the silica transformation, in section 2, an effect of enhanced sintering was seen, but this cannot be allied in this case with

---

*There may be some objection to considering hot pressing as a creep phenomenon; however, it has usually been treated in this way in the past.

TABLE I

Possible Types of Systems Involving Superplastic Deformation

Metals	Ceramic Analogues
1. Monophasic No Transformation	
Pb/Ti [7]	Hot pressing of fine particles single or mixed oxides
Sn/Bi [22]	from alkoxides, chelates, etc. [26]
Ni [23]	Creep of MgO [40, 41]
2. Simple Phase Transformation	
α Fe $\longleftrightarrow \gamma$ Fe [3]	Monoclinic $ZrO_2 \longleftrightarrow$ tetragonal ZrO_2 [27, 31]
α Brass $\longleftrightarrow \beta$ Brass [1]	(hot pressing and bend deformation). Sintering of
Cu/38% Zn	$\alpha SiO_2 \longleftrightarrow \beta SiO_2$ [32]
	Hot pressing of $Zr_{0.85} Ca_{0.15} O_{1.85}$
	Hot pressing of monoclinic $Gd_2O_3 \longleftrightarrow$
	cubic Gd_2O_3 [34]
3. Eutectic or Eutectoid Exsolution	
Pb/Sn & Bi/Sn [4]	MgO/BeO, MgO/CaO, MgO/MgAl$_2$O$_4$
Al/Cu & Others	
4. Order-Disorder	
β Brass, Cu/48% Zn [1]	
5. Solid Solution Exsolution	
Zn/Al & others [1]	Creep of $MgAl_2O_4/Al_2O_3$ [35]
Ferrite/Austenite [28]	UO_2/CaO
6. Decomposition	
$ZrH_2 \longrightarrow Zr + H_2$ [29]	Hot pressing $Mg(OH)_2 \longrightarrow MgO + H_2O$ [36, 37, 38]
$UH_3 \longrightarrow U + H_2$ [30]	Hot pressing $Ca(CO_3) \longrightarrow CaO$
	Hot pressing $[Si(NH)_2]_x \longrightarrow Si_3N_4 + NH_3$
	Hot pressing $BaTiO(C_2O_4)_2 \cdot nH_2O \longrightarrow BaTiO_3$ [29]
7. Decomposition of Single Crystal into Two or More Phases	
Mixed hydrides, e.g.,	Hot pressing $MgCa(CO_3)_2 \longrightarrow MgO + CaO + 2CO_2$ [29]
$ZrNiH_3, Zr_2CuH_2, ZrAgH_3$	Hot pressing clays $\longrightarrow$ silicates + oxides [39]

boundary sliding. Keski [34] has observed anomalously easy hot pressing of Gd_2O_3 at the monoclinic $\longleftrightarrow$ cubic transformation. In section 3, there are some useful ceramic eutectic systems where eutectic exsolution might be expected to produce unusually easy deformational effects. In the case of the MgO/CaO system the effect of producing the two-phase microstructure from a single-precursor crystal during hot pressing is shown in section 7, and here the effect appears to be very real. As mentioned in section 6, Figure 2 illustrates the effects

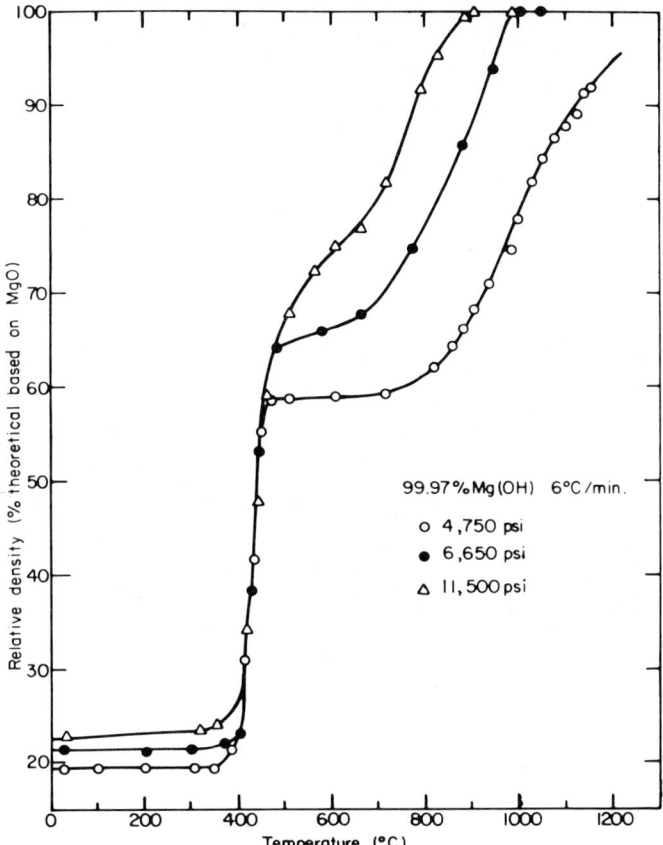

Figure 2. Pressure calcintering of $Mg(OH)_2 \longrightarrow MgO$.

of generating a very fine grain structure by decomposition of $Mg(OH)_2 \longrightarrow MgO$, starting at 350°C, during hot pressing and the very ready densification achieved during this stage. At 600°C in this process the grain size is still only 300 Å [27] while the lattice parameter of the MgO formed at this time is still 0.2 percent oversize in agreement with the work of Glasson [42]. Weiss [5] also has observed that changes in lattice parameter may be significant in some superplastic phenomena. The final microstructure of this material at full density, achieved at 900°C and 10,000 psi, is shown in Figure 3. Strength data of this material as obtained by Rice [43] and compared with other hot pressed magnesia is shown in Figure 4. In Figure 5 is shown a case where a metal is the product, as mentioned in Table I, section 6. Figure 6 illustrates the effect in $Al(OH)_3 \longrightarrow Al_2O_3$. Little is seen [37] at the decomposition temperature of $Al(OH)_3$,

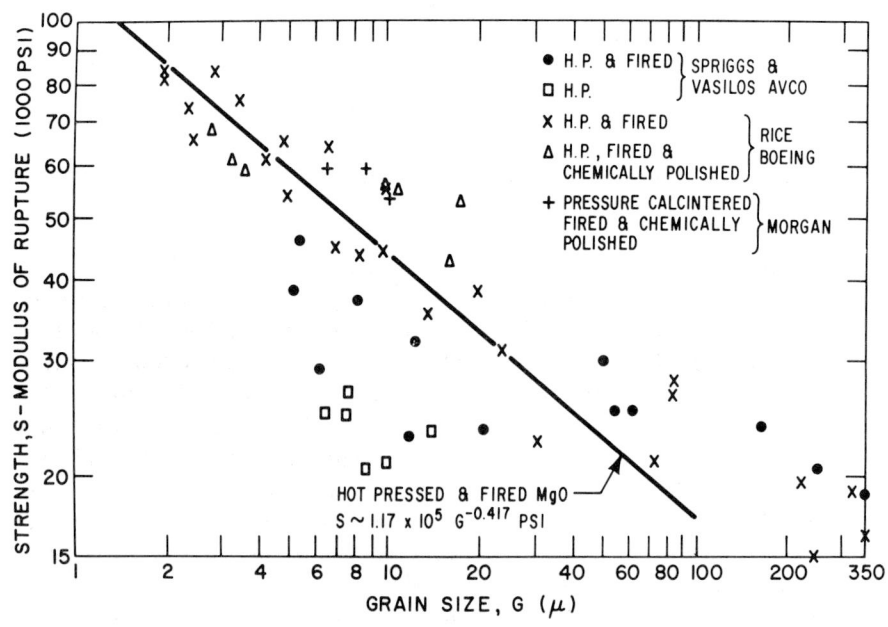

Figure 3. MgO ceramic. 99.8% theoretical density, transparent specimen. 900°C, 11,500 psi. X 32,000.

Figure 4. Strength of pressure calcintered material compared with other hot pressed MgO.

258

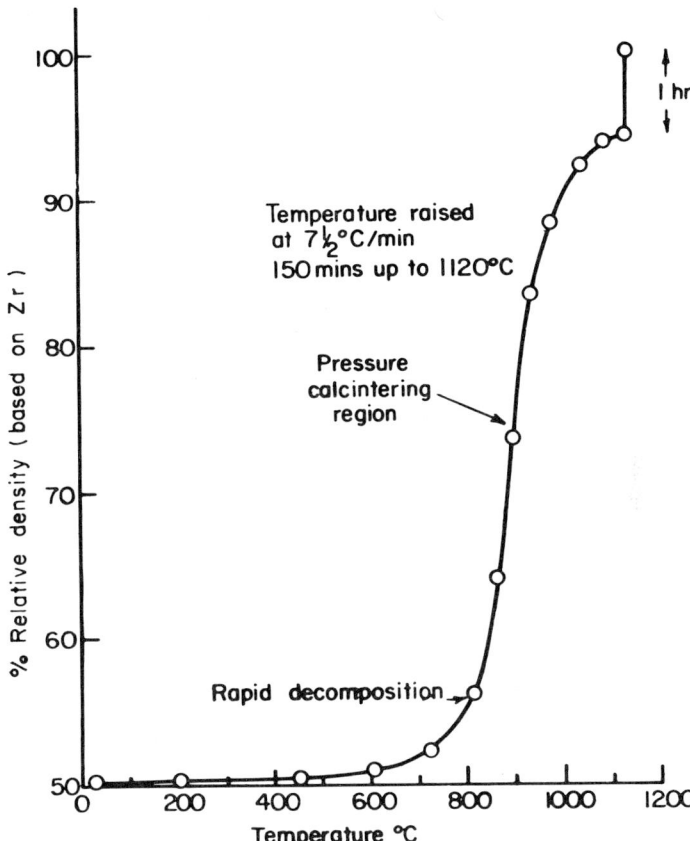

Figure 5. Pressure calcintering of $ZrH_2 \longrightarrow Zr$.

300°C, but a rapid densification sets in at 1000°C. This is now believed to be due probably to the onset of grain boundary sliding permitted by the local deformation of boundary kinks, etc., which can start as dislocations begin to move at this temperature. It is widely thought that dislocations are immobile in aluminum oxide below 1000°C. This result, then, is almost identical with results obtained by starting with very active aluminum oxides prepared by pre-calcination techniques. Figure 7 illustrates a somewhat different behavior in the hot pressing of $Cr(OH)_3 \longrightarrow Cr_2O_3$. $MgAl_2O_4$ generated from an amorphous precipitated hydroxide-basic carbonate mixture, densified as shown in Figure 8; $MgAl_2O_4$ is the only phase detected by X-ray diffraction methods, and begins to appear at about 400°C. Figure 9 shows densification results of hot pressing a mixed crystal so that a two-phase microstructure is achieved. Here, either of the two generated phases can mutually inhibit the grain growth of the other, as

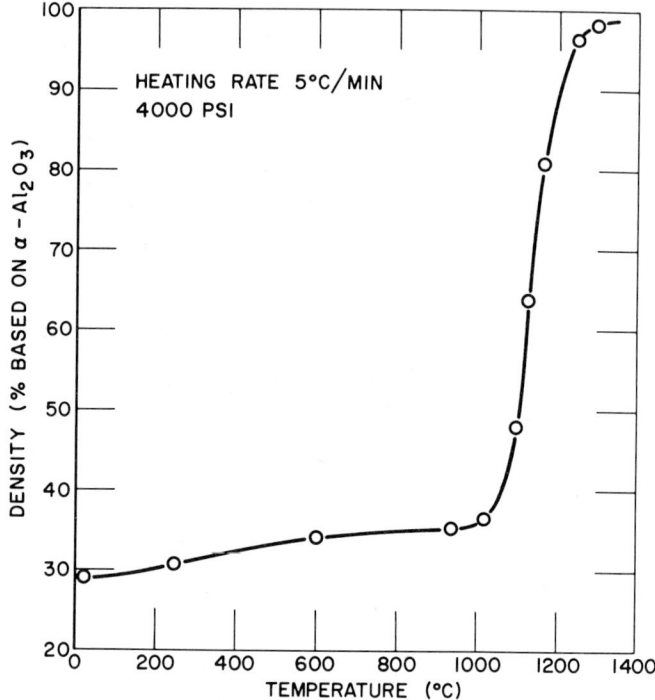

Figure 6. Pressure calcintering of $Al(OH)_3 \longrightarrow Al_2O_3$.

in the case of superplasticity in the quenched eutectics. A striking demonstration of this is shown in Figure 10, illustrating the microstructure of a piece of dolomite rock before hot pressing and the resulting sub-micron grain structure of the generated fully dense ceramic. No densification of this order has been seen in magnesite, brucite, or calcite rocks where only the one-phase oxide product is obtained.

Another two-phase system is generated by hot pressing $MgCrO_4$, magnesium chromate, Figure 11. It is noteworthy that the hot pressing of $MgCr_2O_7$, where only one phase, $MgCr_2O_4$, is generated, does not proceed at all as well as the former case, suggesting that the grain growth inhibition effect is very important, Grain boundary sliding is probably easier at high temperatures between different phases than between like phases because of the lack of the "islands of good fit," and, tentatively, this has also been experimentally observed [44]. Some other support for grain boundary sliding as an important mechanism in both hot pressing [29] and in press forging [45] comes from studies of the effect of sudden changes in stress which certainly do not support the Nabarro-Herring

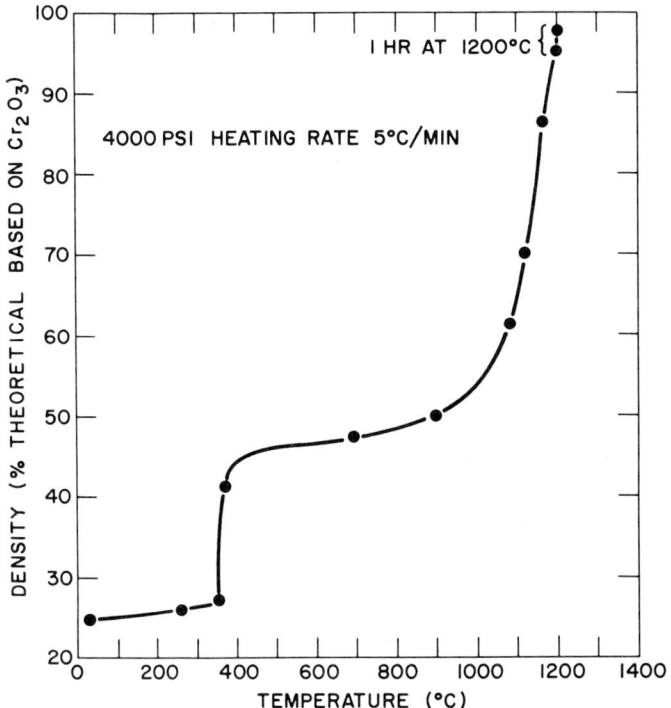

Figure 7. Pressure calcintering of $Cr(OH)_3 \longrightarrow Cr_2O_3$.

mechanism so favored earlier. Sudden small changes in stress can lead to very large changes in rate of deformation consistent with the idea of the sudden breaking down or locking of boundaries.

It has from time to time been proposed, especially by metallurgists, that grain boundaries have viscous liquid-like properties well below the normal melting point of the crystalline material, and there seems to be merit in pursuing this concept. Thus, grain boundary diffusional work [46], internal friction measurements of viscous relaxation in grain boundaries [47, 48], and high-temperature grain boundary separation [49], etc., have been analyzed in these terms, and very recent data on superplastic flow has supported this view [45]. If we say (1) glass is superplastic, (2) glass is amorphous, and then proceed to say, "except in special cases, grain boundaries are amorphous regions" and then move progressively towards solids of smaller and smaller grain size so that the volume percent of grain boundary region as opposed to actual grain content rises, we should not be surprised to find, what is indeed the case, that the superplastic effect increases.

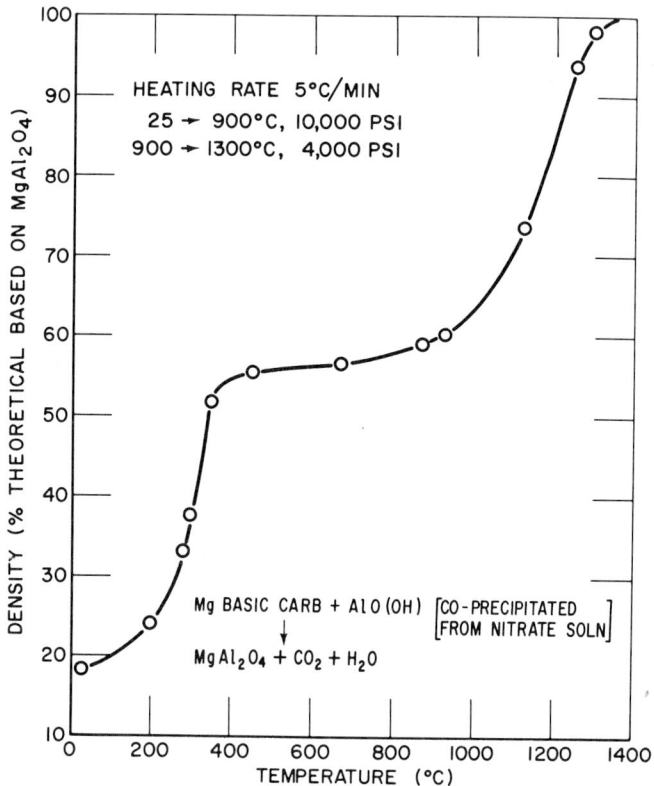

Figure 8. Pressure calcintering to produce $MgAl_2O_4$.

Several observations in ceramics support this boundary view. For example, in the hot pressing of MgO with additions of LiF by Rhodes and co-workers [50], no break was observed in the hot pressing characteristics either at the melting point of LiF or at any other temperature. The truth is probably that LiF markedly reduces the viscosity in boundaries where it forms solid solution layers, and diffusional accommodation is also enhanced. No sudden change occurs in the boundaries at the melting point of LiF or at any other temperature. We have followed up this line of thought by adding LiF during the pressure calcintering of $Mg(OH)_2 \longrightarrow MgO$ with the interesting result shown in Figure 12. The effect of the LiF is to prolong the boundary sliding effects to higher temperatures after the removal of sufficient $(OH)^-$ to cause a temporary pause in this behavior in the pure material. With the addition of the 1/2 w/o LiF, however, the densification/sliding mechanism continues without pause as the temperature rises, so that a fully dense translucent MgO ceramic is produced at only

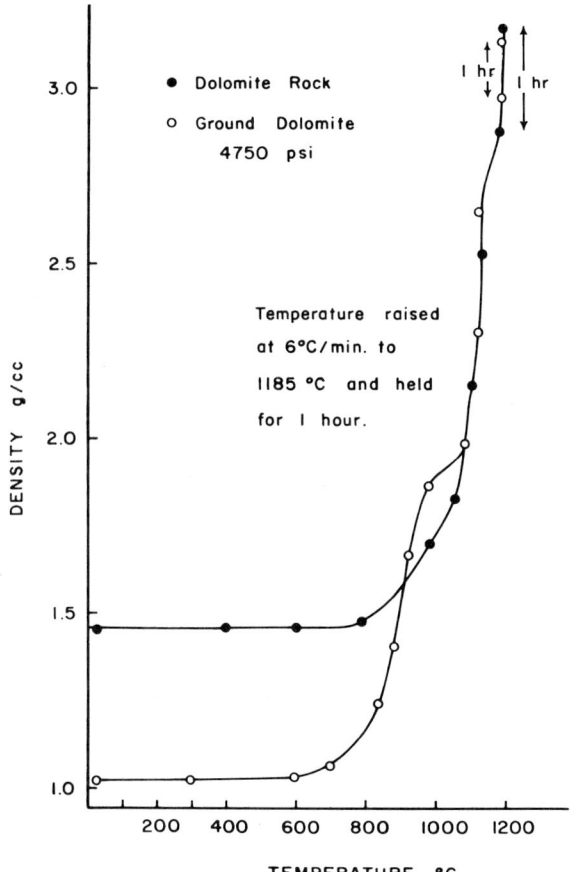

Figure 9. Pressure calcintering of dolomite, $MgCa(CO_3)_3 \longrightarrow MgO + CaO$. In the form of ground powder and also natural whole rock.

$675°C$, below the melting point of LiF; mechanisms invoking melting as necessary may be ruled out. With this addition also (Figure 13), at only **200** psi it is possible to get rapid densification to full density at $900°C$. The synergistic effect of both decomposition and addition of LiF during the pressure calcintering is rather remarkable, but in particular, no other mechanism but viscous grain boundary sliding seems reasonable at such low pressures. However, what in this case would appear to be a grain sliding phenomenon could also be a manifestation of a dynamic solution and reprecipitation of grains in their own boundaries, and some work with boundary diffusion at low temperatures in metal systems could support this view also [51].

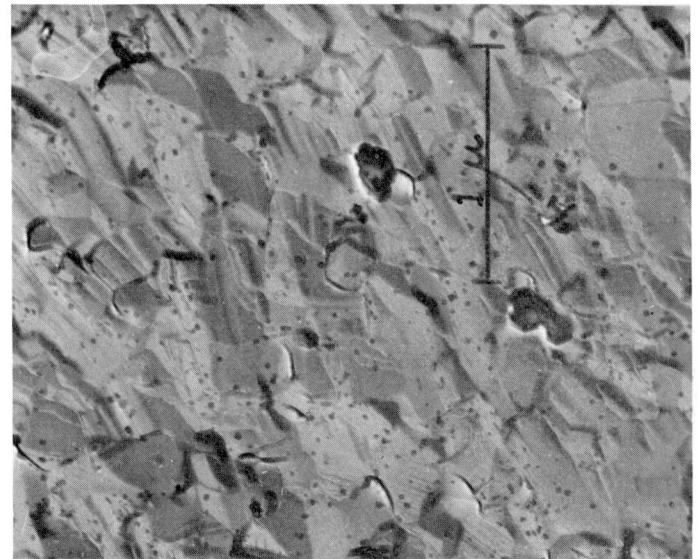

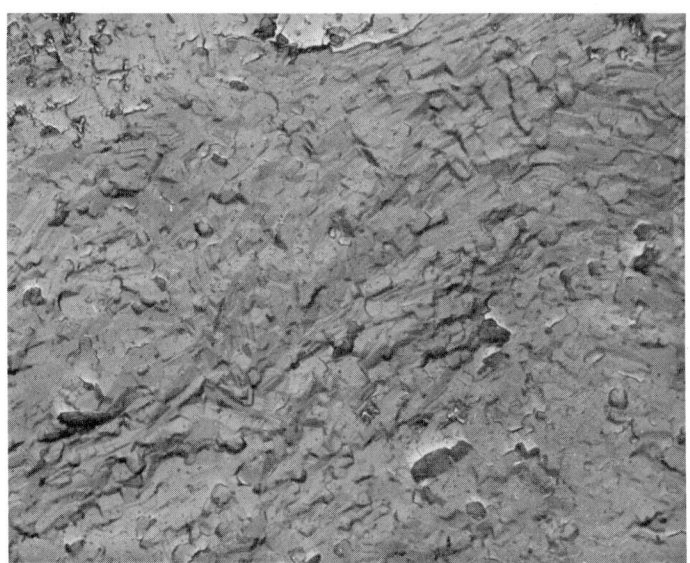

Figure 10. (Facing page) Dolomite rock microstructure. Polarized light micrograph; polished surface. × 100. (Above) Replica electron fractograph after pressure calcintering to 1200°C at 4750 psi. Left, × 500; right × 30,000.

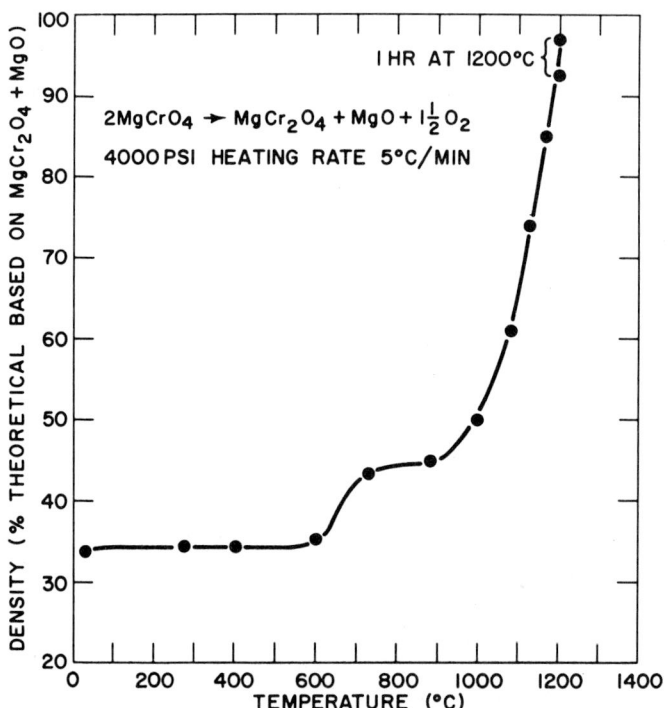

Figure 11. Pressure calcintering of $2MgCrO_4 \longrightarrow MgCr_2O_4 + MgO$.

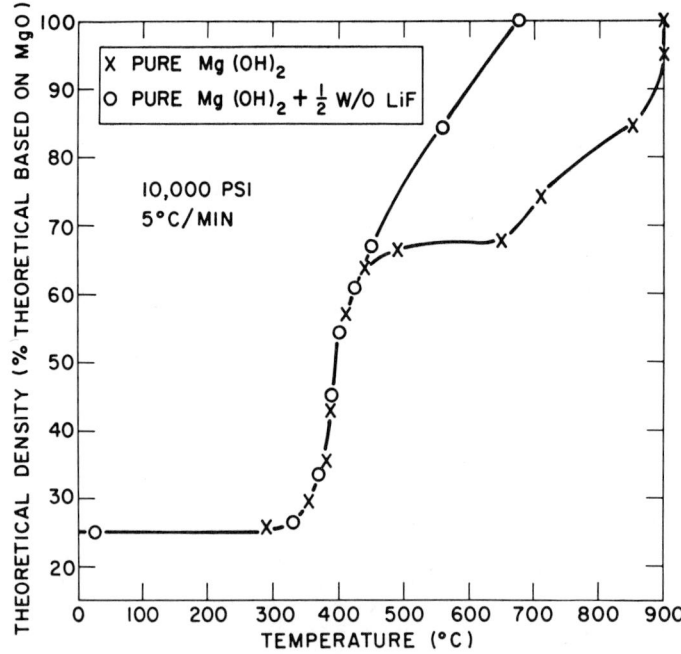

Figure 12. Pressure calcintering of $Mg(OH)_2 \longrightarrow MgO$, with the effect of additions of 1/2 w/o LiF.

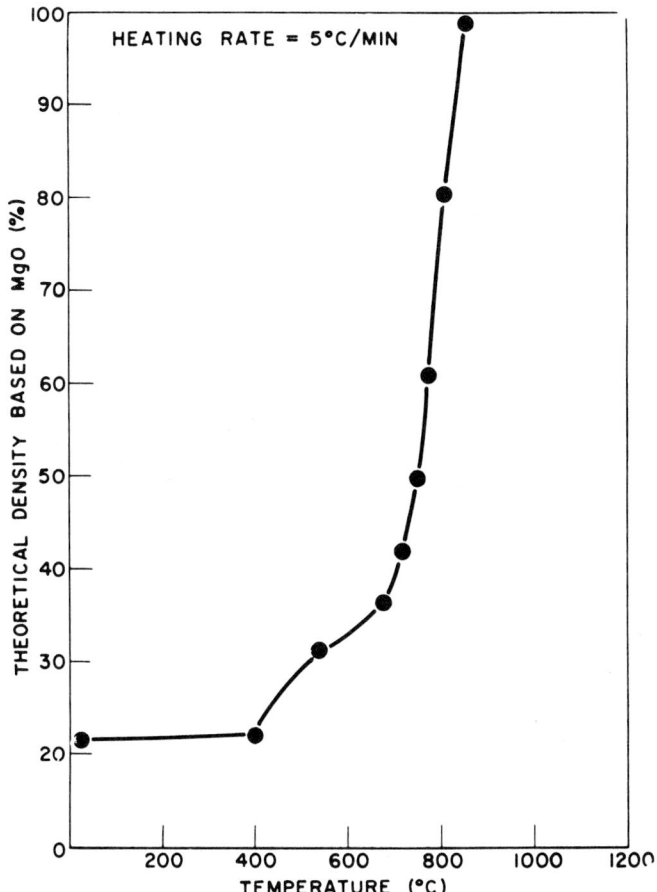

Figure 13. Pressure calcintering of Mg(OH)$_2$-1/2 w/o LiF $\longrightarrow$ MgO. Effect at low pressure, 200 psi.

Much of the data on hot pressing (and perhaps even sintering) can be rationalized in these terms; particularly the effects of impurities which segregate to the boundary in fine-grain materials. It must be stated at this point that whereas grain boundaries are often hardened by impurities at low temperatures they may be softened at high temperatures. There is evidence that if a pure material of fine grain size hot presses at a certain rate at some specified fraction of its melting point, then similar hot pressing may be achieved with certain additives at roughly the same fraction of the eutectic melting point. Some, certainly meager, evidence from the relative fraction of grain boundary sliding versus impurity content in metals also reinforces this view [52].

Hart [53] has recently constructed models of superplastic creep, based on the concept of crystals deforming surrounded by Newtonian viscous films, with some success in reproducing the stress/strain rate curves actually found.

I do not wish, in conclusion, to say that this or that process is superplastic or indeed any different, except in magnitude, from what has been observed before in conventional creep, but only to point out the very interesting effects seen in the deformation of very fine grained metals and to extend the possibilities to various ceramic forming processes also involving very fine grained states. The data being generated in the metal systems suggest also that views of hot pressing and creep have been very much oversimplified.

The properties and behavior of grain boundaries remain the largest uncertainty in attempts to describe deformation in polycrystalline materials, including ceramics, but it appears that viscous sliding properties become increasingly important as the relative amount of such boundaries increases.

References

1. Underwood, E. E., "A Review of Superplasticity and Related Phenomena," *J. Inst. Met.*, 14 (1962), 914–19.
2. Bochvar, A. A. and Sviderskaya, Z. A., *Izv. Akad. Nauk SSSR Otd. Tekhn. Nauk,* 9 (1945), 821.
3. Sauveur, A., "What is Steel?—Another Answer," *Iron Age,* 113 (1924), 581.
4. Pearson, C. E., "The Viscous Properties of Extruded Eutectic Alloys of Lead–Tin and Bismuth–Tin," *J. Inst. Met.,* 54 (1934), 111.
5. Weiss, V. and Kot, R., "Superplasticity," Report Battelle Memorial Inst. (1967), AD 817–836, Clearinghouse for Federal Information.
6. Backofen, W. A., Azzarto, F. J., Murty, G. S. and Zehr, S. W., "Deformation Processing of Anisotropic Metals," Final Report No. N000 19–67–C–0089, Naval Air Systems Command.
7. Weiss, V., Kot, R. and Krause, G., "Investigation of Phenomenon of Superplasticity in Metals," Final Report, BUNW, Contract No. NOw 66–0109–d (1966).
8. Gifkins, R. C., "Grain Movements During Creep," *Nature,* 169 (1952), 238.
9. Alden, T. H., "The Origin of Superplasticity in the Sn–5% Bi Alloy," *Acta Met.,* 15 (1967), 469–80.
10. Cline, H. E. and Alden, T. H., "Rate Sensitive Deformation in Tin–Lead Alloys," *Trans. AIME,* 239 (1967), 710–14.
11. Bell, R. L. and Langdon, T. G., "An Investigation of Grain Boundary Sliding During Creep," *J. Mat. Sci.,* 2 (1967), 313–23.
12. Stevens, R. N., "Grain Boundary Sliding in Metals," *Met Rev.,* 11 (1966), 129–42.
13. Gifkins, R. C., "Structural Studies of the Creep of Lead," *J. Inst. Met.,* 82 (1953–54), 39.
14. Mott, N. F., "Slip at Grain Boundaries and Grain Control in Metals," *Proc. Phys. Soc.,* 60 (1948), 391.
15. Gifkins, R. C., "Superplasticity During Creep," *J. Inst. Met.,* 95 (1967), 373–77.
16. Weinberg, F., "Grain Boundary Shear in Aluminum," *Trans. AIME,* 212 (1958), 808.
17. Umeno, M. and Shinoda, G., "Deformation Behavior of Aluminum Macrocrystals in the Grain Boundary Region," *J. Mat. Sci.,* 3 (1968), 120–26.

18. Lifshitz, I. M., "On the Theory of Diffusion-Viscous Flow of Polycrystalline Bodies," *Soviet Physics JETP*, 17 (1963), 909.
19. Gifkins, R. C., "Diffusional Creep Mechanisms," *J. Am. Ceram. Soc.*, 51 (1968), 69–72.
20. Hensler, J. H. and Cullen, G. V., "Grain Shape Change During Creep in Magnesium Oxide," *J. Am. Cer. Soc.*, 50 (1967), 584–85.
21. Coble, R. L., "A Model for Boundary Diffusion Controlled Creep in Polycrystalline Materials," *J. Appl.·Phys.*, 34 (1963), 1679.
22. Alden, T. H., "The Origin of Superplasticity in the Sn–5% Bi Alloy," General Electric Tech. Inf. Series, Contract No. 66–C–289 (1966).
23. Floreen, S., "Superplasticity in Pure Nickel," *Scripta Met.*, 1 (1967), 19–23.
24. Walter, J. L. and Cline, H. E., "Grain Boundary Sliding, Migration and Deformation in High Purity Aluminum," *Trans. AIME*, 242 (1968), 1825–30.
25. Underwood, E. E., *et al.*, "Mechanism of Superplasticity in Al–78% Zn Alloys," Lockheed-Georgia Co., N000 14 67C–0503 (1968) AD670–800.
26. Mazdiyasni, K. S., Lynch, C. T. and Smith, J. S., "Preparation of Ultra High Purity Sub-Micron Refractory Oxides," *J. Am. Ceram. Soc.*, 48 (1965), 372–75.
27. Hart, J. L. and Chaklader, A. C. D., "Superplasticity in Pure ZrO_2," *Mat. Res. Bull.*, 2 (1967), 521–26.
28. Morrison, W. B., "Superplasticity of Low Alloy Steels," to be published in *Trans. Quarterly*.
29. Morgan, P. E. D. and Schaeffer, N. C., "Chemically Activated Pressure Sintering of Oxides," Report AFML–TR–66–356 (November 1966).
30. Hausner, H. H., "Powder Metallurgical Products," in Vacuum Metallurgy, R. F. Bunshah, ed. (1958).
31. Chaklader, A. C. D. and Baker, V. T., "Reactive Hot Pressing: Fabrication and Densification and Non-Stabilized ZrO_2," *Bull. Am. Ceram. Soc.*, 44 (1965), 258–59.
32. Chaklader, A. C. D., "Deformation of Quartz Crystals at the Transformation Temperature," *Nature*, 197 (1963), 791.
33. Arias, A., "Mechanism by which Metal Additions Improve the Thermal Shock Resistance of Zirconia," *J. Am. Ceram. Soc.*, 49 (1966), 339.
34. Keski, J. R., Private communication.
35. Newey, C. W. A. and Radford, K. C., "Plastic Deformation of Magnesium Aluminate Spinel," presented at "Anisotropy in Single Crystal Refractory Compounds," Conf., Dayton, Ohio (June 1967).
36. Morgan, P. E. D. and Scala, E., "The Formation of Fully Dense Oxides by the Decomposition Pressure Sintering of Hydroxides," presented at the 67th Annual Meeting, Am. Ceram. Soc., Phila., Pa. (May 1965).
37. Morgan, P. E. D. and Scala, E., "High Density Oxides by Decomposition Pressure Sintering of Hydroxides," *Proc. Intl. Conf. on Sintering and Related Phenomena*, Notre Dame, Indiana, June 1965, G. C. Kuczynski, ed., Gordon and Breach, New York (1967).
38. Chaklader, A. C. D., "Theory of Reactive Hot Pressing," presented at the 67th Annual Meeting, Am. Ceram. Soc., Phila., Pa. (May 3, 1965).
39. Chaklader, A. C. D. and McKenzie, L. G., "Reactive Hot Pressing of Clays," *Bull. Am. Ceram. Soc.*, 43 (1964), 892–93.
40. Passmore, E. M., Duff, R. and Vasilos, T., "Mechanisms of Deformation in Polycrystalline Magnesium Oxide," Report AFML–TR–65–122 (June 1965).
41. Vasilos, T., Private communication.
42. Glasson, D. R., "Reactivity of Lime and Related Oxides," *J. Appl. Chem.*, 13 (1963), 111–19.

43. Rice, R. W., Private communication.
44. Zehr, S. W. and Backofen, W. A., "Superplasticity in Lead-Tin Alloys," *Trans. ASM,* 61 (1968), 300–13.
45. Spriggs, R. M., Private communication.
46. Niessen, P. and Winegard, W. C., "The Effect of Solute Distribution Coefficient on Grain Growth and Single Boundary Migration," *J. Inst. Met.*, 94 (1966), 31.
47. Ke, T. S., "Experimental Evidence of the Viscous Behavior of Grain Boundaries in Metals," *Phys. Rev.,* 71 (1947), 533.
48. McLean, D., *Grain Boundaries in Metals,* Oxford University Press, London (1957).
49. Conrad, H., "The Role of Grain Boundaries in Creep and Stress Rupture," in *Mechanical Behavior of Materials at Elevated Temperatures,* J. E. Dorn, ed., McGraw-Hill, New York (1961).
50. Rhodes, W. H. and Sellers, D. J., "Mechanism of Pressure Sintering MgO with LiF Additions," presented at 69th Annual Meeting, Am. Ceram. Soc., New York (1967).
51. Richards, J. L., McCann, W. H. and Gilbert, S. L., "Diffusion in Polycrystalline Thin Film Copper-Nickel Couples," to be published in *J. Appl. Phys.*
52. Harper, S., "Structural Processes in Creep," *Iron and Steel Inst.,* London (1961), 56.
53. Hart, E. W., "A Theory for the Flow of Polycrystals," *Acta Met.,* 15 (1967), 1545–49.

ꞌ Comment

A. H. HEUER
Case Western Reserve University
Cleveland, Ohio

P. E. D. Morgan's contributions in exploiting reactive hot pressing for producing many novel, fine-grain dense ceramics and metals are well known. However, Morgan confuses the issue when he describes as superplastic those systems undergoing a reaction during hot pressing. It is certainly fair to state that even a partial understanding of the enhanced densification observed in such systems and documented in this paper is not yet in hand. The suggestion that this is a manifestation of superplasticity in no way increases our understanding of the mechanisms of densification that may be occurring. Although the metallurgical literature on superplasticity is still somewhat confusing, there now seems to be agreement to use this term to refer to large, neck-free elongations; the phenomenon is enhanced by a fine-stable grain size and/or a phase transformation, and is invariably associated with a high strain rate sensitivity

$$m = \frac{d \ln \sigma}{d \ln \epsilon} > \frac{1}{3}$$

No studies are known to the writer which have shown that the presence or elimination of porosity is important in obtaining superplastic extensions (100-1,000 percent).

Certainly, true superplasticity is possible in ceramics, either of the transformational type or associated with a fine grain size. The point here is that Morgan's plots of relative density versus temperature are completely unrelated to superplastic deformation, as that term is presently understood in the literature.

Author's Reply

The article was intended to be both tendentious and controversial; I think both the format and content indicate this. The article would have had no purpose if there was even a partial understanding of the mechanisms in such processes. Superplasticity itself is a matter of definition and the consensus of experts, but there has been disagreement on this. However, regardless of what terminology we are going to use, there is no reason *not* to assume that the micromechanical processes occurring in an ultrafine-grain ceramic material, even though it contains porosity, may be related to the formerly surprising analogies in superplastic deformation. One must suppose, for example, that in pressure calcintering, even more than in hot pressing, the very fine particle compact will exhibit grain boundary sliding and grain rotation, and this has hitherto been neglected. The presence of porosity does not imply that these processes are not taking place—although, of course, it does complicate them.

Doubtless, anomalous mechanical effects have been seen in ceramic materials near to a phase transformation, as in Chaklader's work on zirconia, and also, as I have shown in this article, it is conceivable that a system could deform largely by particle rearrangement and grain boundary sliding, grain rotation, etc., in the pressure calcintering process. For example, in the case of the conversion of rock Dolomite to MgO-CaO ceramic, we do not know that porosity plays a part. Conceivably, a fully dense rock material may convert to the fully dense ceramic without any real porosity being created at some sufficiently high pressure, the transforming particles always filling the total volume of the compact. To show that this was in fact so, like in all pressure calcintering work, would be extremely—if not impossibly—difficult.

It is only the failure of more simplistic models to describe the facts that justifies the search for analogies. The plots of relative density versus temperature were not intended to confirm the mechanism of superplasticity; clearly, they do not. However, in the absence of hard, cold data, we are surely entitled to use circumstantial evidence of this kind to indicate that unusual processes may be occurring—unusual and anomalous, that is, only in the sense that we think they are so.

13. Ultrafine-Grain Ceramics from Melt Phase

J. HURT and D. J. VIECHNICKI
Army Materials and Mechanics Research Center
Watertown, Massachusetts

ABSTRACT

An ultrafine dispersion of two or more oxide phases can be produced by controlled cooling of a melt. The mechanism by which this dispersion is formed can vary, depending upon whether glass formation is possible in the system. In the case where a glass does not form in the system, coupled growth of oxide eutectics is possible. Planar or rod-like eutectic microstructures result. Conditions for producing such microstructures in selected oxide systems will be discussed. In the case where a glass does form in the system, formation of a crystalline phase may be possible. The formation of crystalline phases may occur by one mechanism or a combination of several mechanisms; metastable liquid-liquid separation, unstable liquid-liquid separation, metastable formation of a crystalline phase, stable formation of a crystalline phase. The particle size and shape of the resulting crystalline phase is dependent upon the mechanism. The effect of the mechanism of formation on the microstructure will be discussed.

Introduction

The bulk of the work today directed towards fabrication of high-strength fine-grain polycrystalline ceramics involves solid-state processing; that is, sintering or pressure sintering. A serious limitation of solid-state processing is that the final product is very much dependent upon the particle shape and size of the starting material. The grain size of the final product will not be finer than the particle size of the starting material. There is much interest in using "active" powders of extremely fine particle size as starting materials, but these "active" powders, by their very nature, can result in rapid grain growth. Fabrication of ultrafine-grain ceramics from the melt phase has the advantage that the microstructure of the final product is insensitive to particle size and shape of the starting materials. Rather, the microstructure is dependent upon the mode of decomposition of the melt to the solid. Oxide ceramics may decompose from the melt by one of two general mechanisms to produce an ultrafine-grain microstructure: (1) eutectic solidification, and (2) recrystallization of a glass.

273

The decomposition of such a melt in a binary or ternary system can lead to a fine eutectic microstructure as a result of eutectic solidification. The crystalline phases in oxide systems where eutectic solidification occurs will generally have high entropies and enthalpies of melting, and large volume changes between solid and liquid. Al_2O_3, for example, has an entropy of melting of 12.3 kcal/mole$^\circ$K, an enthalpy of melting of 28 kcal/mole [1], and an increase in volume of 22 percent when going from solid to liquid [2]. The corresponding values for SiO_2 are 1.03 kcal/mole$^\circ$K, 2 kcal/mole and 5 percent respectively [1]. Formation of a glass or of a highly supercooled melt can occur in many oxide systems, most notably the silicates, borates, and phosphates. The crystalline phases in these systems may be generally classified as having low entropies and enthalpies of melting, and a small volume change between solid and liquid. The decomposition of a glass or a supercooled melt, with its very high viscosity to a crystalline phase, will occur by mechanisms different than the decomposition of a melt which has a low viscosity and where a minimal amount of supercooling is possible.

The object of this paper will be to discuss briefly the mechanisms responsible for the fabrication of ultrafine-grain polycrystalline ceramics by these two different modes of decomposition of a melt to a solid, i.e., recrystallization of a glass and eutectic solidification, and to discuss experimental observations of decomposition of the melt in several Al_2O_3-based systems that were chosen for their high strength potential.

Eutectic Solidification

It is more convenient to describe the solidification of a liquid of low viscosity with two experimentally imposed constraints, temperature gradient in the liquid and growth rate, than to try to obtain more basic parameters from the application of a model describing growth of a solid from a liquid. For the solidification of a binary melt of eutectic composition, assuming a planar liquid-solid interface and unidirectional heat flow, a simple model of the solidification process can be drawn.

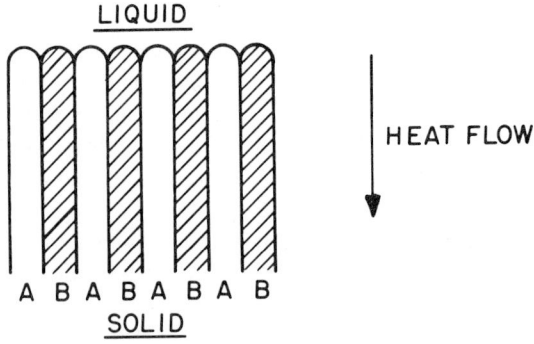

Plates or rods of the two solid phases, **A** and B, form from the liquid. Diffusion in the liquid of the species A and B is short-range and approximately parallel to the liquid-solid interface. For coupled eutectic growth to occur, phases A and B will grow into the melt at the same rate. The experimentally imposed growth rate will control the amount of diffusion occuring in the liquid to the solid and hence the thickness of the plates or rods—in other words, the particle size. A minimal temperature gradient in the liquid ahead of the liquid-solid interface will be needed as a driving force for the reaction [3]. In practice, however, one phase may grow into the melt faster than the other.

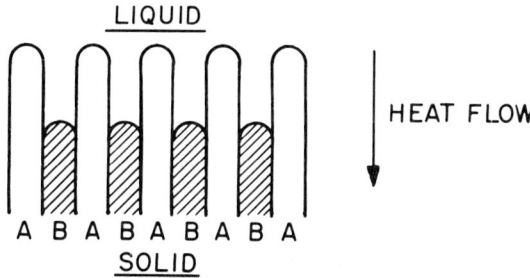

If the difference in growth rate into the melt is great enough, the phases will not grow in a coupled manner. If a large temperature gradient is imposed upon this system, a barrier will be set up so that the one phase cannot grow into the melt much in advance of the other and coupled growth will occur.

The rejection of impurities ahead of the liquid-solid interface leads to non-unidirectional growth by providing a region of constitutionally supercooled liquid for the solid to grow into. A large temperature gradient can prevent this [4]. If the impurity level is high enough, a colony-type microstructure can result because the liquid-solid interface is no longer planar.

Unidirectional heat flow is important to maintain a planar liquid-solid interface. It is more important in the solidification of oxides than metals, because oxides have a large change in the volume between liquid and solid upon solidification. In Al_2O_3 this is 22 percent [2], while in metals it is generally less than 6 percent [5]. Without unidirectional solidification, a very porous microstructure will develop.

This is a brief and incomplete statement of the theory of eutectic solidification. Many review articles are available on the subject to allow the interested party to gain an insight into the problems. The background given here is sufficient for an understanding of the reasons behind the design of our equipment and of the experimental results.

Experimental

The system Al_2O_3-$Y_3Al_5O_{12}$ was chosen for study for two reasons: both Al_2O_3 and Y_2O_3, the starting materials, have low vapor pressures below $2000°C$

[6] and are not easily reduced to sub-oxides [7, 8]; Al_2O_3 has a Young's modulus of 60×10^6 psi [9], and $Y_3Al_5O_{12}$ has a Young's modulus of 41×10^6 psi [10]. These values indicate that this system could be of practical interest for structural ceramics.

The starting materials in this investigation were high-purity alumina powder containing 99.99 percent Al_2O_3 and high-purity yttria powder containing 99.99 percent Y_2O_3. These powders were weighed in proper proportions, mixed in a blender in acetone, dried, and calcined for more than 60 hours at 1000°C. The calcined powders were either put directly into a vacuum furnace for solidification studies or stored in an evacuated desiccator.

Two furnaces were designed and built for the solidification studies to give high-temperature gradients in the material and controlled growth rate. In both cases, vapor-deposited tungsten crucibles were used as the container. The first furnace was simply a gradient furnace, as shown in Figure 1. A graphite resistance furnace was used as the heat source, and heating was accomplished in argon or under a vacuum of 8.0×10^{-2} torr. Melting in a vacuum is preferred because entrapped gases are eliminated upon melting, thereby eliminating gas pores in the ingot. Projecting into the hot zone of this furnace was a tungsten heat exchanger which supported the crucible and melt. Helium was forced into the heat exchanger to draw heat away from the crucible and melt, thereby producing solidification from the bottom of the crucible. The rate of solidification

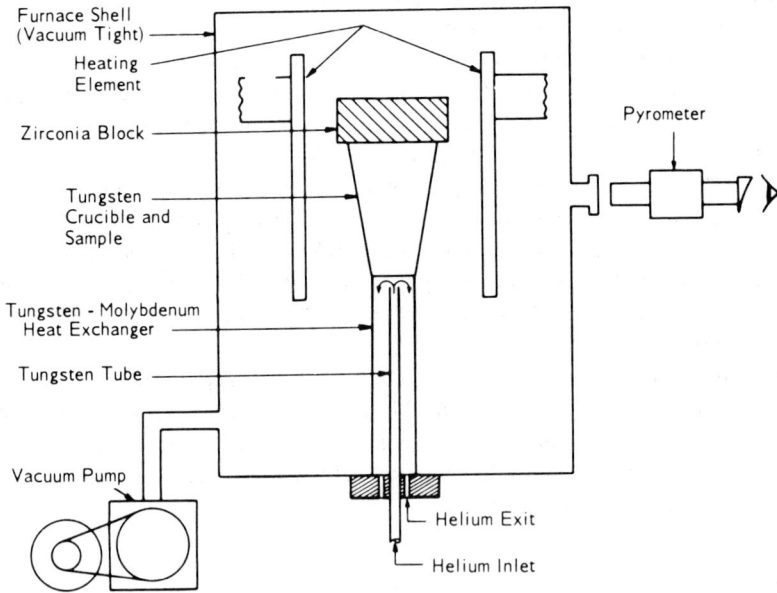

Figure 1. Schematic of gradient furnace.

was controlled by the rate of flow of the helium into the heat exchanger and by the temperature of the melt when the helium was introduced. The temperature gradient in the material was dependent upon the rate of flow of helium into the heat exchanger and the temperature of the melt. A typical operation consisted of melting the material and holding it at some temperature above the melting point. The rate of flow of helium was increased and held for a given amount of time. After this time, the power into the furnace was lowered to drop the whole system below the melting point. In this way, the amount of solidification due to this directional cooling and the growth rate was later measured. The whole system was then slowly cooled so as not to subject the ingot to thermal shock.

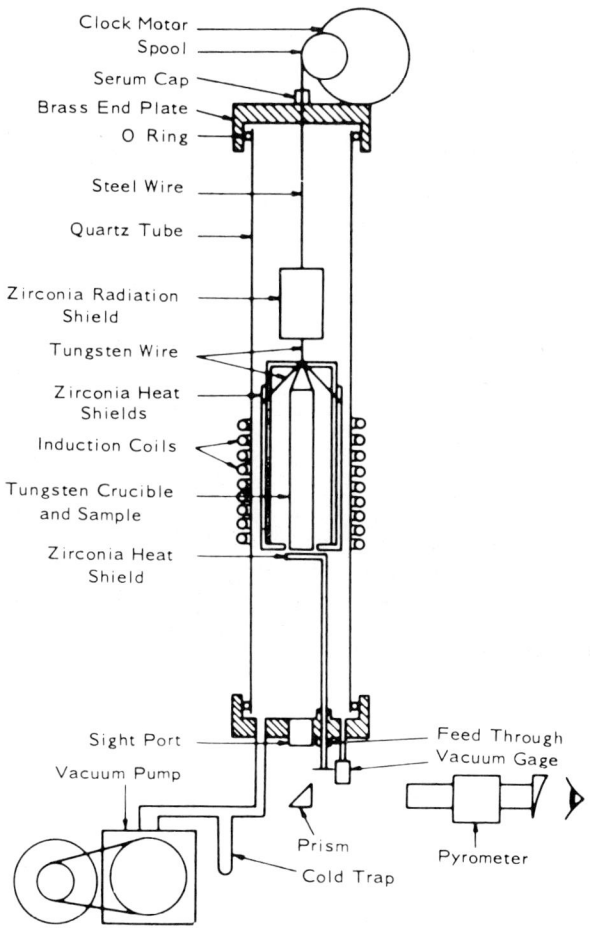

Figure 2. Schematic of Bridgman-type furnace.

The second method for controlled solidification employed a modified Bridgman technique, where the melt was passed through a hot zone at a controlled velocity, as shown in Figure 2. A high-frequency induction generator was the power source and coupling to the tungsten crucible was accomplished. Vacuums below 1.5×10^{-2} torr are employed in this technique. The rate of solidification was controlled by the rate at which the crucible passed through the coils. The temperature gradient was controlled by the configuration of the insulation around the crucible. Heat losses were from radiation.

Results

Figure 3 shows an ingot of the eutectic composition produced in the gradient furnace. Ingots solidified by the Bridgman technique were cylindrical. Large

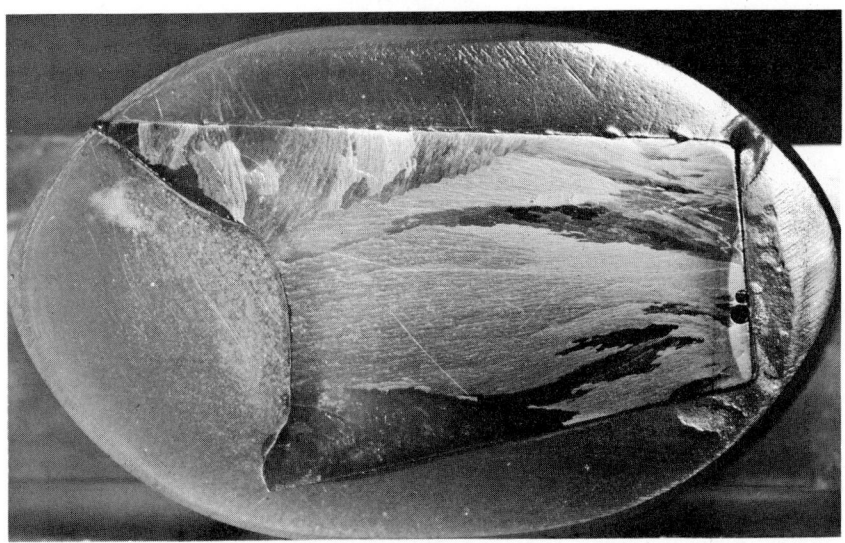

Figure 3. Ingot of Al_2O_3.

columnar grains approximately 0.3 cm wide by 3.0 cm long are in evidence. Near the top of the ingot, solidification has occurred randomly after the furnace power was cut off. In such areas porosity, termed solidification shrinkage by metallurgists, can be noticed in Figure 4. This porosity arose from the difference in volume between liquid and solid, but was eliminated by unidirectional solidi-

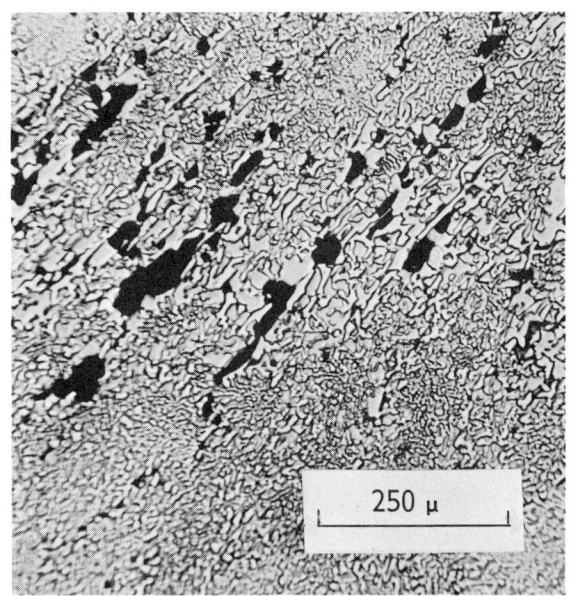

Figure 4. Solidification shrinkage in 77.8 mole% Al_2O_3–22.2 mole% $Y_3Al_5O_{12}$. Original magnification 100 X.

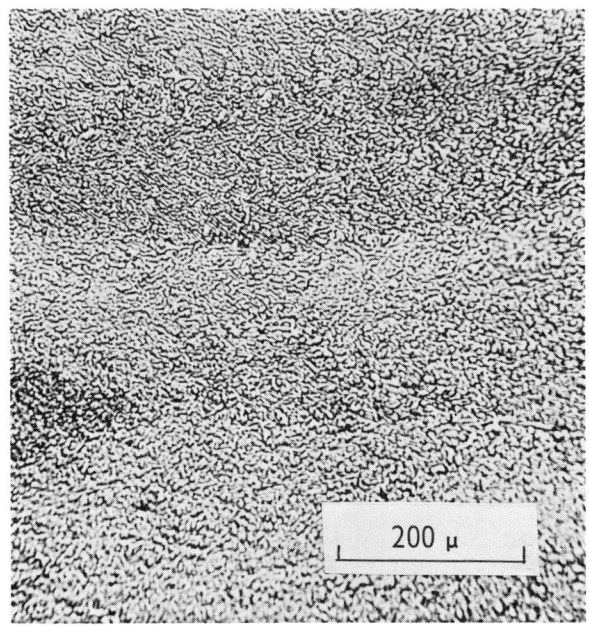

Figure 5. Longitudinal section of microstructure of eutectic composition, 80.1 mole% Al_2O_3–19.9 mole% $Y_3Al_5O_{12}$, after heating to 1850°C. Temperature gradient = 190°C/cm, growth rate = 4 cm/hr. Original magnification 125 X.

fication. No porosity resulting from entrapped gases was observed after melting at pressures of 8.0×10^{-2} torr.

Inside each columnar grain a fine eutectic dispersion was noticed. Figure 5 shows a longitudinal section of an ingot similar to the one in Figure 3. The columnar grains can be noticed, as well as the fine eutectic dispersion. Figure 6

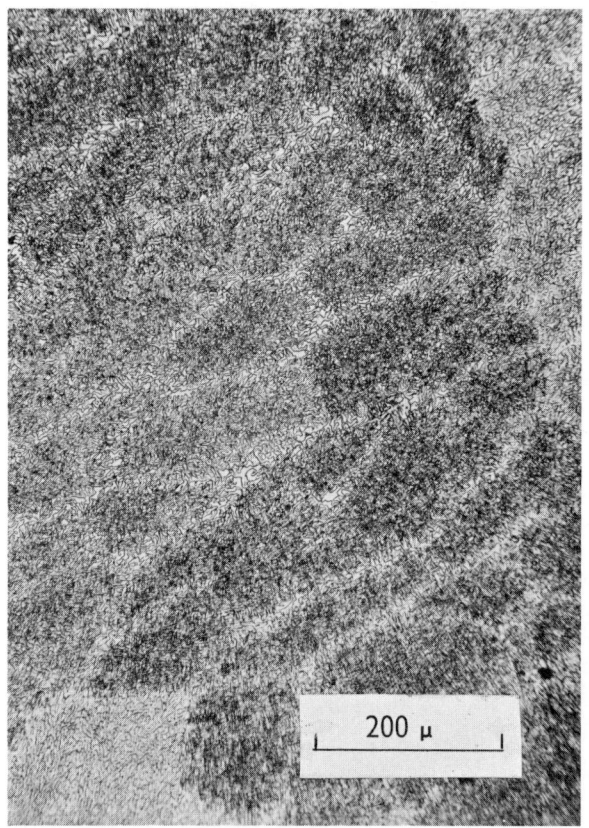

Figure 6. Transverse section of microstructure of eutectic composition, 80.1 mole% Al_2O_3–19.9 mole% $Y_3Al_5O_{12}$, after heating to $1850^{\circ}C$. Temperature gradient = $190^{\circ}C/cm$, growth rate = 4 cm/hr. Original magnification 125 X.

is a transverse section of such an ingot. The formation of these highly dis-oriented regions where the columnar grains meet results from the rejection of impurities. This type of microstructure is called a colony-type, and was the one most commonly observed in this investigation.

At growth rates of 4 cm/hr and above, and with a temperature gradient of 190°C/cm, some of the columnar grains exhibited a highly oriented eutectic microstructure. Figures 7 and 8 are photomicrographs of a longitudinal section of such a grain. Both rods and plates may be seen in Figure 8. The spacing between like rods is 2 μ. Ultrafine-grain ceramics can be produced by unidirec-

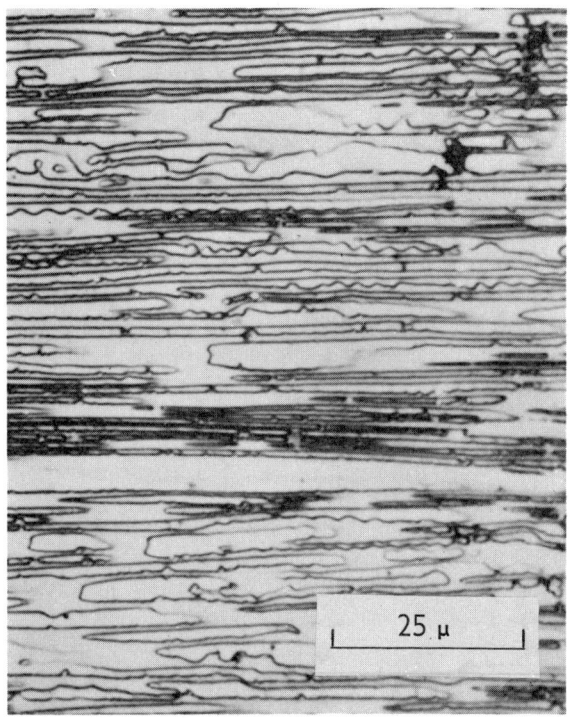

Figure 7. Longitudinal section of highly oriented eutectic microstructure of 80.1 mole% Al_2O_3–19.9 mole% $Y_3Al_5O_{12}$ composition after heating to 1850°C. Temperature gradient = 190°C/cm, growth rate = 4 cm/hr. Original magnification 1,000 X.

tional solidification of eutectic melts. Before the mechanical properties of these microstructures can be investigated, however, the large columnar grain structure found in the ingots must be eliminated. Preliminary mechanical testing of these materials conducted in four-point loading has shown that failure occurs at the interfaces between the large columnar grains at loads of 30,000 psi. The mechanical behavior of the eutectic microstructure could not be obtained. Further work is in progress to control nucleation at the start of the solidification process to eliminate the columnar grain structure.

Figure 8. Transverse section of highly oriented eutectic microstructure of 80.1 mole% Al_2O_3–19.9 mole% $Y_3Al_5O_{12}$ composition after heating to 1850°C. Temperature gradient = 190°C/cm, growth rate = 4 cm/hr. Original magnification 1,000 X.

Fabrication from the Glass Phase

The production of fine-grain ceramics from the glass phase has seen an enormous amount of commercial and academic interest in the past ten years, as evidenced by the large amount of hardware and the volume of literature of recent vintage [11–15]. The long list of obvious benefits of a tailor-made crystalline microstructure need not be repeated here (see [11] for example), but the potential of the process approaches the ultimate dream of the materials scientist.

Ceramics are rather distinct from metals in that some systems have the ability to form glasses. Ceramics can be processed from the melt phase in the same manner as metals, and also offer the advantage of controlled crystallization of a glass phase. This is accomplished by cooling a homogeneous melt to a convenient temperature where the mobility of ions is sufficiently low ($|D| \approx 10^{-17}$ cm^2/sec, or viscosity $\approx 10^{10}$ poises) and the time constant for growth is on the order of $10^3 - 10^4$ sec. This allows the optimum control of grain size.

The fact that some ceramic systems can undercool by an amount equal to their melting temperature allows more paths of decomposition to the final crystalline phase. These systems have the following decomposition mechanisms: (1) sub-liquidus nucleation of a second amorphous phase, (2) sub-liquidus spinodal decomposition, (3) "precrystallization microphase separation," and (4) nucleation of a crystalline phase (homogeneous and heterogeneous). It is the purpose of this portion of this paper to summarize briefly the particle-size dependence of each separation process.

Sub-Liquidus Amorphous Separation

Liquid immiscibility was studied extensively for many glass-forming systems as early as 1927 [16, 17]. Although theory [18] stated that two-liquid equilibrium extends below the liquidus, the experimental evidence was found only recently, e.g., see [19–22]. The two modes of liquid-liquid separation available to the glass are nucleation and spinodal decomposition. These have been described completely elsewhere [23, 24], so extraction of a brief summary will suffice.

Sub-Liquidus Amorphous Nucleation

The one-phase liquid in the two-phase region (yet not within the spinodal) is unstable with respect to a two-phase mixture, but stable to an infinitesimal composition fluctuation. It therefore requires a finite composition fluctuation to become unstable. This is necessary because work must be expended to create a surface between the two phases. It has been shown [25] that this work is inversely proportional to ΔT^2, where ΔT is the undercooling below the liquidus. The work is infinite when ΔT is zero, so a finite undercooling is necessary before nucleation can occur.

Nucleation rates have been shown [25] to be of the form:

$$I_o = A \exp - \frac{a}{\Delta T^2 kT}$$

where

> a is constant
> A is constant $\approx 10^{40}$ nuclei per cm^2 [25]
> k is the Boltzmann constant
> ΔT is the undercooling
> T is the temperature

The presence of both ΔT and T in the exponent indicate a maximum nucleation rate with temperature.

The smallest (critical) nucleus which is stable is proportional to the interfacial free energy and inversely proportional to the free energy of phase change per unit volume of new phase formed [25].

It was pointed out [26] that when ΔT is sufficient to approach the spinodal boundary the finite fluctuation begins to resemble a continuous fluctuation rather than a droplet of second phase. It was also shown that the surface energy effectively goes to zero as the spinodal is reached. Near the spinodal boundary, the work required to form a nucleus is proportional to $(\Delta T_s)^{3/2}$, where ΔT_s is the temperature above the spinodal boundary. In effect, this states that the transition from the metastable state to the unstable state is not discontinuous.

Oxide systems are excellent for nucleation control for several reasons. It is relatively easy to find a system which will undercool. Most "dirt" dissolves in glass so that one is reasonably certain of homogeneous nucleation. It is often easy to quench a system to room temperature with no nucleation. This allows specimens to be subsequently heated to a temperature where $|D|$ is 10^{-16} to 10^{-18} cm^2/sec and experimental times are on the order of minutes or hours. The separation can be followed from the beginning if crystallization can be avoided. Growth of the second phase in glass systems has been shown [27] to be diffusion controlled. Hence, the rate of phase separation and ultimate wavelength will depend on the number of nuclei per unit volume and their distribution. Clearly, a high density of uniformly distributed nuclei will give the smallest grain size and the shortest time to growth completion (shortest diffusion path).

Spinodal Decomposition

In the two-phase region where the molar free energy versus composition curve has negative curvature, a single-phase liquid is unstable to small composition fluctuations. In this case, the single phase is still metastable to an infinitesimal droplet with a large composition difference because of the positive interface energy. However, the single phase is unstable to two phases having large areas with infinitesimal composition difference. The surface energy is not the dominant factor in the activation energy because the system will always select a wavelength sufficiently large that the separation will be spontaneous and continuous. There is no discontinuity to be expected when the two-phase boundary is extrapolated into the solidus region.

Two versions of the kinetics of spinodal decomposition have been reported [28, 29]. The case for a quench from the single liquid region to a temperature below the spinodal and subsequent isothermal treatment for various times shows [28] the wavelength to be proportional to $(\Delta'T_s)^{-1/2}$. $\Delta'T_s$ is the temperature below the spinodal. The wavelength is infinite at the spinodal, but becomes finite a few degrees below the spinodal. The wavelength is typically on the order of 10–100 Å for metals [23]. Because of the exponent, the wavelength falls from infinity to a few angstroms over the first few degrees, and then changes by less than one order of magnitude over the next hundred degrees. The fact that the wavelength spectrum is narrow is due to competition between diffusion distance and surface energy effects. Long wavelengths decrease the surface energy but increase the diffusion distances, and vice versa. The separation rate goes through a temperature maximum much the same as the nucleation rate for the same reasons. It has been shown [24] that the time constant for separation is of the following form:

$$t \approx \frac{10^{-12}}{|D|} \ \text{sec}$$

Solids at the melting point have a value for $|D|$ which is typically 10^{-9} cm^2/sec. Hence t is typically 10^{-3} sec, and there is no hope to quench through the spinodal to room temperature with a homogeneous single phase. Glasses, with their exponential viscosity-temperature relationship and with a sub-liquidus spinodal, can have a value for $|D|$ as low as 10^{-17} cm^2/sec at the spinodal boundary. Hence, the reaction can be suppressed altogether, and has been in several cases [21, 27, 30]. If the spinodal is present at the melting temperature, the prospect of suppression is very slight.

The case of spinodal decomposition during continuous cooling has been described [29] for two cooling rates. The "fast" cooling was defined such that complete decomposition does not occur during the quench. The "slow" cooling was defined as being slow enough for spinodal decomposition to be completed before the end of the quench, but not so slow that a competing process (nucleation, crystallization, coarsening, etc.) can progress. It was shown for the fast rate that the separation wavelength which grows most rapidly and the wavelength spectrum of thermodynamically allowable wavelengths are independent of the quench rate. The "amplification" is inversely proportional to the quench rate. This is to say that the system which will separate, but not completely, during the quench will do so on the wavelength it selects, regardless of the quench rate, and the compositional difference between the two phases (amplification) will vary inversely as the first power of the quench rate.

For the slow quench, the wavelength was shown to be inversely proportional to the quench rate to the one-sixth power, $(\frac{1}{Q^{1/6}})$. The spectrum of wavelengths is a function of the cooling rate, being sharper for the slower cooling rates. However, the spectrum is broader for continuous cooling than for isothermal treatment. The important fact is that it requires a change of six orders of magnitude in the quench rate before the wavelength varies by one order of magnitude.

It must be recognized that the above discussions on the spinodal decomposition were derived for the early stages of separation before coarsening or coalescence occur.

Coarsening theory for metals has been published [31], and the case for isotropic solids is to be published (see reference 8 in [31]). Unfortunately the work by Seward *et al.* [30] is not specifically applicable, but some general trends can be extracted. The driving force for coarsening is surface free energy. The other necessary factor is that the structure be irregular after the latter stages of decomposition.

One major problem in glass production is compositional control because of batching errors, refractory solution, volatility, cords, etc. It has been shown [11] that small variations in composition cause the spinodal temperature to shift by several degrees. Thus wavelength control $[\lambda \alpha (\Delta' T_s)^{-1/2}]$ is extremely difficult because of the uncertainty in the amount of cooling below the spinodal. On the other hand, continuous cooling shows very weak dependence on cooling rate of the wavelength. Once a separated structure has been obtained with the wavelengths ≈ 10–100 Å, coarsening can be readily controlled.

Nucleation of a Crystalline Phase from a Glass

There are few differences between nucleation of a crystalline phase from a glass and nucleation of a second amorphous phase from the glass. The basic equations are of the same form, and only the names and values of the interface energies are different. In fact, it has been reported [32] as being possible for the glass to form amorphous nuclei which will crystallize once beyond the critical nucleus radius. This process is believed to be possible because of a lower surface free energy than that of a crystalline critical nucleus. This step-wise formation was called "precrystallization microphase separation." It was reported that precrystallization microphase separation can occur in a phase diagram of the eutectic type. The major phase is the first to nucleate, whereas the nucleation within the two-phase region has the minor phase nucleate first. Unfortunately, it is very difficult to determine the chemical composition of the precipitate. Only educated guesses can be made based on differential etching rates or electron-density differences. The precrystallization microphase separation was reported to be metastable, and once started is irreversible below the solidus. Amorphous nucleation within the two-phase region is reversible below the solidus. The existence of amorphous nuclei of the precrystallization microphase separation type is especially popular in Russia, where the theories of glass structure were based on just such a picture rather than on one of a random network, which was the most widely accepted theory in America.

The nucleation rate of the crystalline phase from the glass has the same form as that for the amorphous nucleation. Precrystallization microphase separation and formation of a crystalline critical nucleus can occur in any glass, whereas amorphous nucleation can occur only in a two-phase region of the phase diagram. In the two-phase region, the choice between an amorphous and a crystalline critical nucleus is a function of the difference in the work required to form a nucleus for each case.

Once the system has selected its separation mode, the subsequent growth will be by a diffusion process. Diffusion of each species in the glass will be a function of the value of $|D|$, and the flux gradient created by the composition shift as the glass is depleted by the crystallization. It is also possible to have eutectic precipitation from a glass phase.

It is reasonably obvious that the number of nuclei per unit volume will determine the final grain size if the reaction progresses to near-completion without exaggerated growth of a few grains. The ideal case for a grain size of 0.1–1 μ should have 10^{12} to 10^{15} nuclei per cm^3 with a uniform dispersion. It is desirable to nucleate at a temperature where the growth rate is very low to avoid the accidental appearance of a small number of rapidly growing crystals (somewhat similar to the case of exaggerated grain growth in solid-state sintering). Nucleation is often heterogeneous at the glass surface due to the presence of undissolved foreign particles. This can lead to very high nucleation rates at the glass surface and exaggerated growth into the glass is often found. This case is especially a problem if the molar volume difference between glass and crystal is large. It has been shown [33] that elastic stresses often suppress volume nucleation. If this is the case, surface nucleation in the absence of bulk nucleation will lead to extensive surface cracking. Surface nucleation can be reduced by HF etching to remove "dirt," and by avoiding H_2O and O_2 at the surface [34], prior to and during heat treatment, respectively.

Metastable amorphous separation in the two-phase region may complicate the problem of crystallization of a desired phase. One or both of the liquids could crystallize, giving crystals of composition other than that desired.

If nucleation and crystallization do occur in the glass interior, and if the glass-crystal thermal expansion difference is small, the interface energy is probably small. This implies that there will be strong mechanical bonding at the interface. This method of producing a glass-ceramic is superior to the composite method (where ceramic grains are hot pressed with glass powder) for that reason.

Formation of Glass-Ceramics

All of the above discussions are for homogeneous separations in glass. The glass industry today has several other methods to induce nucleation in glass systems:

1. Gold, silver, or copper metal colloids in photosensitive glasses;

2. Platinum colloidal particles acting as nuclei for epitaxial growth;
3. TiO_2;
4. Fluorine (CaF_2);
5. ZrO_2, which is insoluble to a certain extent and acts as nucleation centers for expitaxial growth.

These methods are used for various glass systems with varying degrees of success. The technology for each is well developed. The success in obtaining particle size control is evident from the actual products and from the literature. These methods are of commercial interest, and were of academic interest until they were all explained satisfactorily at various points. However, the homogeneous cases offer more theoretical interest at present.

Practical Applications

It is known that the strengths of glass-ceramics range as high as $0.1–1 \times 10^6$ psi for abraded specimens. In most cases the Young's modulus is higher and the thermal expansion is lower for the recrystallized product than for the corresponding glass. This combination is excellent when the engineering specifications require increased mechanical and thermal stability. If transparency can be maintained, either through precipitating cubic crystals or producing crystals small enough to avoid light scattering, an improvement over both glass and ceramics produced by conventional methods will result.

The system $ZnO–Al_2O_3–SiO_2$ is interesting from both theoretical and practical aspects, and has received attention for commercial applications as a glass-ceramic [35]. However, the ternary without TiO_2 is more interesting because there is the possibility of observing liquid-liquid separation and crystallization independently within the same system. There is stable and metastable liquid-liquid separation extending into the field from the $ZnO–SiO_2$ binary, and a ternary eutectic to provide a variety of separation mechanisms. Also of interest are the following facts: (1) $ZnO \cdot Al_2O_3$ is cubic; (2) $2ZnO \cdot SiO_2$ is tetragonal and hexagonal; (3) both Zn^{2+} and Al^{3+} can have two oxygen coordinations (tetrahedral or octahedral) in the glass; (4) the strength and elastic properties are reasonably high for similar glass-ceramics; (5) a homogeneous glass, a liquid phase separated glass, and a crystalline material have all been obtained experimentally; and (6) the electron-density difference between the ZnO-rich phase and the other phases enhances the use of transmission electron-microscopy.

Figure 9 is a transmission electron-photomicrograph of a quenched ingot of the composition 64% (mole) SiO_2, 10.3% (mole) Al_2O_3, 25.7% (mole) ZnO, which is near the ternary eutectic in the SiO_2 primary field. The specimen was taken from a section of the ingot touching the mold (receiving the highest quench rate). The scale of separation is 25–50 Å. The marker is 1,000 Å for each figure.

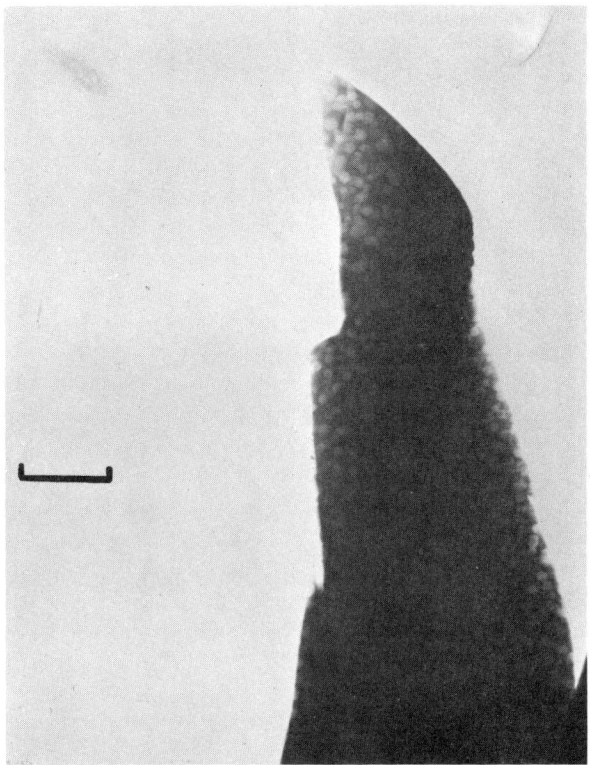

Figure 9. Transmission electron-microscopy of 64 mole% SiO_2–10.3 mole% Al_2O_3–25.7 mole% ZnO, adjacent to mold surface. Separation wavelength is 25–50 Å.

Figure 10 was taken of the same ingot as Figure 9, and was approximately 1/8″ from the mold surface. The separation wavelength is 300–500 Å. Figure 11 is of a specimen taken from 3/8″ away from the mold surface. The separation wavelength is 500–1,000 Å. The estimated minimum quench rate was $10^{2}°C$ per second. The separation is amorphous, as there was no selected area electron diffraction and dark field observation showed diffuse grey areas. The separated minor phase has a lower electron density than the matrix, indicating a ZnO-enriched matrix. The large range in separation wavelength across the ingot (25–1,000 Å) indicates in a preliminary fashion that the effect observed also includes coarsening. Early stages of separation would yield 25–250 Å as a likely range.

Figure 12 is the photomicrograph of an ingot of composition 70% (mole) SiO_2, 8% (mole) Al_2O_3, 22% (mole) ZnO. This composition is near the metastable two-phase region. This glass melt was so fluid that the separation was

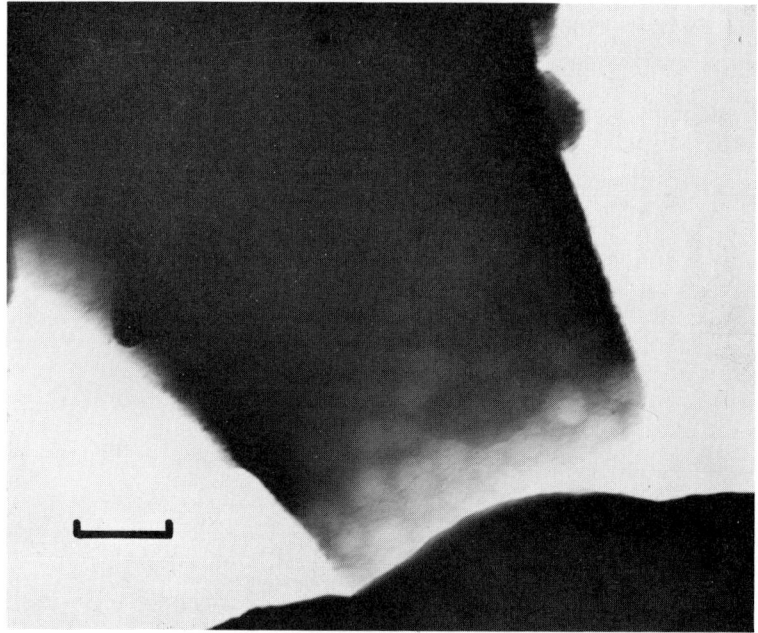

Figure 10. Transmission electron-microscopy of 64 mole% SiO_2–10.3 mole% Al_2O_3–25.7 mole% ZnO, 1/8″ from mold surface. Separation wavelength is 300–500 Å.

similar throughout the ingot. The separation wavelength is on the order of 500–1,000 Å. There was no crystallinity, and the minor phase has a lower electron density than the matrix.

The glass in Figure 13 was taken from the interior of an ingot having the composition 70% (mole) SiO_2, 13% (mole) Al_2O_3, 17% (mole) ZnO. The two-phase region is below the solidus and the melt was less fluid than the previous melts. The amorphous separation wavelength is on the order of 20–30 Å. Of particular interest is the fact that the separation has the appearance of an interconnected structure and that the minor phase appears to have a higher electron density than the major phase. It is possible to conclude that the minor phase is now enriched in ZnO.

It must be emphasized that only the following information can be obtained from the electron-microscopy: (1) the presence of any phase separation; (2) the wavelength and morphology of the separation; (3) the relative electron density of the two phases; and (4) the presence of crystallinity. The phase separation mechanism cannot be determined. The only positive proof that spinodal decomposition has occurred is the fact that the diffusion coefficient is negative within the spinodal.

Another qualitative observation about the morphology of the specimens shown

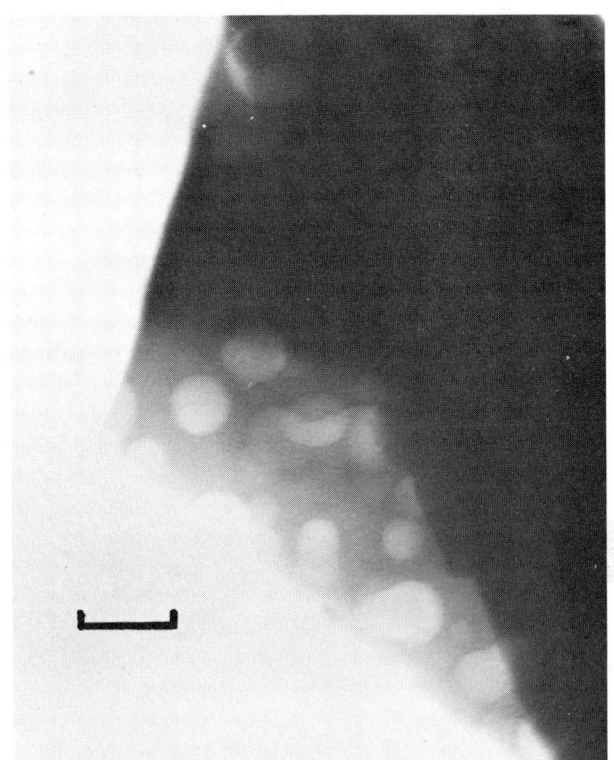

Figure 11. Transmission electron-microscopy of 64 mole% SiO_2–10.3 mole% Al_2O_3–25.7 mole% ZnO, 3/8″ from mold surface. Separation wavelength is 500–1,ooo Å.

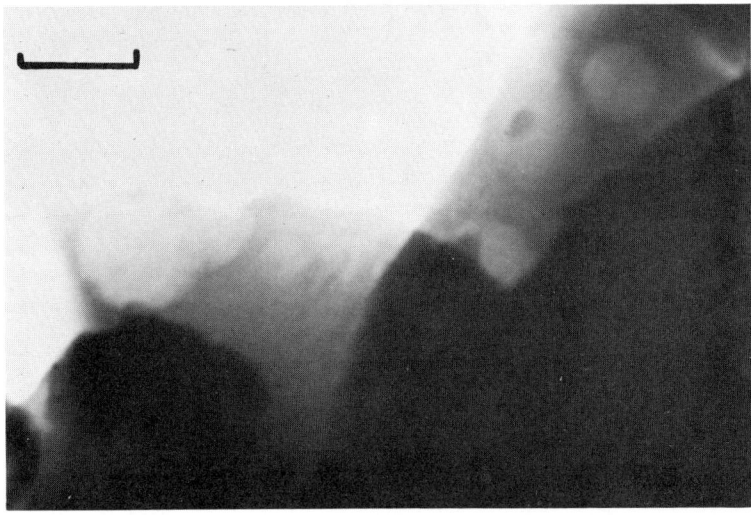

Figure 12. Transmission electron-microscopy of 70 mole% SiO_2–8 mole% Al_2O_3–22 mole% ZnO. Separation wavelength is 500–1,000 Å.

291

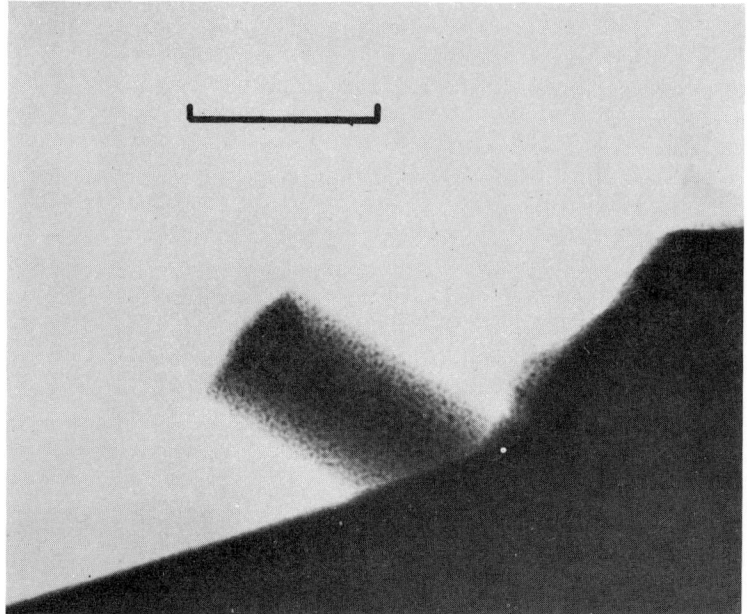

Figure 13. Transmission electron-microscopy of 70 mole% SiO_2–13 mole% Al_2O_3–17 mole% ZnO. Separation wavelength is 20–30 Å.

in the figures is the narrow range of particle sizes and shapes found on each. The range is not more than a factor of 2. The glass in Figures 12 and 13 had the same separation wavelength throughout the respective ingots. This indicates that the separation was uniformly dispersed throughout the specimen and coarsening failed to produce exaggerated growth of a few particles. This condition is essential when the microstructure of the final product must be controlled to a uniform microstructure.

References

1. Janef Thermochemical Tables, The Dow Chemical Co., Midland, Michigan (March 31, 1964).
2. Kirshenbaum, A. D. and Cahill, J. A., "Density of Liquid Aluminum Oxide," *J. Inorg. and Nucl. Chem.*, 14 (1960), 283.
3. Jackson, K. A. and Hunt, J. D., "Lamellar and Rod Eutectic Growth," *Trans. Met. Soc. AIME*, 236 (1966), 1129.
4. Chalmers, B., *Principles of Solidification*, John Wiley and Sons, Inc., New York (1964), 207.
5. Ubbelohde, A. R., *Melting and Crystal Structure*, Clarendon Press, Oxford, England (1965), 171.

6. Alcock, C. B. and Peleg, M., "Vaporization Kinetics of Ceramic Oxides at Temperatures around 2000°C," *Trans. Brit. Ceram. Soc.*, 66 (1967), 217.

7. Yanagida, H. and Kroger, F. A., "Condensed Phases in the System Al_2O_3–Al," *Bull. Am. Ceram. Soc.*, 47 (1968), 366.

8. Carlson, O. N., McMullen, W. D. and Gibson, E. D., U.S. Atomic Energy Commission, IS-351 (1961), 26. See also: Carlson, O. N. and McMullen, W. D., U.S. Atomic Energy Commission IS-193 (1960), 40.

9. Wachtman, J. B. Jr. and Lam. D. G. Jr., "Young's Modulus of Various Refractory Materials as a Function of Temperature," *J. Am. Ceram. Soc.*, 42 (1959), 254. See also: Wachtman, J. B. Jr., Tefft, W. E., Lam, D. G. Jr. and Apstein, C. S., "Exponential Temperature Dependence of Young's Modulus for Several Oxides," *Phys. Rev.*, 122 (1961), 1754.

10. Spencer, E. G., Denton, R. T., Bateman, T. B., Snow, W. B. and Van Uitert, L. G., "Microwave Elastic Properties of Nonmagnetic Garnets," *J. Appl. Phys.*, 34 (1963), 3059.

11. Ohlberg, S. M., Golob, H. R. and Strickler, D. W., *Symposium on Nucleation and Crystallization in Glasses and Melts*, Margie K. Reser, ed., The American Ceramic Society, (1962), 55.

12. Rindone, G. E., "Further Studies of the Crystallization of a Lithium Silicate Glass," *J. Am. Ceram. Soc.*, 45 (1962), 7.

13. Charles, R. J., "Some Structural and Electrical Properties of Lithium Silicate Glasses," *J. Am. Ceram. Soc.*, 46 (1963), 235.

14. Sastry B. S. R. and Hummel, F. A., "Studies in Lithium Oxide Systems: III. Liquid Immiscibility in the System Li_2O–B_2O_3–SiO_2," *J. Am. Ceram. Soc.*, 42 (1959), 81.

15. Roy, R., "Metastable Liquid Immiscibility and Subsolidus Nucleation," *J. Am. Ceram. Soc.*, 43 (1960), 670.

16. Greig, J. W., "Immiscibility of Silicate Melts," *Am. J. Sci.*, 13 (1927), 1, 133.

17. Kracek, F. C., "The Cristobalite Liquidus in the Alkali Oxide–Silica Systems and the Heat of Fusion of Cristobalite," *J. Am. Chem. Soc.*, 52 (1930), 1436.

18. Eitel, W., *Silicate Melt Equilibria*, par. 68, Rutgers University Press, New Brunswick, New Jersey (1951).

19. Ohlberg, S. M., Hammel, J. J. and Golob. H. R., "Phenomenology of Noncrystalline Microphase Separation in Glass," *J. Am. Ceram. Soc.*, 48 (1965), 178.

20. Rockett, T. J., Foster, W. R. and Ferguson, R. G. Jr., "Metastable Liquid Immiscibility in the System Silica–Sodium Tetraborate," *J. Am. Ceram. Soc.*, 48 (1965), 329.

21. Hammel, J. J., paper No. 36, proceedings of VII International Congress on Glass, Brussels, 1965, Vol. I, Institut National de Verre, Charleroi, and Federation de L'Industrie du Verre, Brussels, Belgium, 1966.

22. Charles, R. J., "Metastable Liquid Immiscibility in Alkali Metal Oxide–Silica Systems," *J. Am. Ceram. Soc.*, 49 (1966), 55.

23. Cahn, J. W. and Charles, R. J., "The Initial Stages of Phase Separation in Glasses," *Phys. and Chem. of Glasses*, 6 (1965), 181.

24. Cahn, J. W., "Spinodal Decomposition," *Trans. AIME* (1968), 166.

25. Hillig, W. B., *Symposium on Nucleation and Crystallization in Glasses and Melts*, Margie K. Reser, ed., The American Ceramic Society (1962), 77.

26. Cahn, J. W. and Hilliard, J. E., "Free Energy of a Nonuniform System. III. Nucleation in a Two-Component Incompressible Fluid," *J. Chem. Phys.* 31 (1959), 688.

27. Hammel, J. J., "Direct Measurements of Homogeneous Nucleation Rates in a Glass-Forming System," *J. Chem. Phys.*, 46 (1967), 2234.

28. Cahn, J. W., "Phase Separation by Spinodal Decomposition in Isotropic Systems," *J. Chem. Phys.*, 42 (1965), 93.

29. Houston, E. L., Cahn, J. W. and Hilliard, J. E., "Spinodal Decomposition During Continuous Cooling," *Acta Met.*, 14 (1966), 1053.
30. Seward, T. P. III, Uhlmann, D. R. and Turnbull, D., Division of Engineering and Applied Physics, Harvard University, Cambridge, Mass. (January 1968), Technical Report No. 15, Office of Naval Research Contract NOOO 14-27-A-0298-0009, NR-032-485.
31. Cahn, J. W., "The Later Stages of Spinodal Decomposition and the Beginnings of Particle Coarsening," *Acta Met.*, 14 (1966), 1685.
32. Filipovich, V. N., *The Strucutre of Glass*, Vol. 3, Consultants Bureau, New York, (1964), 11.
33. Cahn, J. W., *Acta Met.* 10 (1962), 907.
34. Ainslie, N. G., Morelock, C. R. and Turnbull, D., *Symposium on Nucleation and Crystallization in Glasses and Melts*, Margie K. Reser, ed., The American Ceramic Society (1962), 97.
35. Stookey, S. D., British Patent No. 829 (1960), 447.

SESSION V

BEHAVIOR OF ULTRAFINE-GRAIN CERAMICS

MODERATOR: S. M. COPLEY
Pratt and Whitney Aircraft
Middletown, Connecticut

14. Proposed Theory for the Static Fatigue Behavior of Brittle Ceramics

D. P. H. HASSELMAN
Allied Chemical Corporation
Morristown, New Jersey

ABSTRACT

A theory is developed for the static fatigue behavior of brittle ceramics on the basis of the growth of microcracks by the stress-enhanced thermally activated formation of vacancies at the crack tip. The driving force for crack growth is derived from the decrease in elastic energy around the microcrack upon an increase in crack length. Above the fatigue limit an applied load results in a thermodynamic non-equilibrium which results in a continuously advancing crack front and eventual gross failure. The role of moisture or other materials adsorbed on the crack surface is to lower the energy required to create vacancies at the crack tip, thereby decreasing the apparent activation energy for crack growth. A modified Griffith equation is derived which contains the original Griffith formulation for the critical fracture stress, but which also predicts the time to failure for values of stress below the critical fracture stress.

The theory is demonstrated for an industrial glass with good agreement with experiment. The effect of a limited rate of transport of adsorbed materials on rates of crack growth is discussed and demonstrated on the basis of experimental results for a polycrystalline alumina. The theory also properly describes all the general phenomena associated with short-time and long-term strength testing. The theory is considered applicable to brittle solids in non-corrosive environments.

Introduction

The decrease in strength of brittle materials with increase of the duration of the applied load is a well-known phenomenon, generally referred to as delayed fracture or static fatigue [1]. Numerous investigations were carried out on glasses and brittle oxides such as those conducted by Baker and Preston [2], Glathart and Preston [3], Mould and Southwick [4], Charles [5, 6], Gurney and Pearson [7], Pearson [8], Williams [9], and Charles and Shaw [10]. The general results of these studies indicate that delayed fracture is strongly de-

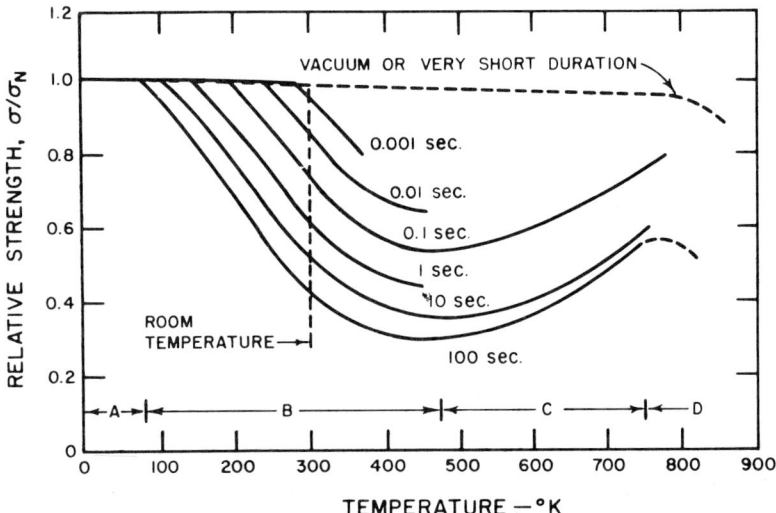

Figure 1. Relative strength of glass tested in air as a function of temperature and load duration. Semi-quantitative composite curves from results of several investigators according to Mould [11].

pendent on atmospheric environment as well as temperature, as illustrated in Figure 1 by the general fatigue curves presented by Mould [11].* The fact that changes in environment or temperatures do not appear to affect observed values of strength at very short loading times, suggests that delayed fracture is governed by a kinetic or transport phenomenon. The results of Charles [6] suggest that this transport phenomenon is governed by a thermally activated process.

The effect of water vapor and other surface-active adsorbed materials on the delayed fracture effect and short-time strength of brittle materials was attributed by Petch [13], Schoenig [14], and Hammond and Ravitz [15] to the decrease in the value of surface energy as it affects the Griffith [16] criterion for brittle fracture, expressed by:

$$S_c \approx (E\gamma/l)^{1/2} \tag{1}$$

which describes the critical fracture stress of a brittle solid, with Young's modulus, E, and surface energy, γ, containing microcracks of length l. As adsorbed water and other materials can lower the surface energy of solids appreciably [17], the decrease in strength appears self-evident.

A number of theories for the time-dependence of strength based on a mechanism for the growth or formation of cracks due to the stress-enhanced thermally activated rupture of the chemical bonds between atoms, were advanced by

---

*Also see W. D. Kingery [12].

Cox [18], Taylor [19], Hodgdon, Stuart, and Bjorklund [20], Poncelet [21], and Stuart and Anderson [22]. Similarly, Zhurkov [23], on the basis of experimental observations, proposed an empirical relationship for the time to fracture as a linear function of stress, which was interpreted in terms of the breaking of chemical bonds at the tip of pre-existing microcracks. Elliot [24] considers a diffusion-controlled reaction to predict the static fatigue behavior of glasses. All of these theories result in general expressions for the tensile strength as a function of time which can be fitted to experimental fatigue curves and, at least over part of the curve, predict the observed fatigue behavior to a good approximation. Gurney [25] suggests a mechanism for the growth of microcracks in glass due to the stress-enhanced attainment of homogeneous equilibrium at the tip of the crack.

More recently Charles and Hillig [26] and Hillig and Charles [27] developed theories for the decrease in strength with time in terms of a stress-enhanced corrosion mechanism which affects the geometry at the tip of the crack. As the stress concentration, σ, at the tip of an elliptical crack of length, l, can be related to the radius of curvature, ρ, at the crack tip by $\sigma = S(1 + 2\sqrt{l/\rho})$, where S is the applied stress, it is clear that a corrosion reaction which decreases the radius of the crack tip will increase the stress concentration, thereby lowering the macroscopic strength. Support for this theory can be found, for instance, in the observations by Hillig [28] for the failure times of vitreous silica in solutions of sodium hydroxide and hydrofluoric acid. Also, the observations by Wiederhorn [29, 30] were interpreted to be in support of the Charles-Hillig stress corrosion theory. Shand [31], however, concludes that, at least at low stress levels, the stress corrosion theories do not appear to be applicable.

It is the purpose of this chapter to propose a fatigue theory based on crack growth due to the stress-enhanced formation of vacancies at the crack tip. The theory properly describes the general phenomena associated with short-time and long-time strength tests and correctly predicts fatigue life and rates of crack growth in agreement with experimental results.

Theory

For the development of the theory, a hypothetical material will be considered which contains flaws in the form of Griffith cracks. The existence of flaws in industrial glasses and ceramics is well established [32–36]. Stress relaxation mechanisms in the form of dislocation motion or other non-linear stress-strain behavior are assumed to be absent. This condition is fulfilled for most brittle solids at low homologous temperatures.

Figure 2 illustrates a cross-section of the tip of an elliptical crack in a crystalline material. The plane of the crack is not necessarily restricted to the orienta-

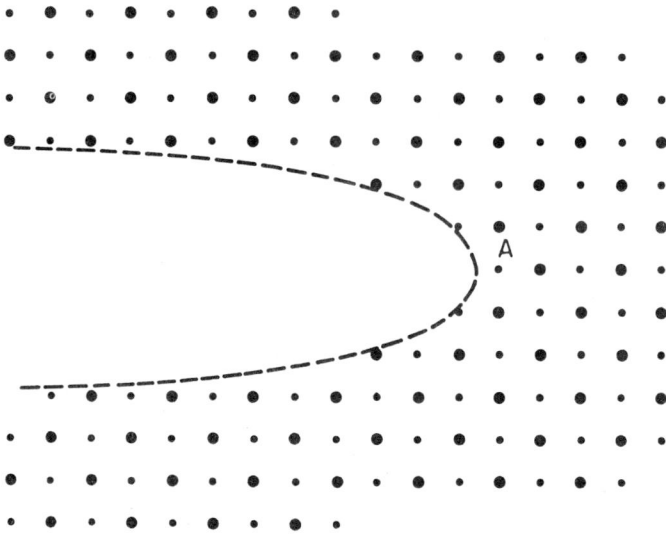

Figure 2. Tip of elliptical crack in hypothetical lattice.

tion of planes with low Miller-indices, nor is the crystallographic arrangement restricted to a simple cubic lattice. The Griffith criterion for brittle fracture is based on the fact that under conditions of stress an increase in the length of the crack results in a decrease in the elastic energy of the stress field around the crack. Catastrophic failure will result if the decrease in elastic energy equals or exceeds the surface energy required to form the new crack surfaces. Although the Griffith criterion was derived for a continuum, analogous considerations apply to crystalline materials. Referring to Figure 2, rupture of an atom bond at the crack tip results in an effective lengthening of the crack equal to the interatomic spacing and a subsequent decrease in elastic energy. Analogous to the continuum material, catastrophic failure will result if this decrease in elastic energy exceeds or equals the energy required to rupture the atomic bonds. The atoms, however, are in continuous thermal motion and occasionally may acquire sufficient energy to cause rupture of bonds, even at stress levels below the critical Griffith stress. A mechanism is thereby provided for the stress-enhanced thermally activated growth of microcracks at stress levels which are insufficient to initiate catastrophic failure. To break single atom bonds, however, requires strains of the order of 100 percent or more. Strains of this magnitude appear unlikely, due to constraints provided by neighboring atoms.*

*Similar conclusions were also reached by W. B. Hillig in: "Sources of Weakness and the Ultimate Strength of Brittle Amorphous Solids," pp. 152–94, in *Modern Aspects of the Vitrous State*, Vol. 2, Butterworth and Co. (London).

Only when a group of adjacent atoms would acquire sufficient energy for the rupture of atom bonds can crack growth take place. The occurrence of an atomistic event of this kind seems highly unlikely. Rates of crack growth which could result from this mechanism are expected to be exceedingly slow.

It is suggested, however, that the above difficulty does not arise if the lengthening of the crack takes place by the creation of vacancies at the crack tip. This would require the simultaneous rupture of a number of bonds for each atom, with an activation energy for the process higher than for single-bond rupture. The rate of vacancy formation, however, would be governed by the motions of the individual atoms only. As a result, rates of crack growth by the formation of vacancies are expected to exceed rates of growth by the rupture of single atom bonds. It is suggested that the actual mechanism by which vacancies are created at the crack tip is by the removal of atoms at the crack tip by means of surface diffusion. The numerical results of Cook and Gordon [37] for the stress concentrations around elliptical cracks suggests that the atoms have to move over a few atomic distances to move from the high-energy crack tip to the adjacent virtually stress-free crack surface. See Gurney [38].

In order to form a quantitative basis for the present theory, Figure 3a illustrates schematically the energy levels for the atoms at the crack tip and the adjacent crack surface in the absence of an applied stress. ΔH represents the difference in energy between an atom at the highly curved crack tip and an atom at the immediately adjacent virtually flat crack surface. ΔH can be related to the surface energy and the radius of curvature, and will be assumed to be a constant independent of crack length or environment.

Figure 3b illustrates the energy levels for the atoms at the crack tip and crack surface under conditions of an applied load. The quantity ΔV represents the decrease in elastic energy of the stress field surrounding the crack when an atom at the crack tip is replaced by a vacancy. The quantity ΔV can be obtained from the known solution [39] of the decrease in elastic energy, V, when a crack of length l is introduced in a plate of unit thickness. V can be expressed:

$$V = \pi l^2 S^2 / 4E \tag{2}$$

where S is the applied stress and E is Young's modulus.

When the length of the crack l is changed by the increment Δl, the change in $V(\Delta V)$ of equation (2) becomes:

$$\Delta V = \pi l S^2 \Delta l / 2E \tag{3}$$

where for the present discussion Δl represents the interatomic distance, d_o. Per lattice site at the crack tip ΔV can be expressed:

$$\Delta V = \pi l S^2 / 2EN \tag{4}$$

where N is the number of lattice sites per unit area.

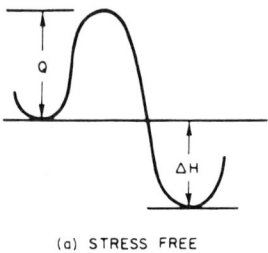

(a) STRESS FREE

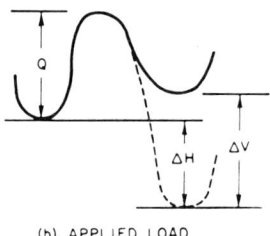

(b) APPLIED LOAD

Figure 3. Schematic diagram for energy levels for atoms at the crack surface (left), and the crack tip (right).

The energy levels of Figure 3b indicate that no net diffusion of atoms from the crack tip to the crack surface will occur, unless the potential energy change V exceeds the quantity H. From equation (4), this occurs at a stress level:

$$S_f = \left\{ 2EN\Delta H/\pi l \right\}^{\frac{1}{2}} \tag{5}$$

The stress S_f represents the value of stress below which no crack growth can occur. Physically, S_f represents the fatigue limit of the material. It may be noted that the mechanism of crack growth by the formation of vacancies at the crack tip leads to a very simple explanation for the existence of a fatigue limit. It should also be noted that at values of stress $S < S_f$, atoms will diffuse towards the crack tip. This leads to a healing of cracks, which at higher temperatures may well occur at an appreciable rate. As a result, increased values of strength should be observed at higher temperatures in agreement with the fatigue curves in Figure 1.

At values of stress, $S > S_f$, atoms at the crack tip will be replaced by vacancies at a net rate:

$$\dot{X}_v = B \exp\left[-Q/RT\right] \left\{ \exp\left[(\Delta V - \Delta H)/RT\right] - 1 \right\} \tag{6}$$

where B is a frequency factor [40, 41, 42], with $B \approx 10^{13}$, R is the gas constant and T is the absolute temperature.

With the aid of equations (4) and (5), the rate of vacancy formation can be expressed by:

$$\dot{X}_v = B \exp\left[-Q/RT\right] \left\{\exp\left[\pi l(S^2 - S_f^2)/2ENRT\right] - 1\right\} \tag{7}$$

The atoms at the crack tip will diffuse away at the above rate in an attempt to reach the equilibrium concentration of vacancies, which corresponds to the heat of formation of surface vacancies at the crack tip decreased by the elastic energy term of equation (4). The fact, however, of replacing an atom at the crack tip by a vacancy advances the crack front by one atomic distance, which automatically produces new atoms at the crack in a high state of energy which in turn can be replaced by vacancies at the rate described by equation (7). As a result, levels of stress in excess of the fatigue limit S_f give rise to a thermodynamic non-equilibrium, which gives rise to a continuous formation of vacancies at the crack tip. Continuous growth of the crack is thereby obtained, until the crack reaches sufficient size to cause catastrophic failure.

From the known rate of vacancy formation the rate of crack growth, dl/dt, can be expressed:

$$dl/dt = 2\dot{X}_v d_o \tag{8}$$

where d_o is the interatomic distance, and the factor 2 is introduced to account for the fact that both sides of the crack grow simultaneously.

For the derivation for the expression for the time to failure, only values of stress $S > S_f$ will be considered. Also, for mathematical simplicity, the stress will be considered high, so that back-diffusion of atoms to the crack tip can be neglected. This eliminates the term equal to unity in equation (7). Equation (8) can then be written:

$$dl/dt = 2Bd_o \exp\left\{\left[-Q_v + \pi l(S^2 - S_f^2)/2EN\right]/RT\right\} \tag{9}$$

Rearrangement of equation (9), followed by integration, using the boundary condition that when $t = 0$, $l = l_o$, the initial crack length, the length of the crack as a function of time is expressed by:

$$\frac{ENRT}{\pi Bd_o(S^2 - S_f^2)} \exp\left[\frac{Q_v}{RT}\right] \left\{\exp - \left[\frac{\pi l_o(S^2 - S_f^2)}{2ENRT}\right] - \exp\left[-\frac{\pi l(S^2 - S_f^2)}{2ENRT}\right]\right\} = t \tag{10}$$

The crack reaches a critical length, l_c, when the rate of energy release, $\Delta V/\Delta l$, of equation (3) is equal to or exceeds the fracture energy, G, required to propagate a crack over a unit area. This condition can be expressed:

$$l_c = 2EG/\pi S^2 \tag{11}$$

Substitution of equation (11) for l in equation (10) results in the length of time, t^*, to catastrophic failure:

$$t^* = \frac{ENRT}{\pi B d_o (S^2 - S_f^2)} \exp \left[\frac{Q_v}{RT} \right] \left\{ \exp \left[-\frac{\pi l_o (S^2 - S_f^2)}{2ENRT} \right] - \exp \left[-\frac{G(S^2 - S_f^2)}{S^2 NRT} \right] \right\}$$

(12)

which is defined only for values of stress $S > S_f$.

When $l_c = l_o$, $t^* = 0$ at a value of stress:

$$S_o = (2GE/\pi l_o)^{1/2}$$

(13)

with S_o hereafter referred to as the "instantaneous fracture stress." Equation (12) can be considered to represent a modified time-dependent Griffith equation, which not only contains the original Griffith formulation, but which also predicts the time to failure for values of stress below the instantaneous fracture stress.

The effect of moisture or other atmospheric constituents on the rate of crack growth and fatigue life can now be examined in terms of the mechanism of crack growth by the creation of vacancies at the crack tip. Hillig and Charles [27] base their fatigue theory on the existence of a stress-enhanced corrosion reaction at the crack tip. For the present theory, an explanation is offered from the field of surface chemistry. Gjostein [43], for instance, calculates that the presence of adsorbed molecules decreases the activation energy for surface diffusion of the underlying host material. This is supported by the observations of Holscher and Sachtler [44], who noted that carbon monoxide adsorbed on tungsten promotes the surface diffusion of the underlying tungsten surface, with the activation energy for the surface diffusion of tungsten decreasing with increasing coverage of CO molecules.

Burton, Cabrera, and Frank [45] (see also [46]) relate the increase of surface vacancies to a decrease of the surface energy of an initially smooth atom plane. The decrease in energy required to create a surface vacancy will be equal to or at least approximately equal to the decrease in surface energy, $\Delta\gamma$. If the activation energy for the creation of surface vacancies is considered to be the sum of the energy required to create the vacancies and the activation energy for the mobility of the diffusing atoms, then a decrease in the energy required to create surface vacancies will manifest itself in an equivalent decrease in the over-all activation energy for the creation of surface vacancies. As a result, in terms of the present theory the presence of moisture decreases the apparent activation energy for crack growth. Taylor [19] came to the same conclusion without specifying the actual mechanism. A net activation energy for crack growth can now be defined by:

$$Q_{cr} = Q_V - \Delta\gamma$$

(14)

Incorporating equation (14) in the fatigue equation (12) shows that the effect of a decrease in surface energy, $\Delta\gamma$, due to adsorbed moisture, reduces the time to fracture at a given stress level by the factor $\exp(-\Delta\gamma)/RT$. Rates of crack growth are increased by the reciprocal of this factor.

Numerical Examples and Discussion

As an illustration of the present theory, static fatigue curves and rates of crack growth were calculated for a hypothetical material with the following property values: interatomic spacing, $d_o = 2 \times 10^{-8}$ cm; Young's modulus, $E = 7 \times 10^{11}$ dynes cm^{-2}; frequency factor, $B = 10^{13}$ sec^{-1}; initial crack length, $l_o = 10^{-3}$ cm; lattice site density, $N = 2.5 \times 10^{15}$ cm^{-2}; instantaneous fracture stresses, $S_o = 10^9$ dynes cm^{-2}; and a fatigue limit, $S_f = 0.3 \times 10^9$ dynes cm^{-2}. Two values, $Q_{cr} = 20$ and 25 kcal/mole were selected so as to be somewhat in excess of the value of 18.8 kcal/mole observed by Charles [6] for a soda-lime glass. The above property data correspond to those for an industrial glass. Calculated fatigue curves can then be directly compared with the fatigue curves for a glass. The value of the fatigue limit, S_f, was selected approximately equal to the fatigue limit observed for a glass annealed after scoring [31]. The values selected for S_o and l_o result in a calculated value for the fracture energy, $G = 2660$ erg cm^{-2}, in good agreement with values measured by Wiederhorn [29]. Similarly, the quantity ΔH can be calculated to be approximately 1.6 kcal/mole.

Figure 4 illustrates the calculated fatigue curves at liquid nitrogen temperature, 300°K and 600°K. At short loading times, strength is nearly independent of temperature in agreement with the data in Figure 1. At liquid nitrogen temperature, the static fatigue effect is minor. Also, at short loading times, strength is nearly independent of the values of the activation energy, Q_{cr} (i.e., environment), again in agreement with several observations outlined in the Introduction. The calculated curve for 20 kcal/mole and 300°K agrees with the results of Figure 5 for the glass annealed after scoring. The actual numerical agreement may be fortuitous, due to the particular values of the constants selected. The relative agreement, however, provides support for the present theory.

Figure 6 shows the the initial rates of advance of the crack front, $0.5dl/dt$, for three values of Q_{cr}, as calculated by means of equation (9). These results may be compared with experimental data obtained by Wiederhorn [29] for a soda-lime glass by means of the split-cantilever beam technique, reproduced in Figure 7. As the stress levels shown on Figures 6 and 7 are approximately the same relative to the fracture stress for each case, the elastic energy driving force for crack growth should have approximately the same value. The present theory predicts rates of crack growth which agree with the experimentally observed values. Figure 6 also indicates that, at a given stress level, a small change in the activa-

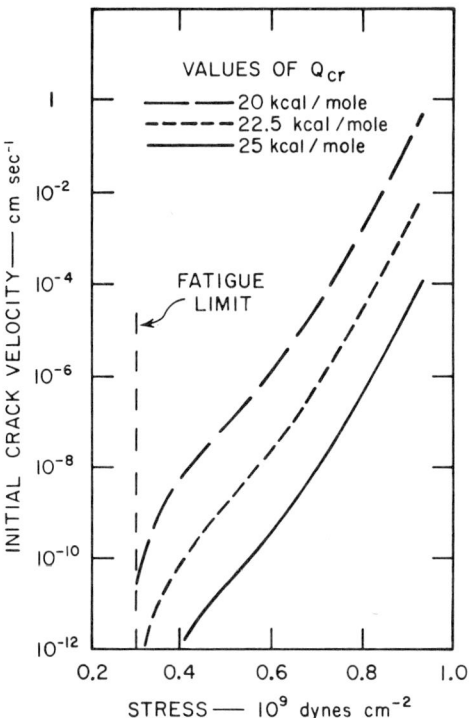

Figure 4. Calculated static fatigue behavior of a material which approximates glass in mechanical behavior.

tion energy can cause large relative changes in the rates of crack growth. The slight upward curvature of the calculated curves is due to the quadratic stress dependence of the elastic driving force for crack growth. At higher stress levels a similar upward curvature was also noticed by Wiederhorn [30] for the two right-hand curves of Figure 7 when plotted on an extended scale. At stress levels near the fatigue limit, the present theory predicts a downward curvature, not indicated on Figure 6. As a result, the curve for the rate of crack growth versus stress will have the general appearance of a sigma-curve, which to a first approximation may appear to be linear.

The data in Figure 7 allow an estimate to be made for the value of $\Delta\gamma$ for crack growth under different environments. Referring to the curves for the crack growth in moist air and nitrogen containing 0.0001 percent water vapor, at an applied load of 0.9 kg, the ratio of rates of growth under the two conditions is of the order of 10^3. Equating this ratio to the factor $\exp(\Delta\gamma/RT)$ yields a value of $\Delta\gamma \approx 4.2$ kcal/mole. This value is of the order of the change in surface energy to be expected due to the presence of adsorbed water on a glass surface.

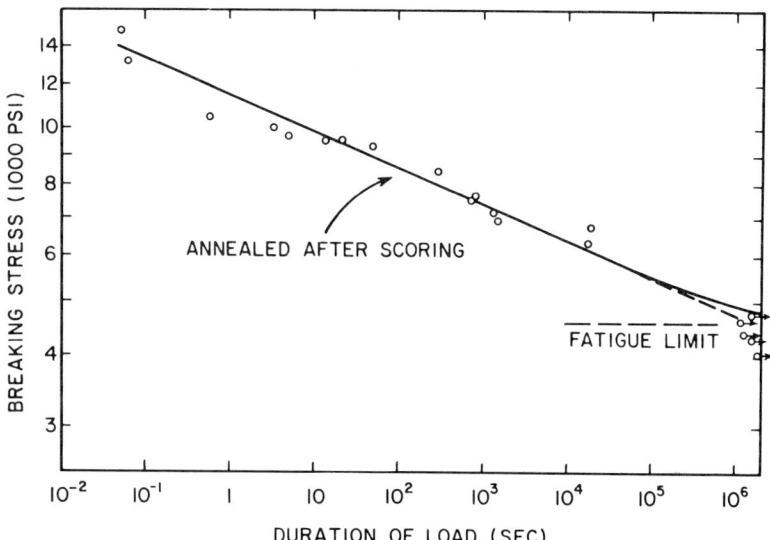

Figure 5. Static fatigue curves for a glass aged 24 hours and when annealed after scoring (after Shand [31]). The data for the samples annealed after scoring agree with calculated curve in Figure 4 for 300°K and Q_{cr} = 20 kcal/mole.

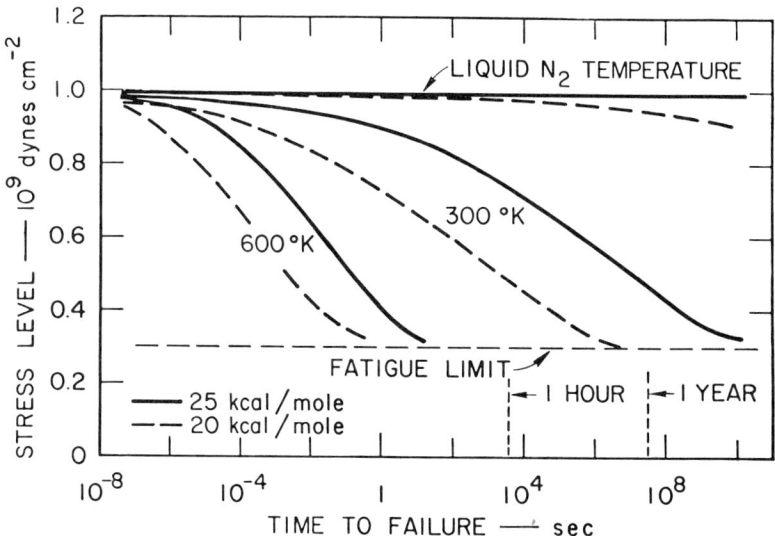

Figure 6. Calculated initial rates of advancement of the crack tip for the glass of Figure 4.

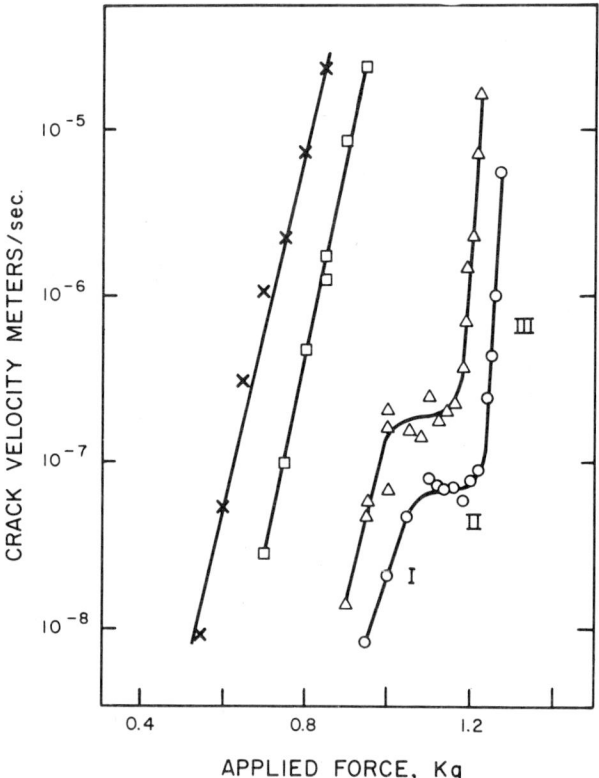

Figure 7. Experimental results (after Wiederhorn [29]) for the crack velocity in soda-lime glass as a function of load applied to the cantilever-beam specimens, in various environments. x = water; □ = moist air containing 1.5 mole % water vapor; △ = nitrogen gas containing 0.001 mole % water vapor; ○ = nitrogen gas dried over liquid nitrogen.

Converse to calculating changes in surface energy from differences in rates of crack growth, from a known value of a change in surface energy, changes in rates of crack growth and fatigue life can also be obtained. For instance, water vapor adsorbed on quartz has been observed [47] to lower the surface energy by 244 ergs cm^{-2}. With a molecular weight of 60 grams per mole and a specific gravity of 2.60 grams per cc, 244 ergs cm^{-2} corresponds to a value of $\Delta\gamma = 5.8$ kcal/mole of surface lattice sites. Substitution of this value in the factor $\exp(\Delta\gamma/RT)$ indicates that water vapor adsorbed on quartz increases rates of crack growth by a factor of 10^4. Also, the time to failure is reduced by the same factor.

The present theory was also applied to predict the fatigue behavior of a polycrystalline alumina (Wesgo AL-995). The experimental data selected were those obtained by Sedlacek [48] for room temperature and a relative humidity of approximately 50 percent. This material exhibited a fatigue limit,

$S_f \approx 1.25 \times 10^9$ dynes cm^{-2}, and an instantaneous fracture stress, $S_o \approx 2.5 \times 10^9$ dynes cm^{-2}. For the calculations, pertinent property data assumed were: frequency factor, $B = 10^{13}$ sec^{-1}; initial crack length, $l = 25\ \mu$ (average grain size); Young's modulus, $E = 54.4 \times 10^6$ psi, which corresponds to the minimum value of Young's modulus of the single crystal [50, 51] of alumina. A stress concentration factor, A, was also introduced to account for the stress non-uniformity [52] within the polycrystalline material, which results from the elastic anisotropy of the individual grains. For alumina, the factor $A \approx 1.23$.

From these data the quantity ΔH and the surface fracture energy, G, were calculated to be $\Delta H \approx 6.2$ kcal/mole and $G \approx 10^4$ dynes cm^{-2}, respectively. The density of lattice sites was taken as $N = 4.09 \times 10^{15}$ dynes cm^{-2}, as calculated from the interatomic spacing $d_o = 2.83 \times 10^{-8}$cm, assuming close-packing. The appropriate values of the activation energy, Q_{cr}, were selected from those values which gave the best fit to the experimental data. Figure 8 shows the results obtained. It is immediately apparent that the experimental data do not correspond to any single value of the activation energy for crack growth, Q_{cr}. Indeed, at stress levels near 1.75×10^9 dynes cm^{-2} fatigue life with decreasing stress levels increases so slowly that no combination of physical property data could possibly describe the experimental fatigue behavior. This suggests that the present theory may not be applicable to polycrystalline alumina.

An explanation, however, is offered by the observations of Wiederhorn [49], which indicate that under certain conditions the rate of advance of the crack tip is affected by a limited rate of transport of water vapor to the crack tip. Over a range of stress the rate of crack growth appears to be independent of stress levels. This phenomenon is indicated for a glass by region II in the right-hand curves of Figure 7. For sapphire at a humidity of 50 percent, the stress-independent region occurs at a rate of advance of the crack tip of approximately $10^{-2.5}$ cm sec^{-2}. For values of the activation energy, Q_{cr}, of 20 and 25 kcal/mole for polycrystalline alumina, this rate of crack growth on the basis of the present theory occurs at stress levels of approximately 1.7×10^9 dynes cm^{-2} and 1.95×10^9 dynes cm^{-2}. Over this range of stress, rates of crack growth will be independent or nearly independent of stress level, and fatigue life is not expected to increase materially with decreasing stress level.

An explanation can be offered for the apparent stress-independent region in terms of the proposed mechanism of crack growth by the formation of vacancies at the crack tip. In agreement with Wiederhorn [49], it is suggested that the stress-independent region arises because the rate of transport of water vapor is insufficient to keep up with the advancing crack front. At values of stress which are below the stress-independent region, the rate of advance of the crack tip is low enough for water molecules to be transported in sufficient numbers such that water is always present at the crack tip. In terms of the present theory, the full effect of the decrease in surface energy on rates of crack propagation

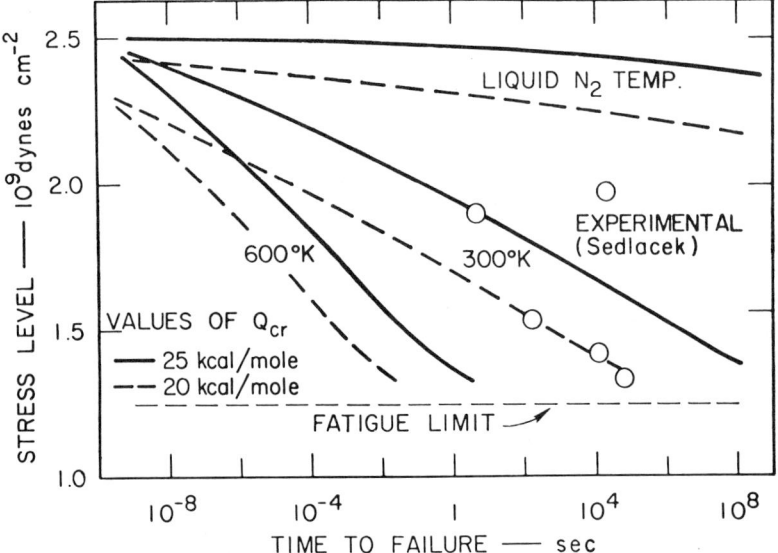

Figure 8. Calculated fatigue curves and experimental data for polycrystalline alumina.

will always be present in this region of stress. Conversely, very high stress levels will result in rates of crack propagation which are too fast for water vapor to be transported to the crack tip. At these levels of stress, the effect of lowering of the surface energy on rates of crack growth will not be felt, and the crack will propagate at a rate which corresponds to the rate of propagation under vacuum conditions. At the rate of propagation at which water vapor can just be transported in sufficient quantity to be present at the crack tip, the transition will be made from the behavior of crack propagation under low stress conditions to the behavior under high stress conditions.

In terms of the present theory, crack propagation in region III corresponds to the activation energy for surface diffusion under vacuum conditions. In region I, crack propagation is governed by the activation energy for surface diffusion in vacuum minus the decrease in surface energy due to the adsorption of water. Region II corresponds to a gradual transition of the activation energy of region I to the activation energy of region III. The mechanism for crack growth, however, is identical in all three regions. At this point, the writer would like to point out that the split-cantilever beam technique for obtaining data for the crack propagation behavior in brittle materials is exceptionally useful, and the data obtained are absolutely essential for the proper interpretation of the mechanical behavior of engineering materials. Further research in this area should be encouraged.

A number of conclusions can be drawn from the fatigue curves of Figures 4 and 8. In Figure 8, the calculated curves for the strength of polycrystalline alumina at very short loading times suggest that strength should be virtually independent of temperature. This conclusion is in agreement with the data of Abbott, Cornish, and Weil [53], who investigated the strength of the identical alumina material as a function of strain rate at 70°F, 1000°F, and 1800°F. These authors attribute the increase in strength at the higher temperatures with increasing strain rate to a strengthening mechanism. In terms of the present theory, higher strength at shorter failure times arises because the weakening mechanism of crack growth has not had sufficient time to occur. The present theory can also be used to calculate fatigue curves for stress conditions which are a function of time. For a constant strain rate, this involves the solution of a differential equation of the form: $dl/dt = \exp(alt^2)$ for which, at the time of this writing, no solution has been found.

Figure 8 suggests that the activation energy for crack growth at high stress levels (i.e., under vacuum conditions) is of the order of 25 kcal/mole. This value is approximately one-quarter to one-fifth of the value for the activation energy for bulk diffusion in alumina, reported [54, 55, 56] to be 130, 114, and 115 kcal/mole, respectively. This ratio of activation energies may be compared with the data for silver with an activation energy for surface diffusion [57] of 10.3 kcal/mole and an activation energy for bulk diffusion [58] of 45.9 kcal/mole. Similarly, the activation energy of 18.3*kcal/mole for the surface diffusion [59] in tungsten is approximately one-sixth of the value of 106.5 kcal/mole for the activation energy for bulk diffusion [60]. As a result, an activation energy of 25 kcal/mole for the surface diffusion of alumina is a reasonable value. As the activation energy for crack growth is of the order of 25 kcal/mole, further support is thereby provided for the hypothesis that crack growth occurs by the formation of vacancies at the crack tip by the mechanism of surface diffusion. As a result, it must also be concluded that the value of Q_{cr} for the glass in Figure 4 must be attributed to the surface diffusion in the Si-O network. The apparent agreement between this value and similar values for the activation energy for the bulk-diffusion of sodium ions in glasses [61] must be considered fortuitous.

Figures 4 and 8 also suggest that the activation energies for crack growth for the glass and the alumina have approximately the same value. The much longer times to failure for the alumina as compared to the times to failure for the glass at the same stress level must be attributed to differences in physical properties other than the activation energies for surface diffusion. Comparison of all the physical properties shows that the largest relative difference occurs for the values of Young's modulus of elasticity. High values of Young's modulus of elasticity reduce the elastic energy driving force for crack propagation. As a result, alumina is far superior to glass from the point of view of load-bearing ability. In general, for high values of short-term strength and long fatigue life, materials

---

*Average for the three values reported.

should be selected with small flaw size and high values of Young's modulus of elasticity, fracture surface energy, bond energy, activation energy for surface diffusion, and high atomic packing. Whenever possible, the effect of adsorbed materials on surface energy should be minimized. The refractory oxides, carbides, nitrides, borides, and silicides should meet most of these requirements. As flaw size generally is related to the grain size, fine-grain bodies of the above materials should prove to be superior to bodies having large grain size.

It is suggested that the present theory is applicable to materials and environments where any surface-absorbed species play a passive role and do not react chemically with the underlying load-bearing material. For environments where the structural element is subjected to gross chemical attack under zero-stress conditions, the theories of Charles and Hillig [27] are considered more applicable.

As strength studies, in general, have been devoted primarily towards short-time strength behavior, a definite lack of design criteria for long-term load applications of brittle materials currently exists. It is suggested that future investigations include experimental studies of long-time fatigue behavior for the purpose of developing realistic data for engineering design.

Summary and Conclusions

A theory is proposed for the static fatigue behavior of brittle materials based on the growth of microcracks by the stress-enhanced thermally activated formation of vacancies at the crack tip. Calculated rates of crack growth and fatigue behavior agree with experimental results.

Acknowledgment

Considerable appreciation is extended to the San Francisco Symphony Orchestra, as the basic concept for the theory presented was conceived in part during a most brilliant performance of Mozart's 27th Concerto in B-Flat for Piano and Orchestra, K 595, with Clifford Curzon as soloist. The author also is indebted to A. M. Wiederhorn for making available unpublished data for crack propagation in sapphire.

References

1. Shand, E. B., "Stress Behavior of Brittle Ceramics," *Bull. Am. Ceram. Soc.,* 38 (1959), 653-60.
2. Baker, T. C. and Preston, F. W., "Fatigue of Glass under Static Loads," *J. Appl. Phys.,* 17 (1946), 170-78.

3. Glathart, J. L. and Preston, F. W., "The Fatigue Modulus of Glass," *J. Appl. Phys.*, 17 (1946), 189-95.
4. Mould, R. E. and Southwick, R. D., "Strength and Static Fatigue of Abraded Glass under Controlled Ambient Conditions," Part I, *J. Am. Ceram. Soc.*, 42 (1959), 542; Part II, *ibid.*, 42 (1959), 582.
5. Charles, R. J., "Static Fatigue of Glass, II," *J. Appl. Phys.*, 29 (1958), 1549-60.
6. Charles, R. J., "The Strength of Silicate Glasses and Some Crystalline Oxides," Chap. 12 in *Fracture*, B. L. Averback *et al.*, eds., John Wiley and Sons, Inc. (1959).
7. Gurney, C. and Pearson, S., "Fatigue of Mineral Glass Under Static and Cyclic Loading," *Proc. Roy. Soc.* (London), 192A (1948), 537-44.
8. Pearson, S., "Delayed Fracture of Sintered Alumina," *Proc. Phys. Soc.* (London), 69 (1956), 1293-96.
9. Williams, L. S., "Stress Endurance of Sintered Alumina," *Trans. Brit. Ceram. Soc.*, 55 (1956), 287-312.
10. Charles, R. J. and Shaw, R. R., "Delayed Failure of Polycrystalline and Single-Crystal Alumina," General Electric Research Laboratory Report 62-RL-3081 M (1962).
11. Mould, R. E., Glastech. Ber.: V. Intern. Glass Congress, Sonderband, 32K, III/18 (1959).
12. Kingery, W. D., "A Review of the Stress-Strain-Time-Temperature Behavior of Ceramics," pp. 19-34 in *Symposium on Stress-Strain-Time-Temperature Relationships in Materials,* ASTM Special Publication No. 325 (1962).
13. Petch, N. J., "The Lowering of Fracture-Stress due to Surface Adsorption," *Phil. Mag.*, 81 (1956), 331-37.
14. Schoenig, R. L., "On the Strength of Glass in Water Vapor," *J. Appl. Phys.*, 31 (1960), 1779.
15. Hammond, M. L. and Ravitz, S. F., "Influence of Environment on Brittle Fracture of Silica," *J. Am. Ceram. Soc.,* 46 (1963), 329-32.
16. Griffith, A. A., "Phenomena of Rupture and Flow in Solids," *Phil. Trans. Roy. Soc.* (London), A221 (1920), 163-98.
17. Wolf, K. L., *"Physik und Chemie der Grenzflaechen,"* Vol. I, Springer-Verlag (1957).
18. Cox, S. M. Jr., "A Kinetic Approach to the Theory of the Strength of Glass," *J. Soc. Glass Technology*, 32 (1948), 127-46.
19. Taylor, N. W., "Mechanism of Fracture of Glass and Similar Solids," *J. Appl. Phys.*, 18 (1947), 943-55.
20. Hodgdon, F. B., Stuart, D. A. and Bjorklund, F. E., "Application of Rate Process to Glass. I. Breaking Strength," *J. Appl. Phys.* (1950), 1156-59.
21. Poncelet, E. F., *Fracturing of Metals*, Am. Soc. Metals. (1951), 201.
22. Stuart, D. A. and Anderson, O. L., "Dependence of Ultimate Strength of Glass under Constant Load on Temperature, Ambient-Atmosphere and Time," *J. Am. Ceram. Soc.,* 36 (1953), 416-24.
23. Zhurkov, S. N., "Kinetic Concepts of the Strength of Solids," *Inter. J. Frac. Mech.,* 1 (1965), 311.
24. Elliot, H. A., "Stress Rupture in Glass," *J. Appl. Phys.,* 29 (1958), 224-25.
25. Gurney, C., "Delayed Fracture in Glass," *Proc. Phys. Soc.*, 59B (1947), 169-84.
26. Charles, R. J. and Hillig, W. B., "The Kinetics of Glass Failure by Stress Corrosion," *Symposium on the Mechanical Strength of Glass and Ways of Improving It,* Union Scientifique Continentale du Verre (1961).
27. Hillig, W. B. and Charles, R. J., "Surfaces, Stress-Dependent Surface Reactions, and Strength," Chap. 17 in *High-Strength Materials*, John Wiley and Sons, Inc. (1965).
28. Hillig, W. B., "The Factors Affecting the Strength of Bulk Fused Silica," *Symposium on the Mechanical Strength of Glass and Ways of Improving It,* Union Scientifique Continentale du Verre (1961).

29. Wiederhorn, S. M., "Fracture Surface Energy of Soda-Lime Glass," Chap. 27 in *Materials Science Research*, Vol. III, Plenum Press, New York (1966).

30. Wiederhorn, S. M., "The Influence of Water Vapor on Crack Propagation in Soda-Lime Glass," to be published.

31. Shand, E. B., "Strength of Glass—The Griffith Method Revised," *J. Am. Ceram. Soc.*, 48 (1965), 43-49.

32. Anderson, O. L., "The Griffith Criterion for Glass Fracture," Chap. 17 in *Fracture*, Averbach *et al.*, eds., John Wiley and Sons, Inc. (1959).

33. Ernsberger, F. M., *Advances in Glass Technology*, Plenum Press, New York (1962), 511-24.

34. Ernsberger, F. M., *Progress in Ceramic Science*, Vol. III, J. E. Burke, ed., Pergamon Press, New York (1963), 57-75.

35. Philips, C. J., "Strength and Weakness of Brittle Materials," *Am. Scientist*, 53 (1965), 20-51.

36. Barenblatt, G. I., "The Mathematical Theory of Equilibrium Cracks," *Advances in Applied Mechanics,* Vol. 7, Academic Press, New York (1962), 55-129.

37. Cook, J. and Gordon, J. E., "A Mechanism for the Control of Crack Propagation in All-Brittle Systems," *Proc. Roy. Soc.*, 282 (1964), 508-20.

38. Gurney, C., "The Effective Stress Concentration at the End of a Crack in Materials Atomic Constitution," *Phil. Mag.*, 39 (1948), 71-76.

39. Timoshenko, S. and Goodier, J. N., *Theory of Elasticity*, 2nd ed., McGraw-Hill, New York (1951), 161.

40. Shewmon, P. G., *Diffusion in Solids,* McGraw-Hill, New York (1963).

41. Glasstone, S., Laidler, K. J. and Eyring, H., "The Theory of Rate Processes," 1st ed., McGraw-Hill, New York (1941).

42. Swalin, R. A., *Thermodynamics of Solids*, John Wiley and Sons, Inc., New York (1962).

43. Gjostein, N. A., "Surface Self-Diffusion," Chap. 4 in *Metal Surfaces, Structure Energetics and Kinetics*, Amer. Soc. for Metals (1963).

44. Holscher, A. A. and Sachtler, W. M. H., "Chemisorption and Surface Corrosion in the Tungsten + Carbon Monoxide System, as Studied by Field Emission and Field Ion Microscopy," *Disc. Faraday Soc.,* 41 (1966), 29-42.

45. Burton, W. K., Cabrera, N. and Frank, F. C., "The Growth of Crystals and the Equilibrium Structure of their Surfaces," *Phil. Trans. Roy. Soc.*, (London), 243A (1951), 299-358.

46. Gjostein, N. A., "Adsorption and Surface Energy (I): The Effect of Adsoprtion on the J-Plot," *Acta Met.* 11 (1963), 957-67.

47. Boyd, G. E. and Livingstone, H. K., "Adsorption and Energy Changes at Crystalline Solid Surfaces," *J. Am. Chem. Soc.*, 64 (1942), 2383-88.

48. Sedlacek, R., to be published.

49. Wiederhorn, S. M., "Effect of Water Vapor on Crack Propagation in Sapphire," presented at the 69th Annual Meeting of the American Ceramic Society (April 30-May 4, 1967). Paper 29-B-67.

50. Nye, J. F., *Physical Properties of Crystals: Their Representation by Tensors and Matrices*, Oxford University Press (1957).

51. Wachtman, J. B., Tefft, W. E., Lam, D. G. and Stinchfield, R. P., "Elastic Constants of Synthetic Single Crystals of Corundum at Room Temperature," *J. of Res., NBS,* 64 (1960), 213-28.

52. Hasselman, D. P. H., "Single Crystal Elastic Anisotropy and the Mechanical Behavior of Polycrystalline Ceramics," presented at the Intern. Symposium on Anisotropy in Single-Crystal Refractory Compounds (June 13-15, 1967), Dayton, Ohio, in print.

53. Abbott, B. W., Cornish, R. H. and Weil, N. A., "Techniques for Studying Strain Rate Effects in Brittle Materials," *J. Appl. Polymer Sc.,* 8 (1964), 151-67.

54. Folweiler, R. C., "Creep Behavior of Pore-Free Polycrystalline Aluminum Oxide," *J. Appl. Phys.,* 32 (1961), 773-78.

55. Paladino, A. E. and Kingery, W. D., "Aluminum-Ion Diffusion in Aluminum Oxide," *J. Chem. Phys.,* 37 (1962), 957–62.

56. Rossi, R. C. and Fulrath, R. M., "Final Stage Densification in Vacuum Hotpressing of Alumina," *J. Am. Ceram. Soc.,* (1965), 558-64.

57. Hoffmann, R. E. and Turnbull, D., "Lattice and Grain Boundary Self-Diffusion in Silver," *J. Appl. Phys.,* 22 (1951), 634-39.

58. Nickerson, R. A. and Parker, E. R., "Surface Diffusion of Radioactive Silver on Silver," *Trans. Am. Soc. Met.,* 42 (1950), 376-86.

59. Mueller, E. W., "Oberflaechewanderung von Wolfram auf dem eigenen Kristallgitter," *Z. Physik,* 126 (1949), 1039-49.

60. Ehrlich, G. and Hudda, F. G., "Atomic View of Surface Self-Diffusion: Tungsten on Tungsten," *J. Chem. Phys.* 44, (1966), 1039-49.

61. Williams, E. L. and Heckman, R. W., "Sodium Diffusion in Soda-Lime-Aluminosilicate Glasses," *Phys. Chem. of Glasses,* 5 (1964), 166-71.

15. Crack Propagation in Polycrystalline Ceramics

S. M. WIEDERHORN

National Bureau of Standards
Washington, D.C.

ABSTRACT

In this chapter, the strength and shock resistance of ceramic materials will be related to microstructural details and the energy necessary to form fracture surfaces. The influence of grain size, grain orientation, crystal anisotropy, and residual stresses on crack propagation will be discussed. It will be argued that the strength of ceramic materials is closely related to the grain boundary or single-crystal fracture surface energy, while shock resistance is related to the surface energy required to propagate large cracks through polycrystalline materials. The use of fracture surface energies as a research technique and as a design parameter will be discussed.

Key words: Fracture, ceramics, strength, fracture surface energy, thermal shock.

Introduction

The recent development of high-strength, fine-grain ceramics has belied the traditional concept that ceramics are inherently weak. Strengths approaching 7×10^8 N/m² (100,000 psi) are now possible, and, as a result, fine-grain ceramics are finding uses in applications not possible with the older ceramics. These advances have come about largely through empirical investigations in which it was observed that the elimination of pores and the reduction of grain size brought about an increase in strength [1-3]. Surprisingly, little use has been made of a basic understanding of the fracture process in these developments, and the functional dependence of strength on microstructural parameters is not entirely understood in these materials.

A fundamental approach to studying the mechanical failure of polycrystalline ceramics depends on a detailed insight into the mechanisms of crack propagation in these materials. The role played by the grains and grain boundaries are important in this approach. Fortunately, techniques are now available for measuring the relative resistance to fracture of grains and grain boundaries, and means of correlating these measurements with polycrystalline microstructure are

317

also available. Consequently, a fundamental approach to the mechanical failure of ceramic materials may now be feasible.

This chapter presents a fundamental approach to the fracture of ceramics. Some of the basic causes of weakness of polycrystalline ceramics are discussed in the hope that a deeper understanding of the fracture process will lead to further improvement of these materials. Mechanical failure will be discussed from the point of view of the flaw theory of solids, in which failure is due only to the presence and propagation of pre-existing flaws in the solid. Techniques will be described to measure the relative resistance of single crystals and grain boundaries to crack propagation. Data obtained by these methods will be discussed with respect to engineering properties such as strength and thermal shock. Ceramics will be considered to be free of plastic deformation and stress corrosion.

Flaw Theory of Fracture

Most ceramics are at least two orders of magnitude weaker than the theoretically predicted strength of 10^6 psi [4, 5]. This great reduction of strength is due to the presence of small cracks and other flaws in the surface and body of ceramics. These flaws act as stress concentrators, greatly magnifying the applied load so that the stress in the immediate vicinity of the flaw is considerably greater than the applied stress. Failure occurs when the stress in the vicinity of the flaw exceeds the cohesive strength of the material.

The flaw concept of material failure was first expressed by Inglis and Griffith [6–8], who showed that the stress at the tip of a long elliptical crack in a two-dimensional elastic medium (Figure 1) is given by the following formula,

$$\sigma = 2 S \sqrt{L/\rho} \qquad (1)$$

where σ is the stress at the crack tip, S is the applied stress, ρ is the radius of curvature of the crack tip, and L is the crack length. The critical load for failure occurs when the crack tip stress, σ, is equal to the cohesive strength of the material. Equation (1) is both a necessary and sufficient condition for failure, in that fracture always occurs when the crack tip stress exceeds the cohesive strength of the material. Unfortunately, equation (1) has no real engineering value since there is no way of measuring the crack tip radius of curvature, ρ.

A more practical approach to materials failure was developed by Griffith [7] from energy considerations. He reasoned that the sole result of propagating a crack through a completely brittle elastic solid was the formation of two new surfaces. Associated with these surfaces is a surface energy which Griffith assumed to be the surface free energy of the solid. A necessary condition for brittle failure is that the work associated with failure be equal to the total energy necessary to form the new surfaces.

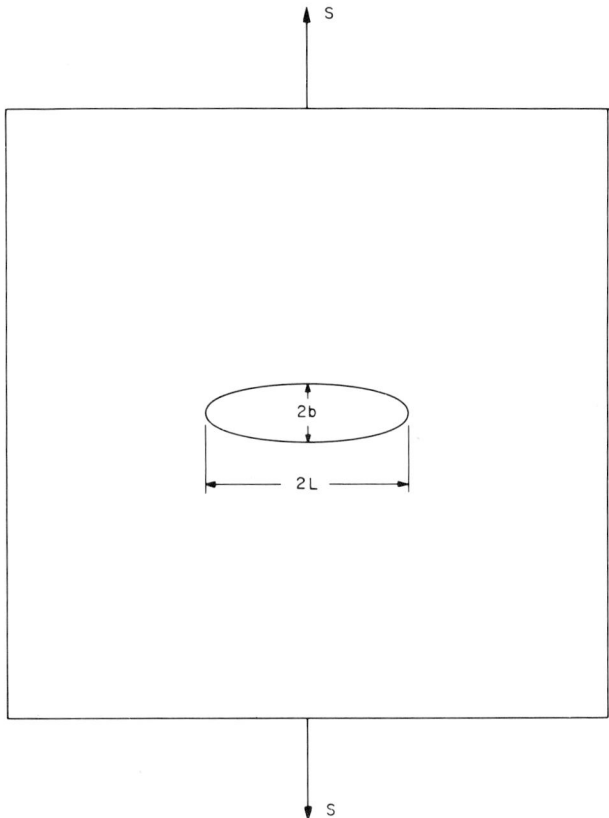

Figure 1. Elliptically shaped crack in an infinite two-dimensional plane.

The energy concept was made quantitative by putting it into differential form. Consider a crack extending an infinitesimal distance, dL. The minimum condition for failure occurs when the work on the body, dW, just equals the sum of the change in surface energy, dS, and elastic strain energy, dV, of the body,

$$dW = dS + dV \qquad (2)$$

dW and dV can be obtained from an elastic solution of the configuration shown in Figure 1 and dS is just equal to $4\,\gamma dL$, where γ is the surface energy per unit area of surface formed (usually referred to as the surface energy). Solving the elastic problem and making the appropriate substitutions, the following equation is obtained as a minimum condition for crack propagation [8],

$$S = \sqrt{4E\gamma/\pi L} \qquad (3)$$

where E is Young's modulus.

Many different crack problems have been solved to determine the critical condition for failure. They are always of the form given by equation (3) in that the fracture load or fracture stress depends on the square root of the surface energy modified by a function of the elastic constants and the crack geometry [9]. The surface energy for fracture is therefore an important variable for characterizing the fracture process. Although the surface energy for fracture was originally thought to be equal to the surface free energy of the solid, it is now recognized that the surface energy for fracture may contain terms that reflect irreversible processes at the crack tip, such as plastic deformation and stress corrosion [3, 10-12]. In general, the fracture surface energy reflects processes occurring at the root of the crack during fracture and fracture is easy or difficult, depending on the magnitude of the fracture surface energy. Consequently, the fracture surface energy has real engineering value since it can be used to measure the fracture resistance of solids.

Modifications to the Flaw Theory of Fracture

The flaw approach to the fracture of solids was originally devised for homogeneous, isotropic solids, and must be modified whenever these conditions are not satisfied. The modifications usually involve altering the fracture surface energy to reflect the new conditions of crack propagation. Some of the conditions for which the fracture surface energy must be modified will now be discussed.

Fracture of Single Crystals

The fracture surface energy of isotropic materials such as glass is a constant of the materials, and as such is independent of the direction of crack propagation in the solid. This is not the case for crystalline materials. The fracture surface energy of crystalline solids depends on the relative orientation of the crack to the crystallographic axes of the solid [10]. Therefore, the determination of the fracture surface energy for single crystals requires the determination of this quantity as a function of direction of crack propagation in the solid. One consequence of this directional dependence in many crystals is the occurrence of cleavage fracture, which results from the fact that one crystallographic plane in the solid has a much lower surface energy than any other [10]. Another consequence is the observation of directional strength dependence in some crystals, as in the case of single-crystal alumina [13].

Fracture Along Grain Boundaries

Crack propagation along grain boundaries has been treated in a straight-forward manner by Gilman [4, 14]. The fracture surface energy is found to consist of two terms, the grain boundary energy, γ_{gb}, which arises from elastic strains involved in the formation of the boundary, and the surface energy of the newly created surfaces, γ_a and γ_b. The fracture surface energy, γ, is equal to the difference of these two quantities, $2\gamma = \gamma_a + \gamma_b - \gamma_{gb}$. For symmetrical grain boundaries, $\gamma_a = \gamma_b$, and $\gamma = \gamma_a - \frac{1}{2}\gamma_{gb}$. The strength of a grain boundary will depend on the relative magnitudes of the grain boundary energy and the surface energy of the newly created surfaces. If the two are of the same order of magnitude, grain boundaries will be considerably weaker than crystals. Gilman has given convincing reasons to expect that grain boundary weakness is a principal characteristic of covalent refractory non-metallic substances. This opinion is supported by the fact that polycrystalline ceramics usually fracture along grain boundaries.

Changes in Direction of Fracture

Crack propagation may be hindered by obstacles such as grain boundaries, which may cause an abrupt change in the direction of crack propagation. The problem of the forked crack has not been solved in its entirety; the problem of crack initiation in a new direction has, however [15]. Consider a crack that has been momentarily arrested at a grain boundary (Figure 2). Crack motion is assumed to re-initiate either at some tilt angle, θ, to the original crack surface, or at a twist angle, φ. The apparent increase in surface energy required for a crack

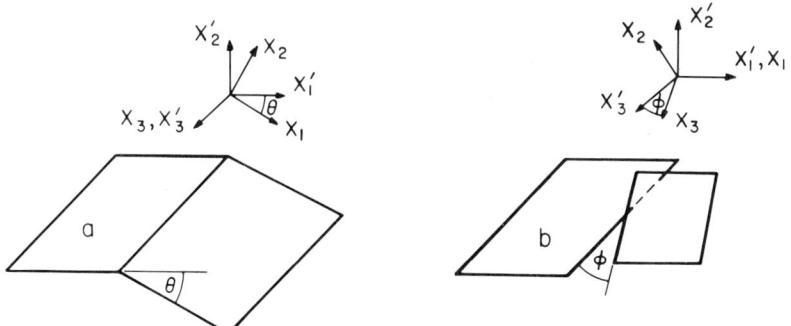

Figure 2. Change of crack plane orientation as crack traverses a grain boundary: (a) tilt propagation, (b) twist propagation (after M. Gell and E. Smith [15]).

to abruptly change directions is given by [15]

$$\gamma_2 = \gamma_1 \ \sec^4 \theta /2 \qquad (4)$$

for the case of tilt propagation, and

$$\gamma_2 = \gamma_1 \ \sec^4 \varphi \qquad (5)$$

for the case of twist propagation, where the subscripts 1 and 2 refer to the fracture surface energy before and after the change in direction of crack propagation.

The apparent increase in fracture energy for crack re-initiation depends on the angle of tilt or twist, as might be expected. For fracture along grain boundaries in polycrystalline ceramics, tilt propagation is most important, since all grain boundaries meet at tilt angles. If a polycrystalline aggregate is assumed to be a nested assembly of truncated octahedra, the angle of tilt will always be $60°$ [16]. In real materials, the angle of tilt scatters around this value. Since the larger angles control the ease of fracture in a polycrystalline ceramic, tilt angles ranging from $60°$ to $90°$ should be important to the fracture process. Tilt angles in this range are equivalent to surface energy increases of from 1.8 to 4 times the fracture surface energy value before the change in propagation direction. Fracture surface energy increases of this magnitude are an important factor in the propagation of cracks in polycrystalline ceramics.

Anisotropic Thermal Contraction

Ceramic materials are produced at high temperatures. When cooled to room temperature, internal stresses may arise due to anisotropic thermal contraction of the crystalline grains that make up the ceramic. These stresses can be considerable, and have the effect of weakening the ceramic body. Internal stresses of this type may also arise from phase transformations, such as occurs in ZrO_2, or from differential expansion due to radiation damage, such as occurs in BeO.

The effect of anisotropic thermal contraction has been treated by Clarke [17], who assumed that the localized stresses were relieved by the passage of the crack and contributed to the energy required for fracture. Fracture is assumed to initiate at grain boundary pores and then to propagate along the grain boundary (Figure 3), absorbing energy from both the external and internal strain fields, which arise respectively from the applied stresses and the expansion anisotropy. A Griffith-type energy balance is used to derive the critical stress for fracture,

$$\sigma = \left\{ E[2\gamma_b - E\epsilon^2(\ell - c_o)/12(1 - \nu^2)] / \pi c_o \right\}^{1/2} \qquad (6)$$

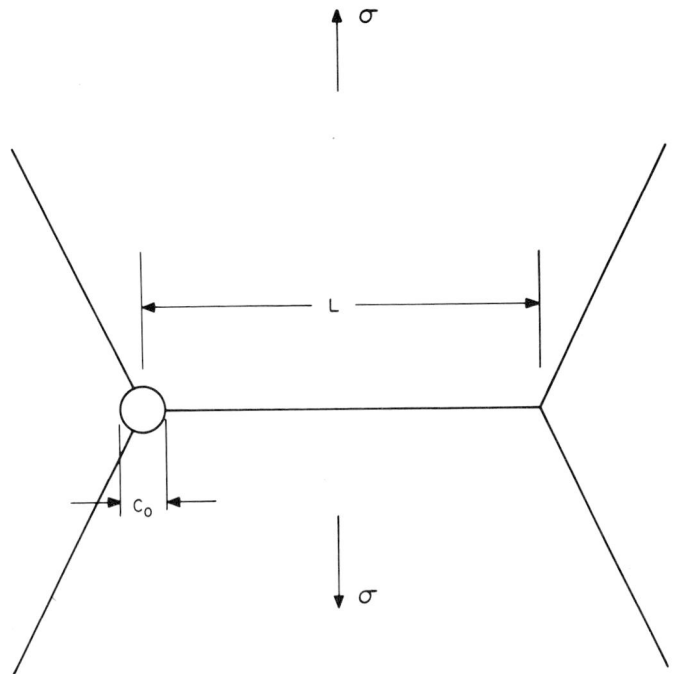

Figure 3. Model for crack initiation at a grain boundary pore due to thermal stresses (after F. J. P. Clarke [17]).

where ϵ is the grain boundary strain, the grain size is 2ℓ, γ_b is the grain boundary fracture surface energy, and c_o is the grain boundary pore diameter. The fracture surface energy of the grain boundary is seen to be modified by a term that accounts for the localized elastic strain. Spontaneous fracture is predicted when the term inside the square brackets equals zero.

$$2\gamma_b = E\epsilon^2(\ell - c_o)/12(1 - \nu^2) \tag{7}$$

or for $c_o \ll \ell$,

$$\epsilon = (24\gamma_b/E\ell)^{1/2} \tag{8}$$

The validity of this last equation has been demonstrated recently by Davidge and Tappin [18] on polycrystalline BeO.

Other Factors

The fracture surface energy of solids may be modified by processes in addition to those described above. The fact that crystalline materials are not elastically isotropic will give rise to localized stress variations in polycrystalline ceramics under imposed loads. These stress variations arise because of the random orientation of the grains in the solid, and the fact that the elastic constants of adjacent grains are not equal in a given direction. Other stress concentrators such as pores, inclusions, and other cracks may also affect the ease of crack propagation.

Experimental Methods of Measuring the Fracture Surface Energy

There are three experimental methods of measuring fracture surface energy. One requires a complete elastic solution of the crack configuration under consideration [9] from which a fracture surface energy equation can be obtained. The fracture surface energy is expressed as a function of crack size, maximum load at failure, and the specimen goemetry. Several crack geometries used in these types of determinations are shown in Figures 4-6. The experiment is performed by loading the specimen to failure and then calculating the fracture surface energy from pertinent experimental parameters. The center-notched plate tension specimen [19] and the edge-cracked bend specimen [19] have been used extensively in the testing of metals, and the dimensions suggested in Figures 4 and 5 are for metallic materials. The double-cantilever specimen [10, 20–22], also known as the infinite crack-line loaded edge-crack or Manjoine specimen, has been used in various forms for both metals and ceramic materials.

The second method of determining the fracture surface energy is called the compliance method [11, 23, 24], since it requires a knowledge of the compliance of the specimen. Consider a crack in a solid, subjected to a constant load, P (Figure 7). As the crack extends, the work, dW, performed on the solid by the external load must equal the change of internal strain energy, dV, plus the change of surface energy, dS; $dW = dV + dS$. The work by the external load is Pdu, where du is the differential displacement of the load, P, as the crack extends a small amount. The total elastic strain energy of the body is $V = \int_o^u Pdu$. For a linear elastic body, the displacement is proportional to the applied load, $u = \lambda P$, where λ is called the compliance of the system. It therefore follows that the total elastic energy is $V = u^2/2\lambda = Pu/2$, and at constant load $dV = \frac{1}{2}Pdu = \frac{1}{2}dW$. The change in surface energy is $2\gamma dA$, where $2dA$ is the amount of surface formed as the crack propagates. Substituting expressions for dW, dV, and dS into the energy balance, the following equation is obtained,

$$\gamma = (P/4)\,(\partial u/\partial A)_P = (P^2/4)\,(\partial\lambda/\partial A)_P \qquad (9)$$

for specimens of constant thickness, t, $dA = tdL$, where dL is the increase in crack length. Equation (9) then becomes

$$\gamma = (P^2/4t)\,(\partial\lambda/\partial L)_P \qquad (10)$$

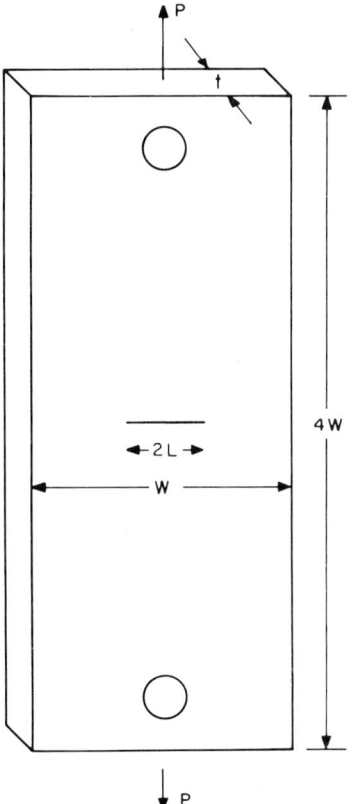

Figure 4. Center-notched plate tension specimen containing a crack of length $2L$ (after J. E. Srawley and W. F. Brown [19]). $\gamma = (P^2/Wt^2 2E) \tan \pi L/W$.

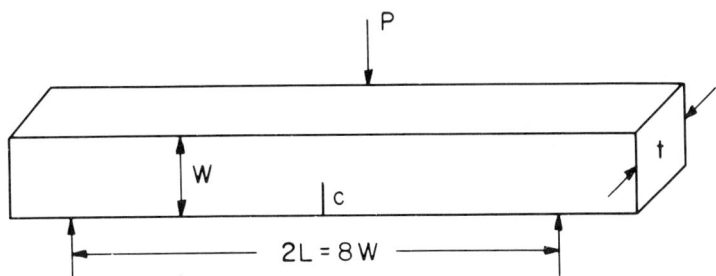

Figure 5. Edge-cracked bend specimen containing a crack of length c. For four-point loading, see [19] (after J. E. Srawley and W. F. Brown [19]). $\gamma = (PL/Wt)^2 [31.7 \, c/W - 64.8 \, (c/W)^2 + 211 \, (c/W)^3]/2E$.

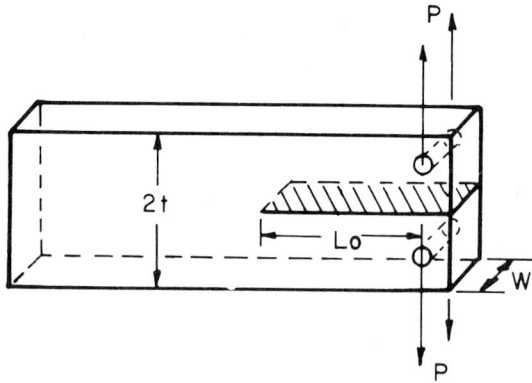

Figure 6. Double-cantilever specimen, $\gamma = 6P^2L^2/Ew^2t^3 \ [1 + 1.335\,t/L + 0.446\,(t/L)^2]$.

Equation (10) is the basis for the compliance method of measuring the fracture surface energy. A series of specimens is prepared containing a range of crack sizes. As a specimen is loaded, the relative displacements of the loading points, u, must be measured as a function of P for each crack length. Since $\lambda = u/P$, λ can be plotted as a function of L and $\partial\lambda/\partial L$ can be determined from

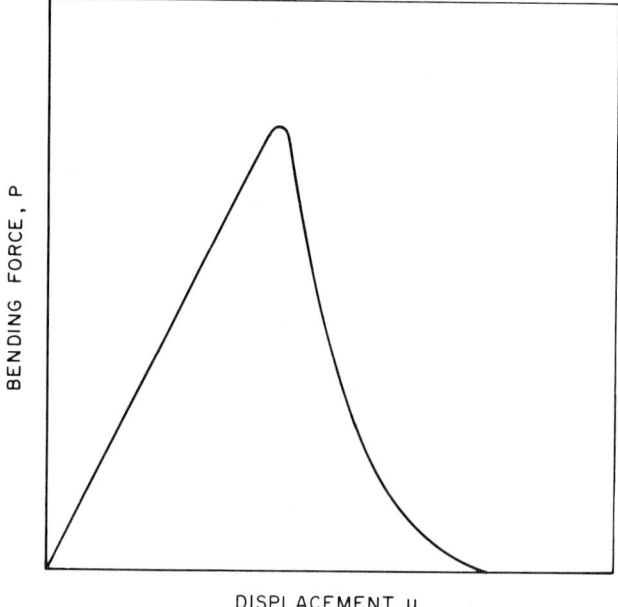

Figure 7. Force displacement curve for the work of fracture method of determining the fracture surface energy.

the slope of the plot for any crack length. Because the derivative of λ is important in the determination of γ, displacements and loads must be measured to a high degree of accuracy. The main limitation of this method is the displacement measurements. The method has the great advantage of being applicable to any specimen shape, and can be used to check elastic solutions of crack problems.

In the third method of measurement, the work to propagate a crack over a complete cross-section is determined [25, 26]. The test is conducted in three-point bending and edge-cracked bend specimens are used (Figure 5). This type of test requires a testing machine with a large spring constant. Specimens are slowly deformed at a continuous, constant rate throughout the test, and the test is continued until specimens have been completely fractured. The method depends on the fact that as the crack propagates, the compliance, λ, decreases, and crack propagation occurs continuously and stably with gradually decreasing load. The force-displacement curve during the test is as depicted in Figure 7, and the area under the curve is equal to the total work performed during the experiment. There is no elastic energy stored in the testing machine at the completion of the test, since the machine is load-free once the crack has passed through the specimen. Consequently, all of the work performed during the experiment has gone into the fracture process, and the fracture surface energy can be determined by dividing the total work by the total fracture surface area.

The first two methods of measurement are made on cracks that are originally stationary and therefore give the fracture surface energy required to initiate the motion of a stationary crack. Fracture surface energies obtained by the third method are related more to crack propagation than to crack initiation, since the surface energy is determined on moving cracks. This difference was recognized recently by Davidge and Tappin [27], who classify fracture surface energy measurements with respect to crack initiation, γ_I, or crack propagation, γ_F. On completely brittle ceramics containing atomically sharp cracks, there should be no difference between the two types of fracture surface energies. Experiments by Davidge and Tappin [27] on graphite, alumina, polymethylmethacrolate, and glass show that γ_I and γ_F are not necessarily equal, but depend on physical processes occurring during fracture.

Experimental Values of the Fracture Surface Energy

Single Crystals

Most fracture surface energy determination on single crystals have been made on crystals that cleave. Thus, the energy values published are characteristic only

TABLE I.

Fracture Surface Energies of Single Crystals

Crystal	Energy, J/m^2	Reference
Mica, vacuum, 25°C	4.5	[28]
Mica, vacuum, 25°C	4.57	[29]
LiF, $N_2(\ell)$, −196°C	0.40	[20]
MgO, $N_2(\ell)$, −196°C	1.5	"
CaF_2, $N_2(\ell)$, −196°C	0.55	"
BaF_2, $N_2(\ell)$, −196°C	0.31	"
$CaCO_3$, $N_2(\ell)$, −196°C	0.35	"
Si, $N_2(\ell)$, −196°C	1.8	"
MgO, Vacuum, −175°C	1.6	[30]
NaCl, $N_2(\ell)$, −196°C	0.28	"
NaCl, Vacuum, −175°C	0.24	"
Sapphire, (11$\bar{2}$3) plane, −196°C	32.2	[31]
Sapphire, (10$\bar{1}$1) plane, −196°C	24	"
Sapphire, (22$\bar{4}$3) plane, −196°C	16.4	"
Sapphire, (11$\bar{2}$3) plane, 20°C	24.4	"
Sapphire, (10$\bar{1}$1) plane, 25°C	6.0	[32]
Sapphire, (10$\bar{1}$0) plane, 25°C	7.3	"

Date from references [20] and [30] were recalculated using the equation in figure 6. The values given represent averages of the three smallest surface energy values in each group. Data from references [31] were obtained on center-notched plate tension specimens. Data from the other references were obtained on double cantilever specimens. Morphological indices are used for sapphire, c/a = 1.365.

of the cleavage plane in the crystal. Since many crystals that cleave exhibit fracture on no other crystal planes, these fracture surface energy values completely characterize fracture in these materials. A summary of some of the fracture surface energy data on crystals that cleave is given in Table I. The data obtained by the double-cantilever technique have been recalculated by the author according to the equation in Figure 6, which was not available at the time of the original publications. The data presented in Table I were obtained on specimens tested in liquid nitrogen or in vacuum. The test conditions were such as to eliminate environmental effects and to reduce plastic flow at crack tips. The results presented in Table I were close to those expected theoretically; therefore, plasticity and environmental effects may have been eliminated.

Fracture surface energy measurements on non-cleavable single crystals have been obtained on sapphire, α-Al_2O_3. The rather meager available data is presented in Table I. Energy values obtained by Petch *et al.* [31] are about ten times that expected theoretically. The authors explain the high values as due either to energy absorption by plastic deformation or the possible fact that the drilled hole did not offer a suitably sharp crack from which fracture could initiate. Considering the low fracture energy values obtained by Wiederhorn [32], it is improbable that plastic deformation plays a role in increasing the surface energy for fracture.

Bicrystals

Bicrystal grain boundary energies are important because they bridge the gap between single-crystal and polycrystal behavior. The fracture surface energy along grain boundaries of various orientation can be obtained and, in principle, can be used to obtain a deeper understanding of the fracture of polycrystalline solids. Fracture surface energies obtained on grain boundaries can be combined with values obtained on single crystals of the appropriate orientation to give the amount of weakening caused by grain boundaries in polycrystalline materials. In theory, the grain boundary fracture surface energy can be obtained on specimen geometries such as those given in Figures 4 through 6, provided the boundary is coplanar with the crack. Despite the importance of grain boundary fracture energies, only one complete experiment of this type has been conducted. Class and Machlin [33] have measured the fracture surface energy on twist boundaries of KCl using the double-cantilever method. Their results (Figure 8)

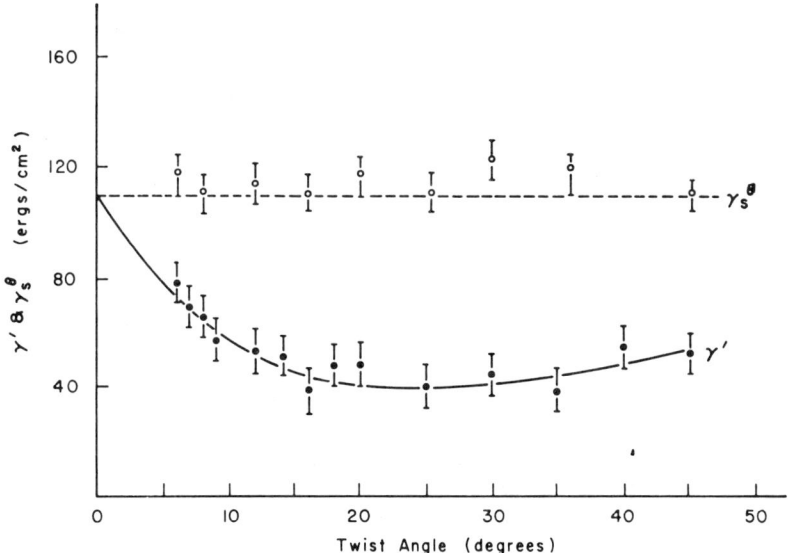

Figure 8. Fracture surface energy of KCl bicrystal twist boundaries. The upper curve gives the fracture surface energy of the (001) surface, and the twist angle refers to the direction of crack propagation on the (001) surface. The lower curve gives the fracture surface energy of the twist boundary. The angle of twist is given by the abscissa (after W. H. Class and E. S. Machlin [33]).

indicate that twist boundaries are weaker than cleavage planes in single crystals, and probably act as preferred fracture paths in polycrystalline ceramics.

Additional work on the fracture of grain boundaries is clearly necessary for a fundamental approach to the fracture of polycrystalline ceramics. In particular, experimental data should be obtained on tilt boundaries to determine if they, too, are easy paths for fracture. Other materials, more brittle than KCl, should also be studied.

Polycrystalline Ceramics

Fracture surface energy values obtained on polycrystalline ceramics, Table II, are an order of magnitude greater than those obtained from single crystals. The reason for the increase in value is the tortuous nature of the crack path in these materials [34] (Figure 9). Cracks must extend through and around many grains,

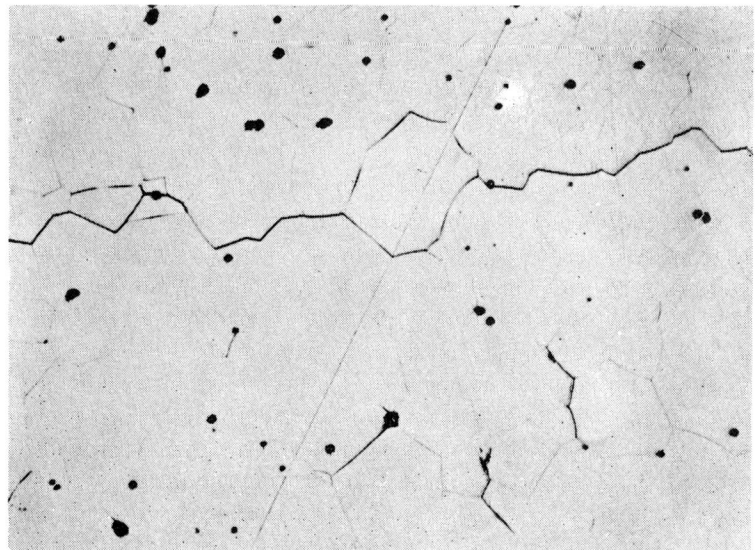

Figure 9. Crack propagating through polycrystalline Al_2O_3. Note the random nature of the crack path (after R. L. Coble [16], p. 623).

and can be pinned at many points along the boundary by high-energy obstacles consisting of poorly aligned grains and grain boundaries; equations (4) and (5). The crack must overcome all of these obstacles to propagate, requiring a greater amount of energy for fracture than is necessary in single crystals. Energy-absorbing processes that make crack propagation difficult have been discussed by Clarke

TABLE II.

Fracture Surface Energy of Polycrystalline Ceramics

Material	Density $\times 10^3 Kg/m^3$	Grain Size meters $\times 10^{-6}$	Energy J/m^2	Reference
Firebrick, Douglas X			30	[34]
MgO	3.48	10	16	"
MgO	3.51	50	19.5	"
MgO	3.56	100	35	"
MgO, Theoretical Density	3.58	7	4.2	"
MgO, Theoretical Density	3.58	13	8.9	"
MgO, Theoretical Density	3.58	23	16	"
MgO, Theoretical Density	3.58	38	17	"
MgO, Theoretical Density	3.58	130	14	"
MgO, Theoretical Density	3.58	150	7.9	"
BeO	2.79	5	15	"
Graphite			100	"
SiN	2.2		20	"
Al_2O_3, Lucalox	3.97	10	18	[35]
Al_2O_3, Lucalox	3.97	30	27	"
Al_2O_3, Lucalox	3.97	45	46	"

et al. [34], and include secondary crack formation, plastic flow, area increase due to fracture surface roughness, and step formation which leads to high-energy tearing fracture.

A second feature of the data given in Table II is the dependence of the fracture surface energy on grain size. For small grain sizes it is observed that the fracture surface energies of polycrystalline MgO and Al_2O_3 decrease with decreasing grain size. No explanation for this behavior is yet available.

Engineering Applications

Design and selection of new materials for particular applications require parameters by which the material can be judged. In this section, the possibility of using the fracture surface energy as an engineering parameter will be discussed with regard to strength and thermal shock behavior.

Strength of Polycrystalline Ceramics

Application of the flaw theory to a determination of strength requires a knowledge of both the flaw size and fracture surface energy of the solid. Depending on the size and location of the flaw, the fracture surface energy can vary by an order of magnitude in polycrystalline solids. Flaws of the same size as the grain diameter will have surface energies characteristic of grain boundaries or single crystals, depending on the location of the flaws. Flaws that are much larger than the grains will have surface energies characteristic of the poly-crystalline fracture surface energy. It is of interest to determine which flaw sizes are critical to the strength of ceramic materials.

The Griffith equation for the penny-shaped crack, $\sigma = \sqrt{\pi E \gamma / D(1 - \nu^2)}$, is probably the most appropriate one for calculating the fracture strength of polycrystalline ceramics. Strengths calculated from the fracture surface energy data for polycrystalline MgO and Al_2O_3, Table II, are in agreement with measured values [34, 2, 36], provided crack nuclei of 10 to 20 grain diameters are assumed for these materials. Since flaw sizes of this magnitude have not been reported in dense ceramics, it is reasonable to conclude that fracture surface energies determined on polycrystalline ceramics are not to be associated with the strength of these materials.

A similar calculation on the strength of MgO and Al_2O_3 using surface energies characteristic of single crystals gives a predicted flaw size of approximately one to three grain diameters, which is in better agreement with that expected. Therefore, the strength of dense ceramic materials is controlled by crack propagation or nucleation from nuclei of the order of the grain size in diameter, as has been concluded previously by Passmore, Spriggs, and Vasilos [3].

Thermal Fracture

Fracture is often observed to occur in ceramic materials after a rapid change in temperature known as thermal shock. The fracture results from stresses that arise from non-linear temperature gradients or restraints to thermal expansion. The thermal stresses causing fracture depend on the rapidity of the temperature change and on the shape of the body being shocked. To determine the onset of fracture, it is necessary to solve the heat transfer equations and obtain the temperature distribution in the structure, from which the stress distribution can be calculated. A review of the results of such analysis has been given by Kingery [37]. Failure is deemed to have occurred once the thermal tensile stresses exceed the tensile strength of the materials, as in the case of static loading. Thus a maximum stress criterion is used to describe thermal failure of materials.

Hasselman [38] and Clarke et al. [34] have suggested an exception to the maximum tensile strength criterion for thermal shock failure. Their criterion for failure is based on the availability of strain energy for crack propagation. During crack propagation, strain energy imparted to the structure by thermal shock is converted into fracture surface energy. The amount of strain energy available for fracture is limited by the severity of the shock. If the shock is not too severe, the conversion of strain energy into fracture surface energy may be complete before structural integrity is lost. Structural materials may crack in a limited way due to thermal shock, but still be able to serve their original design function. Thermal failure therefore contrasts with constant load failure, for which the available elastic strain energy is always sufficient for complete failure once a critical load is exceeded. Mechanical shock and failure under fixed-grip conditions is similar to thermal shock in that the initial elastic strain energy is fixed so that crack propagation may cease before structural usefulness has been destroyed. In the following paragraphs, the approach taken by Clarke *et al.* [34] will be used to illustrate thermal shock.

In engineering practice it is always useful to have a number that can be used to classify materials for a given end-use. For the shock conditions described above, a useful parameter is the ratio of the fracture surface energy to the elastic strain energy available for fracture, V. Since this ratio may be large or small, the logarithm of the ratio will be more manageable for engineering purposes. Thus, the thermal shock resistance can be defined by a parameter, τ, where

$$\tau = \log_{10}(\gamma/V) \tag{11}$$

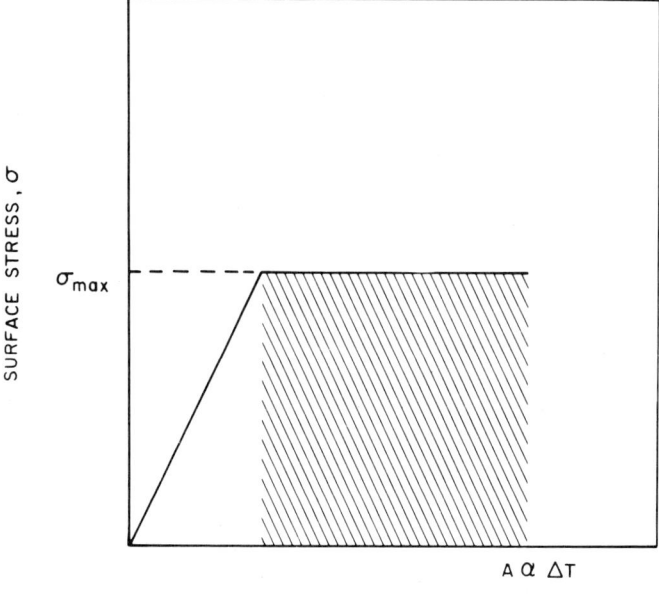

Figure 10. Assumed stress-strain curve in the surface layer of a thermally shocked ceramic.

This definition is essentially the same as the one suggested by Clarke *et al.* [34], differing by an additive and multiplicative constant. Larger values of τ indicate greater resistance to thermal shock.

To evaluate τ, the surface energy, γ, and the excess elastic strain energy must be determined. The fracture surface energies can be determined by methods described earlier in the paper. The excess strain energy can be estimated as follows. The strain imposed by thermal shock at a free surface of a solid is always of the form $\epsilon = A\alpha\Delta T$, where α is the thermal coefficient of expansion, ΔT the temperature difference between the initial body temperature and that of the environment, and A is a constant that depends on the rate of heat transfer [37], thermal conductivity, and the dimensions of the structure. The value of A is usually less than 1. The stress at the free surface of the material is proportional to the thermal strain up to the fracture stress. At higher strains, cracks form so that the surface stress cannot exceed the fracture stress of the material. For purposes of approximation, it is assumed that crack formation releases strain energy in such a manner that the surface stresses remain constant as the material fractures. The stress-strain relationship of a region near the surface of the structure is assumed to be as depicted in Figure 10. The excess

TABLE III.

Thermal Shock of Commercial Alumina

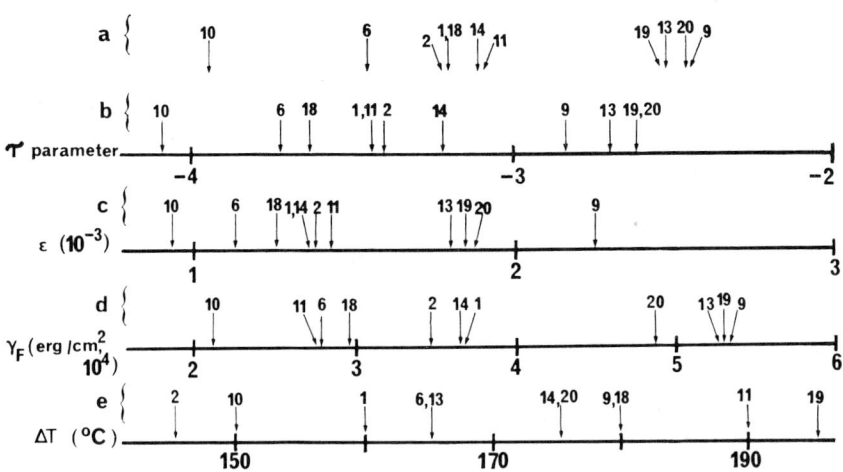

Comparison of a series of commercial aluminas. a) and b) are two different methods of calculating τ. c) represents the strain that can be accommodated before a fresh amount of cracking can occur. d) is the fracture surface energy determined by method c. e) is the British Standard thermal shock test. After F. J. P. Clarke, H. G. Tattersall and G. Tappin [34].

strain energy for fracture is given by the shaded portion of Figure 10. In quantitative terms,

$$V = \int_{\sigma/E}^{A\alpha\Delta T} \sigma d\epsilon = (\sigma/E)\,(EA\alpha\Delta T\text{-}\sigma) \qquad (12)$$

TABLE IV.

Thermal Shock of Ceramics

Material	γF $(10^4\ ergs\ cm^{-2})$	σF $(10^9\ dyn.\ cm^{-2})$	α $(10^{-6}\ {}^\circ C^{-1})$	Young's Modulus $(10^{12}\ dyn.\ cm^{-2})$	$\mathcal{T}$
Firebrick[i]	3	0·1	6·0	0·13	+1
Polycrystalline alumina[ii]					
No. 9	5·3	2·2	7·55	3·5	−2·9
No. 10	2·1	3·45	7·8	3·9	−4·5
Single-crystal alumina[iii]	1·2	3·5	7·0	3·0	−4·8
Polycrystalline magnesia[iv]					
3·48 g.cm^{-3}, 10μm	1·6	0·4	12	3·0	−3·17
3·51 g.cm^{-3}, 50μm	1·95	0·7	12	3·0	−3·5
3·56 g.cm^{-3}, 100μm	3·5	1·0	12	3·0	−3·2
Theoretical density, 7μm[v]	0·42	2·5	12	3·0	−6·9
Single-crystal magnesia[vi]	<0·2	1·0	12	3·0	−6·8
Polycrystalline beryllia[viii]	1·5	2·0	7·0	3·4	−4·2

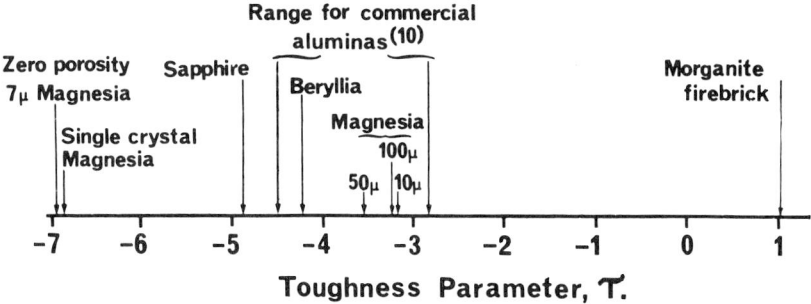

Toughness Parameter, $\mathcal{T}$.

Comparison of some ceramics in respect to their resistance to thermal shock. Specimens are the same as those given in table III. After F. J. P. Clarke, H. G. Tattersall and G. Tappin [34].

where σ is the strength of the material, and E is Young's modulus. From equation (11) it may be concluded that fracture does not initiate unless $\sigma = EA\alpha\Delta T$, which is Kingery's maximum stress criterion for thermal failure [37]. Therefore, the assumptions used in the derivation of equation (12) are consistent with the treatment given by Kingery. Substituting equation (12) into equation (11), the following equation is obtained for thermal shock resistance:

$$\tau = \log_{10}\left[\gamma E/\sigma(EA\alpha\Delta T\text{-}\sigma)\right] \tag{13}$$

For identical amounts of overstressing, $EA\alpha\Delta T\text{-}\sigma =$ constant $> O$, the resistance to thermal shock increases with increasing Young's modulus and surface energy, but decreases with increasing strength. The increased thermal shock resistance with decreasing strength has been noted previously [38].

Clark et al. [34] have shown that a thermal shock equation similar in form to equation (13) compares favorably with other methods of determining shock resistance. Data presented by Clarke et al. are shown in Tables III and IV. A comparison of τ with the British Standard Thermal Shock Test (Table III) for commercial alumina ceramics shows that the thermal shock resistance of the various ceramics occupy similar positions on both scales. The fracture surface energy can also be used to measure the thermal shock resistance for these ceramics (Table III). For a broader range of materials, the fracture surface energy cannot be used as the only criteria for fracture, as can be seen from Table IV. The high thermal shock resistance of firebrick (Table IV) is due to its low strength.

Summary

Mechanical failure of ceramic materials has been discussed with regard to crack propagation. Fracture surface energy is shown to be the important parameter to be used to describe the resistance of solids to fracture. A basic viewpoint is taken and the effect of grain boundaries, anisotropic thermal contraction, crystallographic direction of crack propagation, and the effect of directional changes on the effective fracture surface energy are discussed. Possible engineering applications of fracture surface energies are given. The fracture surface energy is shown to be a useful parameter in the description of thermal shock, and can also be used to determine the weakest links in the structure of polycrystalline ceramics. Additional research in the measurement of fracture surface energies is clearly indicated. It is hoped that fracture surface energy measurements will become a useful adjunct to the usual strength measurements for determining the resistance of ceramic materials to mechanical failure.

Acknowledgment

This work was supported by the U.S. Army Research Office, Durham, North Carolina.

References

1. Knudsen, F. P., "Dependence of Mechanical Strength of Brittle Polycrystalline Specimens on Porosity and Grain Size," *J. Am. Ceram. Soc.*, 42 (1959), 376–87.
2. Vasilos, T., Mitchell, J. B. and Spriggs, R. M., "Mechanical Properties of Pure, Dense Magnesium Oxide as a Function of Temperature and Grain Size," *J. Am. Ceram. Soc.*, 47 (1964), 606–10.
3. Passmore, E. M., Spriggs, R. M. and Vasilos, T., "Strength-Grain Size-Porosity Relations in Alumina," *J. Am. Ceram. Soc.*, 48 (1965), 1–7.
4. Gilman, J. J., "The Strength of Ceramic Crystals," in *The Physics and Chemistry of Ceramics*, C. Klingsberg, ed., Gordon and Breach, New York (1963), 240.
5. Orowan, E., "Fracture and Strength of Solids," *Repts. Prog. Phys.*, 12 (1948), 185–232.
6. Inglis, C. E., "Stresses in a Plate Due to the Presence of Cracks and Sharp Corners," *Trans. Naval Arch.*, 60 (1913), 219–30.
7. Griffith, A. A., "Phenomena of Rupture and Flow in Solids," *Phil. Trans. Roy. Soc.*, (London), 221A (1920), 163–98.
8. Griffith, A. A., "Theory of Rupture," *Proc. First International Contr. Appl. Mechanics*, Delft (1924), 55–63.
9. Paris, P. C. and Sih, G. C., "Stress Analysis of Cracks," pp. 30–81, in *Fracture Toughness Testing and its Applications*, ASTM Special Technical Publication No. 381.
10. Gilman, J. J., "Cleavage, Ductility and Tenacity in Crystals," in *Fracture*, B. L. Averbach, D. K. Felbeck, G. T. Hahn and D. A. Thomas, eds., John Wiley and Sons, Inc., New York (1959), 193–222.
11. Irwin, G. R., "Fracture," in *Encyclopedia of Physics*, Vol. VI, Springer Heidelberg (1958).
12. Orowan, E., "Energy Criteria of Fracture," *Weld. J.*, London, 34 (1955), 157s–160s.
13. Wachtman, J. B., Jr. and Maxwell, L. H., "Plastic Deformation of Ceramic-Oxide Single Crystals," *J. Am. Ceram. Soc.*, 37 (1954), 291–99.
14. Gilman, J. J., "Strength of Ceramic Crystals," in *Mechanical Behavior of Crystalline Solids*, National Bureau of Standards Monograph 59 (1963).
15. Gell, M. and Smith, E., "The Propagation of Cracks Through Grain Boundaries in Polycrystalline 3% Silicon–Iron," *Acta Met.*, 15 (1967), 253–58.
16. Kingery, W. D., *Introduction to Ceramics*, John Wiley and Sons, Inc., New York (1960), **410.**
17. Clarke, F. J. P., "Residual Strain and the Fracture Stress–Grain Size Relationship in Brittle Solids," *Acta Met.* (1964), 139–43.
18. Davidge, R. W. and Tappin, G., "Internal Strain and the Strength of Brittle Materials," *J. Nat. Sci.*, 3 (1968), 297–301.
19. Srawley, J. E. and Brown, W. F., "Fracture Toughness Testing Methods," in *Fracture Toughness Testing and Its Applications*, ASTM Special Technical Publication No. 381, 133–96.

20. Gilman, J. J., "Direct Measurements of the Surface Energies of Crystals," *J. Appl. Phys.,* 31 (1960), 2208–18.
21. Srawley, J. E. and Gross, B., "Stress Intensity Factors for Crackline-Loaded Edge-Crack Specimens," *Nat. Res. and Stand.*, 7 (1967), 155–62.
22. Wiederhorn, S. M., Shorb, A. M. and Moses, R. L., "A Critical Analysis of the Theory of the Double Cantilever Method of Measuring Fracture Surface Energies," *J. Appl. Phys.*, 39 (1968), 1569–72.
23. Irwin, G. R., "Fracture Mechanics," in *Structural Mechanics*, J. N. Goodier and N. J. Hoff, eds., Pergamon Press, New York (1960), 557–92.
24. Irwin, G. R. and Kies, J. A., "Critical Energy Rate Analysis of Fracture Strength," *Welding Res. Suppl.*, 33 (1954), 193s–198s.
25. Nakayama, J., "Direct Measurement of Fracture Energies of Brittle Heterogeneous Materials," *J. Am. Ceram. Soc.*, 48 (1965), 583–87.
26. Tattersall, H. G. and Tappin, G., "The Work of Fracture and Its Measurement in Metals, Ceramics and Other Materials," *J. Mat. Sci.*, 1 (1966), 296–301.
27. Davidge, R. W. and Tappin, G., "The Effective Surface Energy of Brittle Materials." *J. Mat. Sci.*, 3 (1968), 162–73.
28. Orowan, E., "The Tensile Strength of Mica and the Problem of the Technical Strength," *Z. Physik.*, 82 (1933), 235–66.
29. Bryant, P. J., Taylor, L. H. and Gutshall, P. L., "Cleavage Studies of Lamellar Solids in Various Gas Environments," in *Transactions of the Tenth National Vacuum Symposium*, American Vacuum Society, (1963), 21–26.
30. Gutshall, P. L. and Gross, G. E., "Cleavage Surface Energies of NaCl and MgO in Vacuum," *J. Appl. Phys.*, 36 (1965), 2459–60.
31. Petch, N. J., Congleton, J., Hardie, D and Perkins, R. N., "Effect of Surface Energy," in *Study of the Brittle Behavior of Ceramic Materials,* Technical Documentary Report No. ASD–TR–61–628, Parts I, II, III.
32. Wiederhorn, S. M., *J. Am. Ceram. Soc.,* September 1969.
33. Class, W. C. and Machlin, E. S., "Crack Propagation Method for Measuring Grain Boundary Energies in Brittle Materials," *J. Am. Ceram. Soc.,* 49 (1966), 306.
34. Clarke, F. J. P., Tattersall, H. G. and Tappin, G., "Toughness of Ceramics and Their Work of Fracture," *Proc. Brit. Ceram. Soc.*, 6 (1966), 163–72.
35. Gutshall, P. L. and Gross, G. E., "Observations and Mechanisms of Fracture in Polycrystalline Alumina," *Eng. Fract. Mech.*, 1 (1968), 463–72.
36. Charles, R. J. and Shaw, R. R., "Delayed Failure of Polycrystalline and Single-Crystal Alumina," General Electric Research Laboratory Report No. 62–RL–3081 M (July 1962).
37. Kingery, W. D., "Factors Affecting Thermal Stress Resistance of Ceramic Materials," *J. Am. Ceram. Soc.,* 38 (1955), 3–15.
38. Hasselman, D. P. H., "Elastic Energy at Fracture and Surface Energy as Design Criteria for Thermal Shock," *J. Am. Ceram. Soc.*, 46 (1963), 535–40.

16. Plastic Deformation in Fine-Grain Ceramics

A. H. HEUER

Case Western Reserve University
Cleveland, Ohio

R. M. CANNON

Avco Advanced Technology Division
Lowell, Massachusetts

N. J. TIGHE*

National Bureau of Standards
Washington, D.C.

ABSTRACT

Plastic deformation in fine-grain (i.e., $\leqslant 10\,\mu$) ceramics is discussed. It is shown that fine-grain polycrystals can be exceptionally ductile, the fine grain size enhancing diffusional deformation and grain boundary sliding processes. The deformation is sensitive to both grain size and temperature.

The influence of grain size ($1-10\,\mu$), strain-rate ($2 \times 10^{-6} - 3 \times 10^{-4}$/sec), and temperature ($1100-1700°C$) on the deformation of fine-grain alumina has been studied. It is suggested that the predominant deformation mechanism in the larger grained polycrystals is diffusional creep, and that grain boundary sliding makes an increasingly important contribution as the grain size is decreased; in addition, deformation twinning can also be important. These results are shown to be consistent with previous work on deformation in polycrystalline alumina. A brief review of the literature on plastic deformation in fine-grain magnesia, beryllia, thoria, and urania indicates that grain boundary sliding may be important for each of these materials as well.

Introduction

This chapter will be concerned with those features of the plastic deformation of fine-grain ceramic polycrystals which are a direct manifestation of the fine grain size, i.e., features which cannot be readily predicted from the plastic behavior of single crystals. Foremost among these is the phenomenon, recently pointed out by Gilman [1], that under certain conditions, polycrystals can be more ductile than single-crystals.

*Presently at Dept. of Metallurgy, Imperial College of Science and Technology, London, England.

339

The chapter will be organized in the following manner. After a discussion of possible deformation mechanisms, a series of experiments on the deformation of fine-grain alumina will be described, where the deformation characteristics were determined for a number of materials of different grain size as a function of strain rate and temperature. These results were then interpreted, after determining the microstructure and substructure of deformed specimens using transmission electron-microscopy. The conclusions from this study are felt to have general applicability for the deformation of other fine-grain ceramic polycrystals. The literature for other ceramic materials will then be briefly reviewed, with these suggested features being uppermost in the comparisons.

Deformation Mechanisms

Four possible mechanisms of plastic deformation are possible in a polycrystalline body, and all have been identified in one or the other of the ceramic polycrystals which will be discussed. These include conventional dislocation slip, deformation twinning, diffusional deformation (either through the lattice or at grain boundary regions), and grain boundary sliding (including shearing, migration, etc.).

Of these, dislocation slip is expected to be affected least by the fine grain size. It is well known that for a polycrystal to undergo homogeneous plastic deformation without the formation of internal voids or cracks, five *independent* slip systems are necessary [2].*   This insures that when slip bands intersect grain boundaries, stress concentrations are not built up in the neighboring grain but are relieved by additional slip. Thus, for a ceramic polycrystal which has sufficient interpenetrating slip systems, the deformation dynamics should depend on those slip systems which are most difficult to activate. This notion has been confirmed in the plastic yielding behavior of MgO [4] and UO_2 [5]. However, dislocations arriving at a grain boundary may contribute to grain boundary sliding [6], and direct evidence for this has been obtained in creep-tested iron and iron alloys [7]. No similar examples in the ceramic literature are known to the authors.

For the purpose of this chapter, therefore, we shall exclude those cases where the deformation arises mainly from slip and concentrate on the other possible deformation mechanisms.

The importance of deformation twinning in enhancing ductility of ceramic polycrystals is a subject about which little is known. While such twinning has been shown to nucleate fracture in single crystals of sapphire [8,9], twinning

---

*It should be noted that although five slip systems are necessary, they may not be sufficient to allow polycrystalline ductility.   The slip systems must also be able to *interpenetrate* [3].

is a common occurrence when sapphire is compressively deformed above 1100°C [9]. There is no a priori reason why twinning cannot contribute to polycrystalline ductility, and evidence will be presented to show that this is indeed the case in fine-grain alumina polycrystals. The influence of fine grain size on the ease or incidence of twinning is not known and cannot be estimated with certainty. It is the feeling of the present authors, however, that twinning may be less likely to lead to fracture in fine-grain—as opposed to coarse-grain—polycrystals, as the total length of the twin is restricted by the fine grain size.

Diffusional deformation on the other hand, is extremely sensitive to the grain size. In the two presently accepted models leading to Newtonian viscosity ($\dot{\epsilon} \propto \sigma$)—that due to Nabarro and Herring [10] for lattice diffusion, equation (1), and that due to Coble [11] where the rate-controlling process is enhanced diffusion through grain boundary regions, equation (2)—the strain rate varies as the grain size^{-2} and the grain size^{-3}, respectively:

$$\dot{\epsilon} = \left(\frac{13.3 D_1 \, \Omega}{kT (\text{G.S.})^2}\right) \sigma \tag{1}$$

and

$$\dot{\epsilon} = \left(\frac{47.1 \, w D_b \, \Omega}{kT (\text{G.S.})^3}\right) \sigma \tag{2}$$

Here $\dot{\epsilon}$ is the strain rate, σ the stress, Ω the volume of the rate-limiting diffusing species, G.S. the average grain size, D_1 and D_b the self-diffusion and grain boundary diffusion coefficients of the rate-limiting species, w the width of the grain boundary where enhanced diffusion occurs, and k and T have their usual significance.

In both models, application of stress results in an excess of vacancies (above the equilibrium zero stress concentration) at grain boundaries under tension, and a corresponding decrease in the concentration of vacancies at boundaries under compression. The flux of vacancies down this concentration gradient (or, more rigorously, down the chemical potential gradient) results in macroscopic deformation and should lead to elongation of grains.

The enhanced ductility observed in many fine-grain ceramics results primarily from this strong dependence on the grain size. A transition from a 10 μ grain size polycrystal to a 1 μ grain size material will result in an increase in the ductility by a factor of 100 or 1000, according to equations (1) or (2). These diffusional deformation mechanisms have been invoked for a number of ceramic polycrystals, where the deformation was Newtonian-viscous and where the activation energies of the deformation was similar to that for self-diffusion of one or the other atomic species. However, in no case was the required grain elongation observed; furthermore, calculated self-diffusion coefficients from deformation tests are usually several orders of magnitude higher than even the faster self-diffusion coefficients in the system.

Recently, Gifkins [12] has proposed a viscous grain boundary sliding deformation mechanism involving diffusion of grain boundary protrusions, equation (3), as well as a viscous deformation mechanism to accommodate the sliding at triple points also involving grain boundary diffusion, equation (4):

$$\dot{s} \simeq \left(\frac{2D_b \, \Omega}{LkT}\right) \sigma \tag{3}$$

$$\dot{s} \simeq \left(\frac{40 \, wD_b \, \Omega}{kT \, (G.S.)^2}\right) \sigma \tag{4}$$

Here, $\dot{s}$ is the rate of sliding of a boundary at $45°$ to the applied stress, L the average length of the grain boundary protrusions (shown to be present in tungsten by field ion microscopy [13]) and the remaining symbols are defined as for equations (1) and (2). The strain rate resulting from these sliding processes can be obtained from the relation

$$\epsilon \simeq ns \tag{5}$$

where $n = \dfrac{1}{G.S.}$,the number of grains per centimeter. For equation (3), this leads to a grain size dependence $\dot{\epsilon} \, \alpha \, G.S.^{-1}$ (if L is a function of G.S., to $\dot{\epsilon} \, \alpha \, G.S.^{-2}$), while for equation (4), $\dot{\epsilon} \, \alpha \, G.S.^{-3}$.

The virtue of these mechanisms is that they allow for extensive deformation without change of shape of individual grains. However, the grain size dependence may be similar to Nabarro-Herring or Coble creep, which makes differentiation of the possible mechanisms difficult. Furthermore, quantitative agreement between deformation rates and diffusion data is not expected to be unambiguous because of the number of difficult-to-measure parameters in the equations (i.e., width of the boundary region of enhanced diffusion, length of the grain boundary protrusions, etc.)

Nevertheless, it may be significant that Hensler and Cullen [14] have found negligible change of grain shape in polycrystalline MgO deformed 44 percent in compression, and interpreted this result in terms of Gifkins' mechanism(s).

It was first pointed out by Lifshitz [15], and recently emphasized by Gibbs [16], that when solids deform by diffusional deformation, equations (1) or (2), grain boundary sliding (GBS) must occur along with the change in grain shape, to maintain specimen coherency. However, the strain due to such GBS is included in equations (1) and (2). It is, of course, possible for GBS to be a deformation process which is not Newtonian-viscous, i.e., $\dot{\epsilon} \, \alpha \, \sigma^y, \, y > 1$, and thus contribute to the over-all deformation of a fine-grain ceramic polycrystal in a manner not included in equations (1) to (4).* Although the metallurgical litera-

---

*The type of anelastic GBS responsible for internal friction peaks (e.g., [17]) is excluded from the discussion, as this chapter is concerned solely with macroscopic plastic deformation.

ture on GBS is often contradictory (see [18] for a recent review), it now appears [19] that GBS is controlled by the resolved shear stress in the boundary plane, and that the stress sensitivity exponent for sliding (equivalent to the stress exponent in the creep law $\dot{\epsilon} \propto \sigma^y$) has values > 1 for both bicrystals and polycrystals.* It is usually assumed that since GBS depends on the total area of grain boundaries present, the grain size dependence should be as $\dot{\epsilon} \propto$ G.S.$^{-1}$; thus, γ, the fraction of total strain due to GBS, has been found to increase with decreasing grain size. Further support for the notion that GBS is enhanced in fine-grain polycrystals can be obtained from the spate of literature concerned with superplasticity in fine-grain metallic systems (the references are too numerous to cite here completely; the literature has been reviewed at this conference by Morgan†). Although no theory is yet available encompassing the copious amounts of experimental data, it is clear that very fine grain sizes ($< 10\,\mu$) and a high strain rate sensitivity ($dln\ \sigma/dln\ \dot{\epsilon} \equiv 1/y$) are both necessary to obtain superplastic extensions. Alden [22] has advanced the theory that GBS involving climb-glide motion of dislocations near grain boundaries may be the common mechanism for understanding the phenomena.

It is our intention in this chapter to show that non-Newtonian GBS, similar to that occurring in metals, is also important in ceramic systems and is very much enhanced in fine-grain ceramic polycrystals; furthermore, the lack of recognition of this phenomenon may be responsible for much of the confusion in the creep literature between diffusion coefficients calculated from one of the diffusional creep models and directly measured self-diffusion coefficients.

Present Work

Materials and Procedure

Materials used in this investigation were single-phase alumina, either pure or with an addition of $\frac{1}{4}\%$ MgO (to retard grain growth), which were prepared by sintering or hot pressing to high density (> 99 percent of theoretical density, T.D.).

As in many other investigations of ceramic materials, the four-point bend test was used. It is well known that although the elastic analysis of the stress distribution in slender beams is completely understood, the same cannot be said

---

*For lead bicrystals and polycrystals [20], y for GBS was 1 at low stresses (< 150 psi), but was ~ 3 for stresses between ~ 150 and ~ 250 psi (tests at room temperature and at 50°C). Similarly, the stress exponent for GBS in polycrystalline Magnox AL80 (a Mg + 0.78 wt % Al alloy) was found to be ~ 4 over a range of stresses from 2,000 to 8,000 psi at a test temperature of 200°C [21].

†P. E. D. Morgan, "Superplasticity in Ceramics," this volume.

for the stress distribution in plastically deforming beams.* In particular, the stresses in the bent beam can fall to some value which, at a minimum, will be two-thirds of the elastic stresses. However, in the cases of solids deforming by a pure Newtonian-viscous deformation process, equations (1)-(4), the stress distribution is exactly the same as in the elastic case, i.e., the stress varies linearly through the thickness of the beam.

In a previous paper [23], a method has been described whereby the bend test can be used to obtain true stresses in plastically deforming beams. In the general case, the flow stress may be a function of both strain and strain rate, and is not necessarily linear throughout the beam. It is assumed in the derivation that the flow stress is not a function of deformation history. It is necessary to perform a series of constant strain rate bending tests at constant temperature for identical specimens; the desired σ, ϵ, $\dot{\epsilon}$ relationships can then be computed for the outer fibers or for any point within the beam, using a graphical technique. If one desires only the maximum, outer-fiber stress, it is only necessary to know the strain hardening coefficient, n_b, and the strain rate sensitivity, m_b,

$$n_b = \left(\frac{\partial ln\, M}{\partial ln\, \epsilon} \right)_{\dot{\epsilon}} \qquad (6)$$

and

$$m_b = \left(\frac{\partial ln\, M}{\partial ln\, \dot{\epsilon}} \right)_{\epsilon} \qquad (7)$$

where M is the bending moment of the beam. In a four-point test in which there is only a bending moment in the gauge length, M is equal to $Pa/2$, where P is the applied load and a the moment arm (the distance between the inner and outer knife edges). The maximum stress in the outer fibers is then given by

$$\sigma_{max} = \frac{Pa}{bh^2} (2 + n_b + m_b) \qquad (8)$$

where b and h are the depth and width of the beam, respectively.

In the present work, the flow stress was essentially independent of strain, at least to the ~ 3 percent strain limit used in this work (Figure 1, 1350°C sample). Thus, n_b was zero and the rate sensitivity could be determined from discrete rate changes on a single specimen during isothermal tests, as shown in Figure 2. These results indicated no dependence on deformation history (Figure 3).

Experimental Results and Discussion

Spriggs, Mitchell, and Vasilos [24] first reported that fine-grain $(1-2\,\mu)$ alumina exhibited considerable ductility at temperatures as low as 1500°C. Sub-

---

*The strain and strain rate can be determined directly from measured deflections using geometrical considerations.

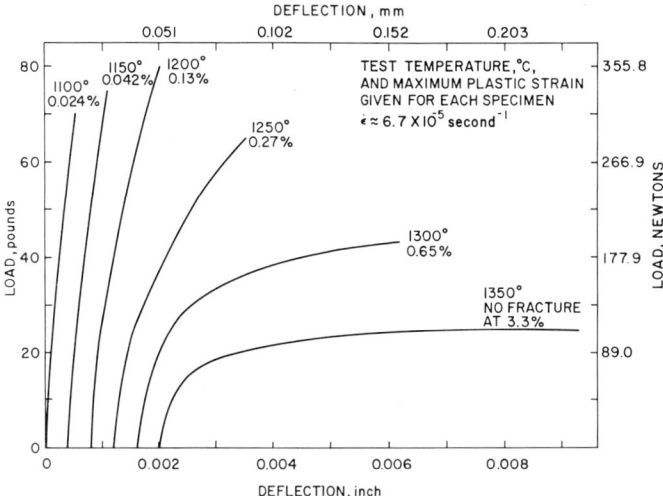

Figure 1. Load-deflection curves for a number of dense, fine-grain (∿1 μ) hot pressed alumina specimens as a function of temperature, indicating a brittle-ductile transition.

sequently, Passmore, Moschetti, and Vasilos [25] determined a brittle-to-ductile transition in 2–3 μ grain size alumina at 1350°C. The understanding of the ductility in these fine-grain materials was the impetus for the present investigation.

The change from brittle to ductile behavior is shown clearly in the load-deflection curves reproduced in Figure 1 for samples of hot pressed pure alumina

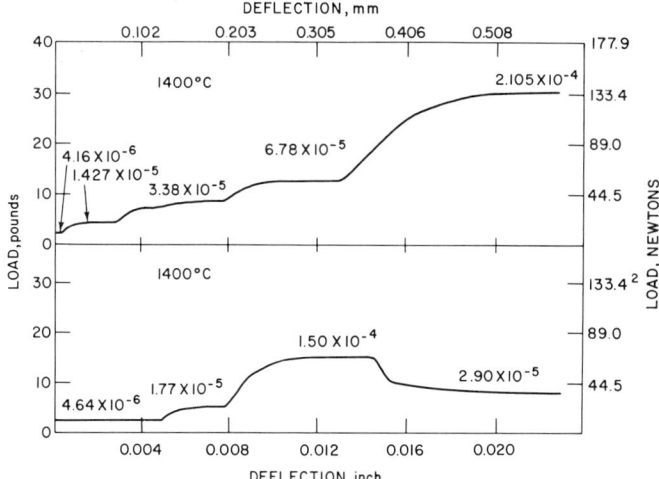

Figure 2. Load-deflection curves for dense, 1.2 μ grain size alumina plus 14% MgO sample, showing change-of-strain-rate tests used to determine the strain rate sensitivity, m. The numbers above the flat portions of the curve are the strain rate per second.

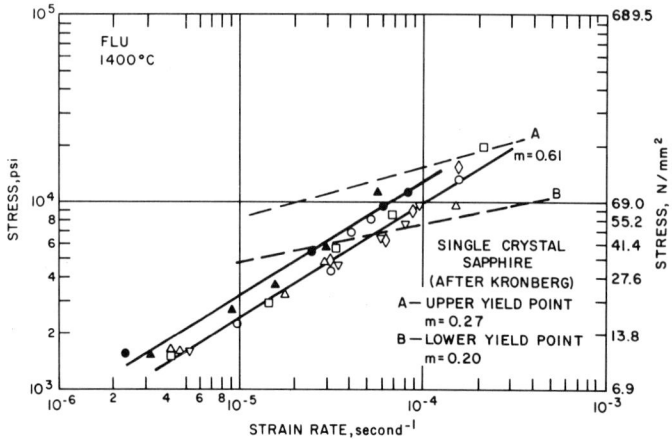

Figure 3. Stress-strain rate relationship for a number of samples identical with those shown in Figure 2. Data from Figure 2 are included in this plot. The dashed line represents the yield stress data of Kronberg [27]. The filled symbols are for slightly larger grained materials (1.3 μ). The slope of the curve, m, is the strain rate sensitivity.

(1.2 μ G.S., > 99 percent T.D.). At 1100°C, completely brittle behavior was observed, while the onset of measurable macroscopic plasticity was just detectable in the sample deformed at 1150°C. The amount of plastic strain at fracture increased with increasing temperature until, at 1350°C, the sample could not be fractured at the true strain rate employed, $\sim 6.7 \times 10^{-5}$ sec, but deformed to an outer fiber plastic strain of 3.3 percent, which was the limit of the apparatus.

Although the 1250°C and 1300°C load-deflection curves suggested an apparent work hardening, this was not so for the 1350°C curve, which exhibited uniform flatness above $\sim$ 1 percent strain. It is thought that in all cases, the samples were deforming as visco-elastic solids; at the lower temperatures, fracture occurred before the approximately exponential build-up of load to a steady state had occurred.

Further information about the deformation was obtained from the change-of-rate tests shown in Figure 2, which were used to determine σ–$\dot{\epsilon}$ relationships. Results on a number of hot pressed specimens (> 99.5% T.D.) of the composition alumina plus 1/4% MgO are shown in Figure 3, all tested at 1400°C. The open symbols were specimens cut from a billet whose average grain size was 1.2 μ, while the full symbols represent samples from a 1.3 μ billet. (Although Hewson and Kingery [26] indicated recently that the creep deformation rates for this composition were lower than for pure alumina, this has not been the case in the present investigation. For equivalent grain sizes and porosities, the ductility of pure alumina and of magnesia-doped alumina specimens were identical. For this reason, most of the research utilized magnesia-containing samples, because grain growth was severely limited, and testing at a constant grain size could be carried out over a range of temperatures.)

It can be seen from Figure 3 that a straight line with a slope of 0.61 (the slope of this log/log plot is m, the strain rate sensitivity) adequately describes the flow behavior over two orders of magnitude of strain rate. Moreover, the specimen-to-specimen reproducibility was good. Also included in this figure is the yield stress as a function of strain rate for single-crystal sapphire at 1400°C, after Kronberg [27]. It can be seen that below a strain rate of about 3×10^{-5} sec^{-1}, the polycrystals are more ductile than the single crystals. The low strain rate sensitivity of the single crystals is expected [28], inasmuch as the deformation occurred by basal slip [27].

Results of tests in the temperature range of 1300–1550°C are shown in Figure 4. It can be seen that the behavior at all temperatures was virtually the same,

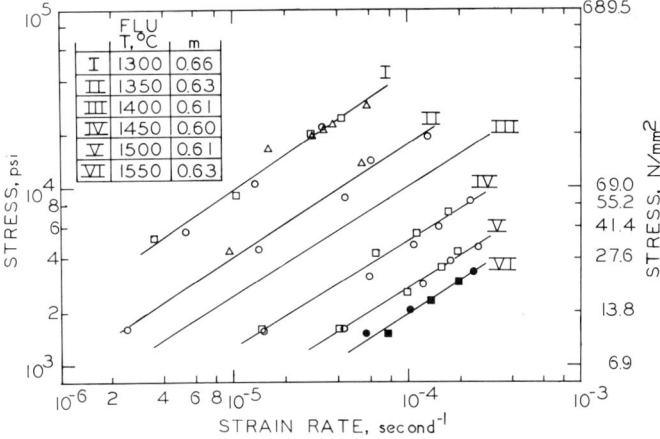

Figure 4. Stress-strain rate data as a function of temperature for samples identical to those in Figures 2 and 3. The curve marked III is the lower curve of Figure 3. The data is all of the form $\sigma \alpha \dot{\epsilon}^m$.

and the strain rate exponent of approximately 0.6 was also independent of temperature. Although the fact that the deformation could be described by $\sigma \alpha \dot{\epsilon}^m$, $m = 0.6$–0.7 was at first surprising, a review of the literature (below) shows that this is a common occurrence in fine-grain ceramics, and the manifestations of this will be further discussed below.

Before proceeding to discuss possible deformation mechanisms, it is of interest to examine the deformation behavior for somewhat larger grained polycrystals. The results for samples of an undoped hot pressed alumina, annealed to give a grain size of $\sim 4.5\,\mu$, are shown in Figure 5, and results for sintered alumina of $10\,\mu$ average G.S. are shown in Figure 6.

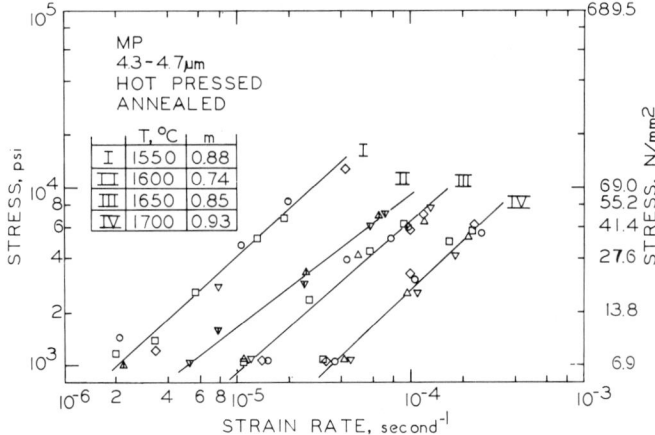

Figure 5. Stress-strain rate data as a function of temperature for a hot pressed alumina annealed so that the average grain size is ∿4.5 μ. The strain-rate sensitivity is given below each temperature.

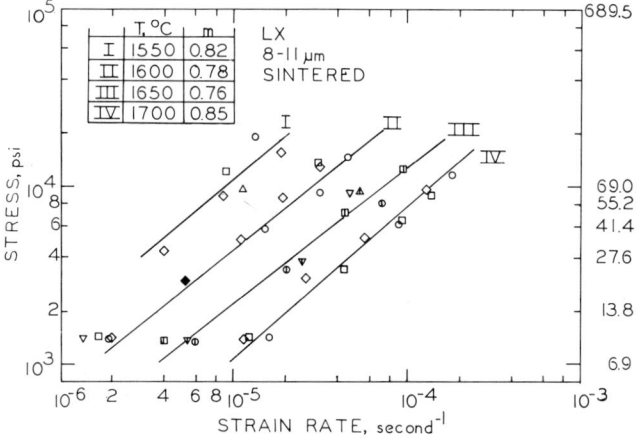

Figure 6. Stress-strain rate data for sintered alumina of average grain size ∿10 μ. The filled symbol denotes a test where the strain rate was decreased rather than increased (this was not so distinguished in Figure 3).

As expected, higher temperatures must be used for the larger grain size materials to get equivalent deformation rates. However, in all cases, non-Newtonian behavior was observed, and there was a hint that the slope of the stress-strain rate curves (i.e., m, the strain rate sensitivity) increased slightly with increasing grain size.

From the type of data already presented, it is possible to determine an "activation energy" for the deformation, and further, to get an estimate of the de-

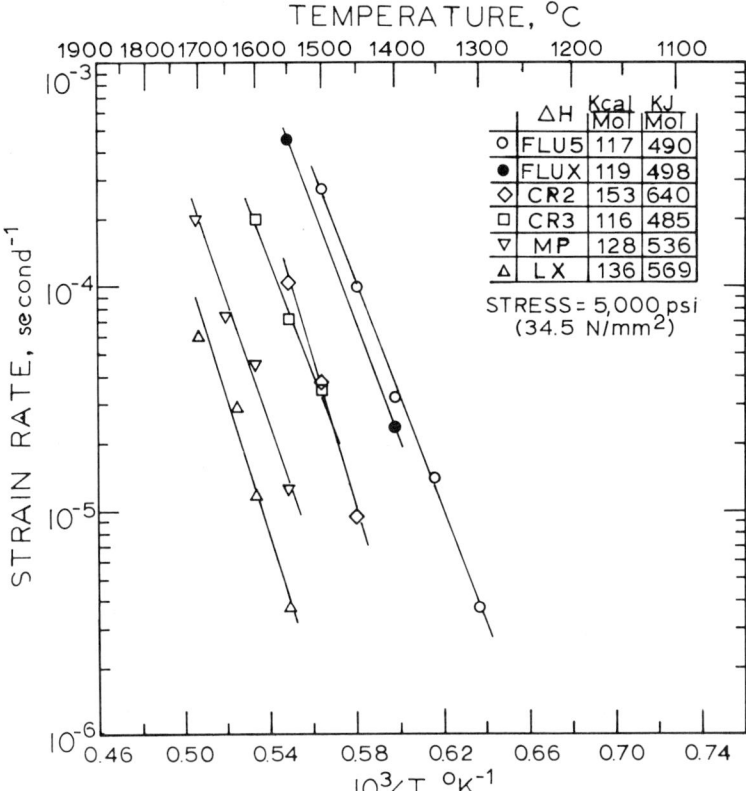

TEMPERATURE, °C

Figure 7. Strain rate (for a stress of 5,000 psi) as a function of temperature for several materials. This data is a cross-plot of Figures 3–6.

pendence of the deformation process(es) on grain size. For the present data, however, the variation in the strain rate sensitivity renders such analyses some-what arbitrary, and the values obtained will depend on the stress level at which strain rates will be compared as a function of temperature and grain size. In this work, we have arbitrarily chosen a stress of 5,000 psi (34.5 MN/m^2), and the strain rate corresponding to this stress is plotted as a function of temperature for a number of samples in Figure 7.* It can be seen that the activation energy falls in the range of 116–136 kcal/mole (485–569 kj/mole), which is in good agree-ment with previous work on deformation of polycrystalline alumina. However, until more is known about the deformation, it is fruitless to attempt to compare

*Data marked CR–2 and CR–3 are for sintered alumina of reported grain size of 2–3 μ. Discussion of this data has not been included in the paper because the specimens were ap-parently atypical, having a duplex microstructure. It is planned to test additional samples of this material.

this value with those of other studies of aluminum oxide—sintering or self-diffusion, for example.

The grain size dependence can be obtained from the slope of a log/log plot of strain rate (at a single temperature) versus grain size. (This is thought to be preferable to the Zener-Hollomon approach [29], because of the variable "activation energies" and m values observed in the range of grain sizes that were available.) The Lucalox and the hot pressed and annealed alumina data are for temperatures of $1550°C$ and above, while data for the $1.2\,\mu$ hot pressed alumina is for $1500°C$ and below. Therefore, the strain rate versus l/T plots were extrapolated to $1530°$, and plotted versus grain size (Figure 8). A straight line with a slope of -2.5 adequately describes this data.

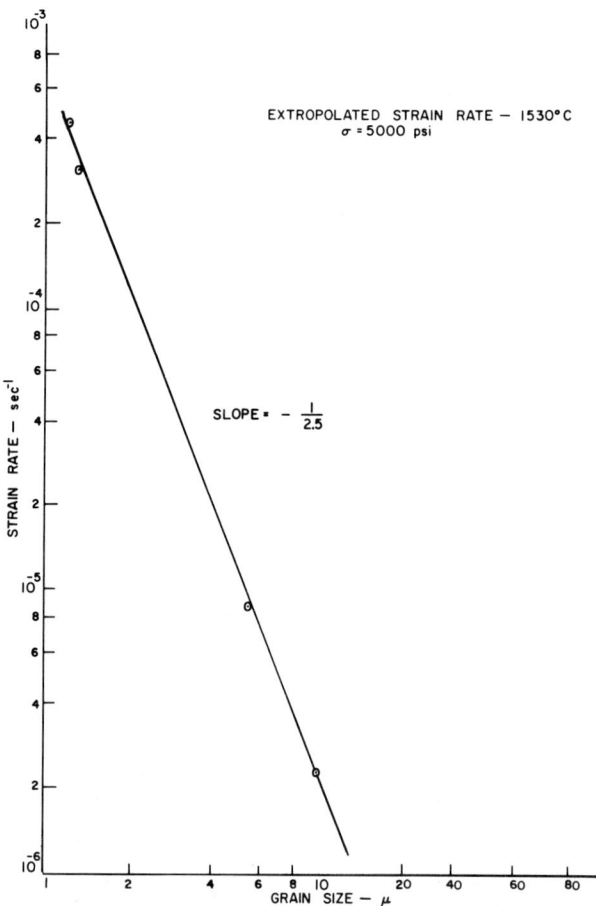

Figure 8. Strain rate (extrapolated from Figure 7 to $1530°C$) at 5,000 psi as a function of temperature. The data obeys the relationship $\dot{\epsilon} \propto G.S.^{-2.5}$.

These results, particularly the variable strain rate sensitivity, m, and the non-integral exponent for the grain size dependence, indicate that more than one deformation process must be contributing over the range of stresses, temperatures, and grain sizes employed. In particular, it appears that a transition is occurring from a viscous deformation mechanism—$m = 1$, equations (1)–(4)—at larger grain sizes to a deformation mechanism of low strain rate sensitivity as the grain size is reduced. It is pertinent, therefore, to consider the possible low m mechanisms:

1. Deformation by dislocation glide (slip) is the best known of the possible mechanisms, and Weertman has considered several possible rate-controlling mechanisms [28] which predict m values between 0.2 and 0.4. Steady state creep of metals at moderate stresses often follows a creep law, $\sigma \propto \dot{\epsilon}^{0.2}$, and have usually been interpreted using Weertman's models [30]. However, Figure 3 indicated that polycrystals of Al_2O_3 can flow at stresses well below the yield stress of single crystals. Furthermore, basal slip is the only slip system which is believed to have sufficiently low critical resolved shear stresses to be important under the experimental conditions employed in this work; therefore, a randomly oriented polycrystal would have very little ductility from this deformation mode [2]. (A more complete discussion of possible deformation modes in alumina can be found in [31].) On the other hand, it is perhaps conceivable that stress concentrations were large enough, for example at triple points, to induce sufficient slip in favorably oriented grains to alter the creep laws. It must be stated, however, that there appears to be no justification for arguing that basal slip would become more important as the grain size is decreased.

2. As has been mentioned, strain rate sensitivities of 0.33 and 0.25 for GBS have been reported for polycrystalline Pb and Mg, respectively [20,21]. Furthermore, GBS can be expected to depend linearly on the area of grain boundaries present; this leads to the often quoted grain size dependence, $\dot{\epsilon}_{GBS} \propto G.S.^{-1}$. Thus, GBS should be more important at the finer grain sizes. However, the assumption that GBS is the low m process important in the present work is contradictory to the strongly held, but apparently incorrect notion [18] that GBS is not a primary deformation process but simply a means of accommodating discontinuities in volume strains across grain boundaries. In view of the argument [19] that GBS depends on the resolved shear stress in the boundary plane, and that GBS can contribute up to 75 percent of the total strain in some cases [21], it seems clear that GBS can be more than a manifestation of deformation processes within the bulk of the grains.

How does GBS occur? To quote Bell and Langdon [21]: "It is clear that GBS is not a simple process to be thought of in terms of the relative motion of two, perfectly flat, smooth surfaces. Real grain boundaries are rugged, both on an atomic and on a microscopic scale, and sliding must involve migration and/or slip." This view that sliding is controlled by deformation of grain boundary irregularities is considered [18] the most plausible of all theories of macroscopic sliding, and considerable evidence exists for its correctness [18].

3. As pointed out previously, deformation twinning is rarely thought of as contributing to polycrystalline ductility, and it is difficult to predict either a strain rate or a grain size dependence for the process. It is doubtless possible to produce phenomenological arguments to explain either low or high m values, or marked or little grain size dependence. It seems best for the present time, therefore, to leave these as unanswered questions.

The three deformation processes considered—slip, GBS, and twinning—all give rise to distinctive microstructures or sub-structures after deformation. One of the samples with the lowest value of m (shown in Figure 3) was studied using transmission electron-microscopy (foils were taken near the tension surface of the bent beam, using the procedure described at this conference by Miss Tighe[*]) and compared with the structure of an undeformed specimen taken from the same hot pressed billet. It is convenient to describe first the observations on the undeformed specimens.

The main feature of interest was a notable lack of dislocations within the bulk of individual grains, but a very large number of low-angle boundaries, whose individual dislocations could be discerned when specimens were tilted into the correct diffraction contrast[*] [32]. In addition, these particular specimens had a slight amount of residual porosity, always less than 1 percent. Pores were generally hexagonal in appearance (Figure 9)[†], and when intergranular, were free of any further substructure. However, intragranular pores usually had dislocations associated with them (Figure 9). The ability of pores to pin dislocations appears to be a general phenomenon and is discussed elsewhere [31].

The structure of the deformed specimen was generally similar. There was no evidence to suggest that either the dislocations which made up the resolvable low-angle boundaries, or the dislocations associated with pores, in any way contributed to the deformation.

On the other hand, evidence for grain boundary sliding is shown in Figures 9, 10, and 11. The hexagonal pores pinning dislocations in Figure 9 have already been discussed. However, the smaller, triangular voids at triple points (arrowed) represent almost classical evidence of GBS. They were present in large numbers in the deformed samples, but were completely absent in undeformed samples. Further evidence for GBS can be found in Figure 10. where a grain boundary triple point (at A) has been displaced. This displacement of the triple point is probably associated with the small void which has formed at B.

What are the possible mechanisms of GBS? The dislocations in the grain boundaries of Figure 11, adjacent to two voids, suggest that the climb and/or glide of dislocations may be important, and the fine structure in boundary C of

---

[*]N. J. Tighe, "Microstructure of Fine-Grain Ceramics," this volume.

[†]Figure 9 was taken from a deformed specimen. However, the hexagonal pores resulting from incomplete densification were identical in both undeformed and deformed specimens, and it was thought unnecessary to include a micrograph of an undeformed specimen.

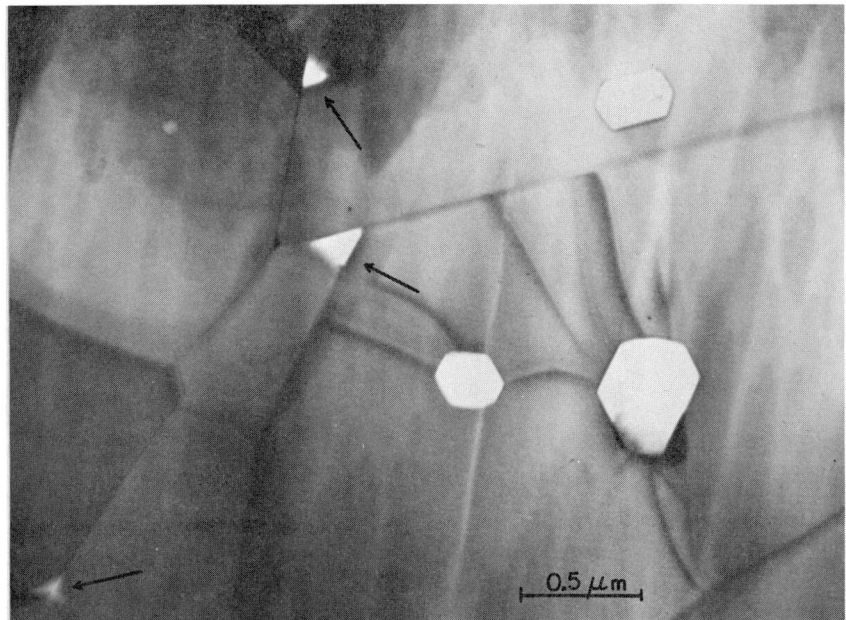

Figure 9. Transmission electron micrograph of one of the deformed specimens shown in Figure 3. The triple-point voids (arrowed) are clear evidence for grain boundary sliding (GBS). The other features of the micrograph are discussed in the text.

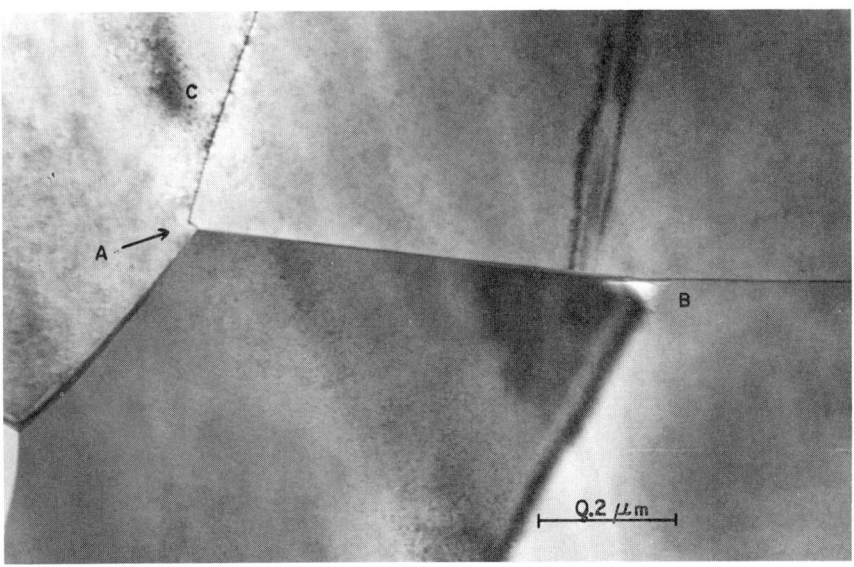

Figure 10. Displaced triple point at A and void at B are further evidence for GBS. The fine structure in grain boundary C may be grain boundary dislocations. (Transmission electron micrograph.)

353

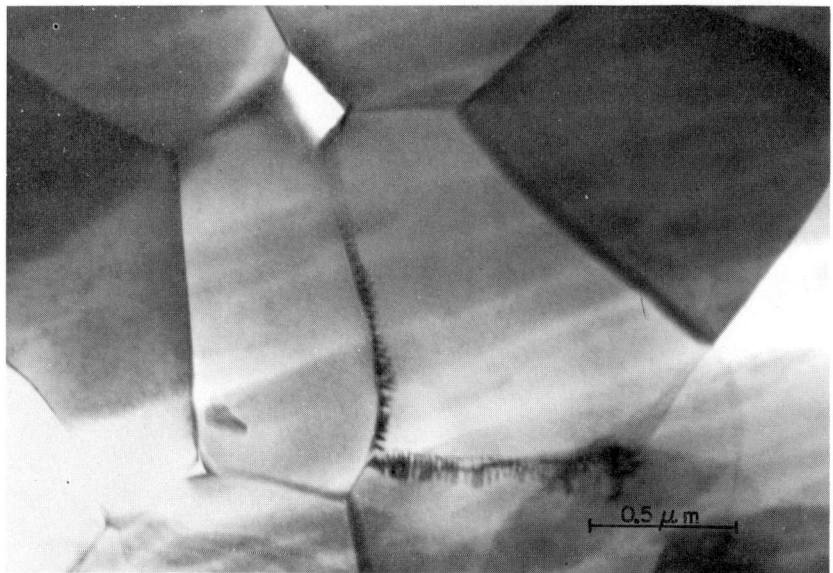

Figure 11. Grain boundary dislocations associated with triple-point voids. (Transmission electron micrograph.)

Figure 10 supports this notion. However, it should be pointed out that most boundaries adjacent to triple-point voids were featureless, and such grain boundary dislocations were difficult to find under the electron microscope.

It is also thought that grain boundary migration is an important mechanism of GBS. This notion is supported by the displaced triple point shown in Figure 10, which must represent a very high energy region and would undoubtedly soon be eliminated by grain boundary migration. This is probably the reason why it was also difficult to find additional observations of such displaced triple points.

How is deformation due to GBS accommodated within the individual grains? In the first place, the copious number of voids at triple points is evidence that the grains could not deform sufficiently to accommodate GBS. Thus, no evidence for massive slip within grains could be detected. (This was not, however, particularly surprising, in view of the relatively low temperatures and stresses employed.) On the other hand, Figure 12 shows two examples of deformation twinning. Although a number of such twinned grains were found in examining the deformed sample, they were absent in the undeformed sample. Analysis of electron diffraction patterns showed unambiguously that these were twins on the rhombohedral system, similar to those recently described by Heuer [8]. These twins were usually associated with triple-point voids, as can be seen in the right-hand micrograph of Figure 12, and undoubtedly formed in response to the high stresses generated by the GBS. In the majority of cases, however, the

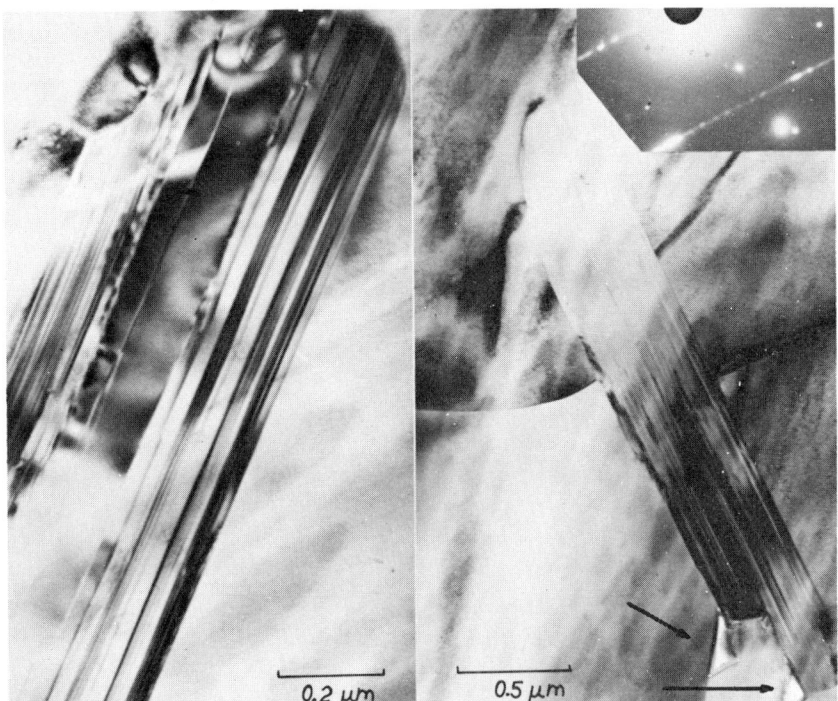

Figure 12. Deformation twins, also associated with triple-point voids (arrowed). The electron diffraction inset showed unambiguously that these were rhombohedral twins.

diffusional deformation within grains must have increased in response to these high stresses sufficiently to avoid twinning.

Thus, it may be concluded from the electron-microscopy that of the possible low m deformation process, grain boundary sliding by grain boundary dislocations and boundary migration is dominant, but is supplemented by deformation twinning. What can be said of the diffusional portion ($m = 1$) of the deformation, which is "in parallel" with the GBS in this transitional region of grain size? The grain size dependence of the deformation ($\dot{\epsilon} \propto$ G.S.$^{-2.5}$) is intermediate between Coble or Gifkins creep—$\dot{\epsilon} \propto$ G.S.$^{-3}$, equations (2) or (4)*— and Nabarro–Herring creep—$\dot{\epsilon} \propto$ G.S.$^{-2}$, equation (1). If the diffusional deformation dominates the grain size dependence, a transition from a G.S.$^{-2}$ to a G.S.$^{-3}$ dependence is expected as the grain size is decreased. This will occur, if equations (1) and (2) are applicable, when

$$3.5 \, w \, D_b \gg \text{(G.S.)} \, D_l \tag{9}$$

---

*It should be pointed out that when equation (4), describing sliding rate, is modified by equation (5) to give strain rate, the results is almost indistinguishable from equation (2). (The constant, 47.1, is changed to 40.)

Equations (3) and (4), on the other hand, describe a situation where sliding must precede accommodation—the processes cannot occur in parallel as for Coble and Nabarro–Herring creep. In this case, accommodation, equation (4), will become more difficult as the grain size increases and will be rate-controlling above a critical grain size given by

$$\frac{(G.S.^2)}{Lx} = 20\,w \tag{10}$$

where x is a factor introduced by Gifkins [12] to account for pauses in the rate of sliding other than that due to triple-point accommodation.* Thus, $\dot{\epsilon} \propto G.S.^{-3}$ will only be observed in a certain range of grain sizes, and inequality (9) will still describe the transition to a $G.S.^{-2}$ process at larger grain sizes. However, the triple-point voids observed in the present work suggested that diffusional accommodation, as described by equation (4), could not occur sufficiently rapidly for the GBS rate.

On the other hand, the $G.S.^{-2.5}$ dependence may arise from a transition from a Newtonian-viscous grain boundary diffusion mechanism ($\dot{\epsilon} \propto G.S.^{-3}$) to a non-Newtonian grain boundary sliding process ($\dot{\epsilon} \propto G.S.^{-1}$). Further support for this notion will be given in the next section, where "diffusion coefficients" are computed from the deformation data and compared with the literature.

This latter interpretation, however, leads immediately to a paradox. If two independent deformation processes are possible, sliding with $\dot{\epsilon} \propto G.S.^{-1}$ and diffusional deformation with $G.S.^{-3}$, then one would expect the transition to GBS to occur with *increasing* rather than with *decreasing* grain size. One possible explanation is that GBS is a more complex function of the grain size than the relationship, $\dot{\epsilon} \propto G.S.^{-1}$. Some evidence that this may be so is available from studies of creep of pure metals. Barrett, Lytton, and Sherby [33], for example, found that the creep rate of pure copper for constant stress and temperature, was constant for grain sizes > 0.1 mm, but increased with finer grain sizes. In addition, γ, the fraction of total strain due to GBS, was greater at the finer grain sizes, which was similar to the results quoted above on Magnox [21]. These results were attributed to the increasing importance of GBS with decreasing grain sizes below 0.1 mm; at larger grain sizes, GBS was thought to make a constant, but smaller, contribution to the total strain. This situation might come about if coarser grained materials gave rise to a geometrical restriction to GBS, which is not so severe in finer grained materials. GBS in bicrystals is of course not subject to such restrictions.

Comparison With the Literature

A perusal of the relevant literature on deformation (including creep) of fine-grain single-phase ceramic polycrystals reveals that the occurrence of non-

*This was felt to be necessary because equation (3) predicted sliding rates ~ 100 times faster than observed in the one case in which it was applied [20].

Newtonian behavior is fairly common and is, in fact, the one factor of commonality among all the studies which will be reviewed. Indeed, the cases where pure Newtonian-diffusional creep has been observed are rare.

Folweiler [34] was the first investigator who studied the effects of grain size, strain rate, and temperature on the plastic deformation of single-phase, dense polycrystalline alumina. Although his results were generally for material whose grain size was larger than 10 μ, he showed that the conditions of Nabarro–Herring creep were fairly well satisfied, namely, that the strain rate was linearly proportional to the stress and inversely proportional to the grain size squared. Analysis of Folweiler's data by the present authors, however, shows m values for 7 μ and 13 μ grain size materials between 0.86 and 0.94 for samples tested between 1500°C and 1700°C.

Warshaw and Norton [35], also working with polycrystalline alumina, found a transition in m in going from small to large grain size. Analysis of their data indicates m values of 0.6 for a 3 μ grain size material, m's of 0.86 for samples with grain size between 7 μ and 15 μ, and an m value of 0.33 for 50 μ to 100 μ grain size specimens. The values of 0.6 and 0.86 were interpreted as agreeing with Nabarro–Herring creep, while the value of 0.3 was interpreted as indicating creep controlled by dislocation climb. (Subsequently, Coble and Guerrard [36] suggested that the small value of m in the large-grain samples investigated by Warshaw and Norton resulted from extensive intergranular separation. This was also observed in their own work, and had previously been reported by Folweiler.) Finally, Passmore and Vasilos [37] found a transition in m from 1.0 to 0.5 when the stress was increased above 2,000 psi for 2 μ grain size material. It is now suggested that in all these cases, the values of $m < 1$ for fine-grain polycrystals arises from non-Newtonian GBS.†

It is common for workers in this field to compare their data by calculating diffusion coefficients, assuming equation (1) is applicable. Figure 13, after Coble and Guerrard [36], shows data from the present investigation,* compared with Folweiler's and Warshaw and Norton's data, as well as literature values [39, 40] for self-diffusion coefficients of aluminum and oxygen. The agreement of our data with Folweiler's is striking, but all "diffusion coefficients" calculated from deformation data are orders of magnitude higher than measured self-diffusion coefficients. This is to be expected, because the present interpretation of the deformation behavior

$$\sigma \propto \dot{\epsilon}^m, 0.6 < m < 0.9$$

implies that only a portion of the deformation can be attributed to the diffusional mechanisms. Thus, the computed diffusion coefficients must be larger

---

*The data marked "press forged" are for hot worked alumina, which will be published elsewhere [38].

†Fryer and Roberts [53] have already invoked GBS to explain m values of 0.53 and 0.33 found in two different commercial alumina bodies of considerable porosity ($\sim$ 8 percent porosity for the case of $m = 0.53$, and $\sim$ 34 percent where $m = 0.33$). The present argument is that GBS is also important in dense bodies.

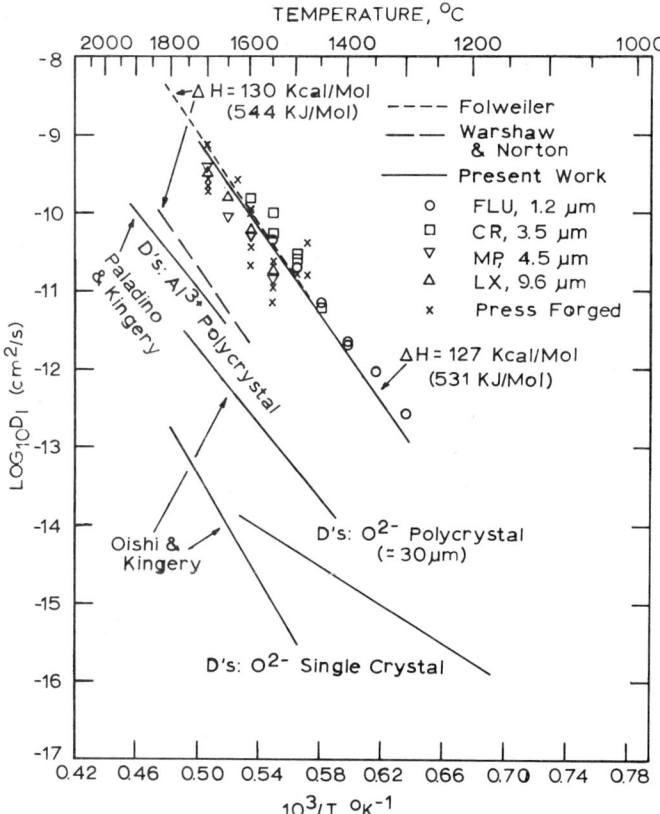

Figure 13. Plot of diffusion coefficients versus $1/T$ for several alumina polycrystals, calculated from deformation data using equation (1). The calculated D's must be larger than measured D's, inasmuch as $\sigma \propto \dot{\epsilon}^m$, $m < 1$. References [34], [35], [39], and [40].

than the actual measured diffusion data; equation (1) is transformed to

$$D_{1\,meas} < D_{1\,computed} = \frac{\dot{\epsilon}\,kT\,(\text{G.S.})^2}{13.3\,\Omega\,\sigma} \qquad (11)$$

A similar inequality exists from equation (2)

$$wD_{b_{meas}} < wD_{b_{computed}} = \frac{\epsilon\,kT(\text{G.S.})^3}{47.1\,\Omega\,\sigma} \qquad (12)$$

The wD_b product, calculated from equation (2), is shown in Figure 14 and compared with similar data calculated by Mistler [41] from secondary grain

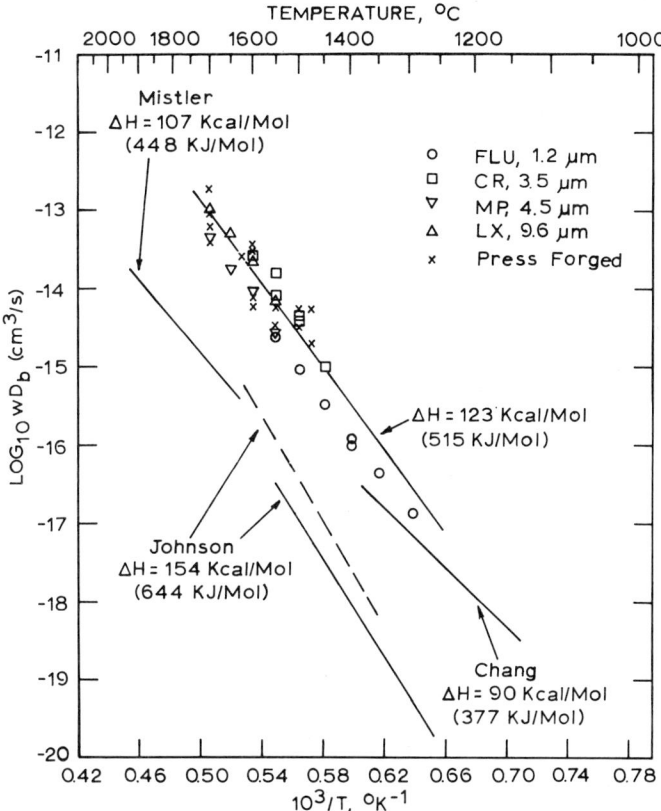

Figure 14. Product of wD_b plotted versus $1/T$. wD_b was calculated using equation (2), and is compared with secondary grain growth data (Mistler [41]), data from initial sintering kinetics (Johnson [42]), and creep data (Chang [43]). Compare with Figure 13.

growth studies, by Johnson [42] from initial sintering kinetics,* and by Chang [43] from creep data.

The discrepancies between inferred wD_b data [41, 42] and those calculated from the deformation data is rather less than for the lattice diffusion data (Figure 13), and this may be further support for arguing that the diffusional deformation occurred by grain boundary diffusion.

Data for fine-grain MgO, UO_{2+x}, BeO, and ThO_2 are collected in Table I. Although experimental conditions varied widely in these cited papers, the

*Johnson (private communication) now believes that his derived wD_b data, as well as the activation energies shown in Figure 14, are too high, in that they were computed neglecting any effect of surface diffusion during sintering. It is now felt that surface diffusion does make a significant contribution in the initial sintering of alumina.

TABLE I

Reference	Grain Size (μ)	Temp. (°C)	m values	Comments
			MgO	
Vasilos, Mitchell, and Spriggs [44]	1.0–3.0	1180–1260	0.59	4-point bending
Passmore, Duff, and Vasilos [45]	2.0–5.5 5.5–20	1007–1527 1007–1527	0.67 1.0	4-point bending—dead weight loading. Dislocation substructure and intergranular porosity aligned along compression axis present after deformation.
Copley and Pask [4]	<15	1400		Initially transparent specimens became opaque after compressive deformation due to grain boundary shearing.
			UO$_{2+x}$	
Scott, Hall, and Williams [46]	2–10 ($x = 0.16$)	800–1000		$\dot{\epsilon} = A \sinh \dfrac{\sigma}{\sigma_0} \exp \dfrac{Q}{RT}$ Samples ~98% T.D. 3-point bending.
Armstrong, Irvine, and Martinson [47]	4–40 ($x = 0$)	1250–1400	0.77	m decreased for $\sigma > 10,000$ psi. Samples 93–98% T.D. 3-point bending. Grain boundary sliding observed.
Armstrong and Irvine [48]	6 ($x > 0.02$)	975–1400	1.0 ($0.02 \leq x < 0.08$) <1 ($x > 0.08$)	Grain boundary sliding observed. 3-point bending. Samples 96% T.D.
			BeO	
Fryxwell and Chandler [49]	<10	1200	<1	Compressive creep testing on samples of variable porosity
			ThO$_2$	
Poteat and Just [50]	10	1430 1535 1600 1666 1790	0.96 0.84 0.76 0.66 0.63	Compression testing of samples with ~2.5% porosity. Intergranular aligned along compression axis noted after deformation. Grain boundary shearing observed.

evidence for m values < 1 is profuse and may be taken as evidence for non-Newtonian GBS in each of these cases. The possibility of GBS has been recognized by several of the cited authors [46, 47, 48, 50].

Effect of Grain Growth

This review would be incomplete without considering the effects of grain growth on plastic deformation, inasmuch as such grain growth may be quite prevalent in fine-grain polycrystals. (Care was taken in our own experiments to exclude the possibility of grain growth, e.g., by the addition of MgO as a grain growth inhibitor or by annealing at elevated temperatures.) Tagai and Zisner [51] were apparently the first to modify equations (1) and (2) to include the effects of grain growth. Isothermal grain growth can usually be described by

$$\text{(G.S.)}^n - \text{(G.S.}_0)^n \, \alpha \, t \tag{13}$$

where t is time, n a constant (usually between 2 and 3), and G.S. and G.S.$_0$ are the grain sizes at time t and time zero respectively.

For G.S.$_0 \ll$ G.S.,

$$\text{G.S.} \, \alpha \, t^{1/n}$$

and

$$\dot{\epsilon} \, \alpha \, \frac{\sigma}{t^q} \tag{14}$$

where q can have values from $3/2$ to $2/3$, depending on the value of n and the grain size dependence. Tagai and Zisner used an analytical graphical method to correct their creep data for the effects of grain growth. Similarly, Gordon and co-workers [52] considered (apparently independently) "time hardening" during isothermal creep and obtained an expression for the creep rate at constant load

$$\dot{\epsilon} = \frac{C_1}{t + C_2} \tag{15}$$

in which $q = \frac{p}{n}$. p is defined by equation (16)

$$\dot{\epsilon} = \frac{k_1}{\text{(G.S.)}^P} \tag{16}$$

and n is defined by equation (13). The constants C_1 and C_2 are given by

$$C_1 = \frac{k_1}{k_n^n}; C_2 = \frac{\text{G.S.}_o^n}{k_n} \tag{17}$$

where k_n is the proportionally constant appropriate for equation (13). Their approach is to find constants C_1 and C_2 which best fit the data, and in this way evaluate the creep process, using measured grain growth kinetics as a basis.

It may be that the transient creep behavior shown in the data of Passmore and co-workers [37, 45] resulted from grain growth during testing.

Summary and Conclusions

Dense, single-phase polycrystalline alumina was deformed in bending from 1100–1700°C, and at strain rates from 2×10^{-6} to 3×10^{-4} per second. The load-deflection curves showed no work hardening, at least up to ~ 3 percent outer fiber strain, and the data could be expressed as

$$\sigma \alpha \dot{\epsilon}^m, 0.6 < m < 0.9$$

Although the variation in the strain rate sensitivity, m, made rigorous interpretation difficult, the deformation was very sensitive to temperature, with an apparent activation energy of 126 ± 10 kcal/mole.

A range of grain sizes, $\sim 1\ \mu$ to $\sim 10\ \mu$, was studied. The deformation was very sensitive to the grain size, varying approximately as G.S.$^{-2.5}$. It appeared that m decreased with decreasing grain size.

Transmission electron-microscopy of deformed samples disclosed extensive evidence for grain boundary sliding, which appeared not to be a Newtonian-viscous process, in addition to some rhombohedral twinning. There was no evidence for bulk slip.

It was suggested that in the range of grain sizes employed, a transition was occurring from a diffusional ($m = 1$) deformation process, possibly involving enhanced grain boundary diffusion, to a grain boundary sliding process involving grain boundary dislocations and boundary migration. The individual grains were unable to deform sufficiently to keep up with the GBS, and voids at triple points formed during deformation.

A brief review of the literature indicated that non-Newtonian GBS may also be important in MgO, UO_2, BeO, and ThO_2.

Finally, the effect of grain growth during deformation was considered. This can lead to an apparent "hardening" with time, and may be responsible for the transient creep behavior occasionally observed.

This study has indicated that non-Newtonian GBS may be important in the deformation of fine-grain ceramic polycrystals. However, little is known about such a process, and knowledge about the rate-controlling mechanism of GBS, as well as the strain-rate and grain size sensitivities of the process, is very much needed.

Acknowledgment

The portions of this work performed at Avco were supported by the U.S. Naval Air Systems Command under Contract N000 19–67–C–0336. The authors thank the Contract Monitor, Mr. Charles Bersch, and Dr. T. Vasilos for their support and encouragement. The authors also acknowledge useful discussions with Dr. W. H. Rhodes and Dr. R. C. Gifkins, a critical reading of the manuscript by Professor A. R. Cooper, and thank Dr. J. S. Nadeau for allowing his results to be quoted prior to publication.

References

1. Gilman, J. J., "Monocrystals in Mechanical Technology," *Trans. ASM*, 59 (1966), 596.
2. Groves, G. W. and Kelly, A., "Independent Slip Systems in Crystals," *Phil. Mag.*, 8 (1963), 877.
3. Day, R. B. and Stokes, R. J., "Mechanical Behavior of MgO at High Temperatures," *J. Am. Ceram. Soc.*, 47 (1964), 493. See also: *ibid.*, "Effect of Crystal Orientation on the Mechanical Behavior of MgO at High Temperatures," *J. Am. Ceram. Soc.*, 49 (1966), 72.
4. Copley. S. M. and Pask, J. A., "Deformation of Polycrystalline MgO at Elevated Temperatures," *J. Am. Ceram. Soc.*, 48 (1965), 636.
5. Nadeau, J. S., "The Strength of Oxygen-rich UO_2 at High Temperatures," to be published in *J. Am. Ceram. Soc.*
6. Crussard, C. and Friedel, J., *Creep and Fracture of Metals*, London, H. M. Stationery Office (1956), 243.
7. Ishida, Y. and Henderson-Brown, N., "Dislocations in Grain Boundaries and Grain Boundary Sliding," *Acta Met.*, 15 (1967), 857.
8. Heuer, A. H., "Deformation Twinning in Corundum," *Phil. Mag.*, 3 (1966), 379.
9. Stofel, E. and Conrad, H., "Fracture and Twinning in Sapphire (α-Al_2O_3) Crystals," *Trans. AIME*, 227 (1963), 1053. See also: Conrad, H., Janowski, J. and Stofel, E., "Additional Observations on Twinning in Sapphire (α-Al_2O_3) During Compression," *ibid.*, 2333 (1965), 255.
10. Nabarro, F. R. N., *Report of a Conference on the Strength of Solids*, Physical Society, London (1948), 75. See also: Herring, C., "Diffusional Viscosity of a Polycrystalline Solid," *J. Appl. Phys.*, 21 (1950), 437.
11. Coble, R. L., "Model for Boundary Diffusion Controlled Creep," *ibid.*, 34 (1963), 1679.
12. Gifkins, R. C., "Diffusional Creep Mechanisms," *J. Am. Ceram. Soc.*, 51 (1968), 69.
13. Ryan, H. F. and Suiter, J. W., "Grain Boundary Topography in Tungsten," *Phil. Mag.*, 10 (1964), 727.
14. Hensler, J. H. and Cullen, G. V., "Grain Shape Change During Creep in MgO," *J. Am. Ceram. Soc.*, 50 (1967), 584.
15. Lifshitz, I. M., "On the Theory of Diffusion-Viscous Flow of Polycrystalline Bodies," *Soviet Physics JETP*, 17 (1963), 909.
16. Gibbs, G. B., "The Role of Grain-Boundary Sliding in High-Temperature Creep," *Mat. Sci. and Eng.*, 2 (1967–68), 269.

17. Turnbaugh, J. E. and Norton, F. H., "Low-Frequency Grain-Boundary Relaxation in Alumina," *J. Am. Ceram. Soc.,* 51 (1968), 344.
18. Stevens, R. N., "Grain Boundary Sliding in Metals," *Met. Rev.,* 11 (1966), 129.
19. Gifkins, R. C., Gittins, A., Bell, R. L. and Langdon, T. G., "The Dependence of Grain Boundary Sliding on Shear Stress," *J. Mat. Sci.,* 3 (1968), 306.
20. Gifkins, R. C. and Snowden, K. U., "The Stress Sensitivity of Creep of Pb at Low Stresses," *Trans. AIME,* 239 (1967), 910.
21. Bell, R. L. and Langdon, T. G., "An Investigation of Grain-Boundary Sliding During Creep," *J. Mat. Sci.,* 2 (1967), 313.
22. Alden, T. H., "The Origin of Superplasticity in the Sn–5% Bi Alloy," *Acta Met.,* 15 (1967), 469. See also: Alden, T. H., *J. of Metals,* 20 (1968), 39A.
23. Heuer, A. H. and Cannon, R. M., "Plastic Deformation of Fine-Grained Aluminum Oxide," presented at Symposium on Mechanical Testing Procedures for Brittle Materials, March 28–30, 1967, IIT Research Inst., Chicago, Ill., to be published.
24. Spriggs, R. M., Mitchell, J. B. and Vasilos, T., "Mechanical Properties of Pure, Dense Al_2O_3 as a Function of Temperature and Grain Size," *J. Am. Ceram. Soc.,* 47 (1964), 323.
25. Passmore, E. M., Moschetti, A. and Vasilos, T., "Brittle-Ductile Transition in Polycrystalline Al_2O_3," *Phil. Mag.,* 13 (1966), 1157.
26. Hewson, C. W. and Kingery, W. D., "Effect of MgO and Mg TiO_3 Doping on Diffusion Controlled Creep of Polycrystalline Al_2O_3," *J. Am. Ceram. Soc.,* 50 (1967), 218.
27. Kronberg, M. L., "Dynamical Flow Properties of Single Crystals of Sapphire, I," *J. Am. Ceram. Soc.,* 45 (1962), 274.
28. Weertman, J., "Theory of Steady State Creep Based on Dislocation Climb," *J. Appl. Phys.,* 26 (1955), 1213. See also: *ibid.,* "Steady State Creep Through Dislocation Climb," *J. Appl. Phys.,* 28 (1957), 362; and *ibid.,* "Steady State Creep of Crystals," *J. Appl. Phys.,* 28 (1957), 1185.
29. Zener, C. and Holloman, J. H., "Plastic Flow and Rupture of Metals," *Trans. ASM,* 33 (1944), 163. See also: *ibid.,* "Effect of Strain Rate Upon Plastic Flow of Steel," *J. Appl. Phys.,* 15 (1944), 22.
30. Sherby, O. D and Burke, P. M., "Mechanical Behavior of Crystalline Solids at Elevated Temperatures," *Prog. Mat. Sci.,* 13 (1967), 325.
31. Heuer, A. H., Sellers, D. and Rhodes, W. H., "Hot Working in Aluminum Oxide: Primary Recrystallization and Texture," to be published. See also: Heuer A. H., "Plastic Deformation in Polycrystalline Alumina," *Proc. Brit. Ceram. Soc.*, 15, in press.
32. Tighe, N. J. and Heuer, A. H., "Substructure of Hot-Pressed Polycrystalline Al_2O_3," *Bull. Am. Ceram. Soc.,* 47 (1968), 349.
33. Barrett, C. R., Lytton, J. L. and Sherby, D., "Effect of Grain Size and Annealing Treatment on Steady State Creep of Copper," *Trans. AIME,* 239 (1967), 170.
34. Folweiler, R. C., "Creep Behavior of Pore-Free Polycrystalline Aluminum Oxide," *J. Appl. Phys.,* 32 (1961), 773.
35. Warshaw, S. I. and Norton, F. H., "Deformation Behavior of Polycrystalline Aluminum Oxide," *J. Am. Ceram. Soc.,* 45 (1962), 479.
36. Coble. R. L. and Guerard, Y. H., "Creep of Polycrystalline Aluminum Oxide," *ibid.,* 46 (1963), 353.
37. Passmore, E. M. and Vasilos, T., "Creep of Dense, Pure, Fine-Grained Aluminum Oxide," *ibid.,* 49 (1966), 166.
38. Rhodes, W. H., Sellers, D. and Heuer, A. H., to be published.
39. Oishi, Y. and Kingery, W. D., "Self-Diffusion of Oxygen in Single-Crystal and Polycrystalline Aluminum Oxide," *J. Chem. Phys.,* 33 (1960), 480.

40. Paladino, A. E. and Kingery, W. D., "Aluminum Ion Diffusion in Aluminum Oxide," *ibid.*, 37 (1962), 957.

41. Mistler, R. E., "Grain Boundary Diffusion and Boundary Migration Kinetics in Aluminum Oxide, Sodium Chloride, and Silver," Sc.D. Thesis, Massachussets Institue of Technology (1967).

42. Johnson, D. L. and Berrin, L., "Grain Boundary Diffusion in the Sintering of Oxides," in *Sintering and Related Phenomena,* G. C. Kuczynski, N. A. Hooton, and C. G. Gibbon, eds., Gordon and Breach, Science Publishers, New York (1967).

43. Chang, R., "Diffusion-Controlled Deformation and Shape Changes in Nonfissionable Ceramics," in *Proceedings of the Conference on Nuclear Application of Nonfissionable Ceramics,* A. Boltax and J. H. Handwerk, eds., American Nuclear Society, Hinsdale, Ill. (1966).

44. Vasilos, T., Mitchell, J. B. and Spriggs, R. M., "Creep of Polycrystalline Magnesia," *J. Am. Ceram. Soc.,* 47 (1964), 203.

45. Passmore, E., Duff, R. H. and Vasilos, T., "Creep of Dense, Polycrystalline Magnesia," *ibid.,* 49 (1966), 594.

46. Scott, R., Hall, A. R. and Williams, J., "The Plastic Deformation of Uranium Oxides Above 800°C," *J. Nucl. Mat.,* 1 (1959), 39.

47. Armstrong, W. M., Irvine, W. R. and Martinson, R. H., "Creep Deformation of Stoichiometric UO_2," *ibid.,* 7 (1962), 133.

48. Armstrong, W. M. and Irvine, W. R., "Creep Deformation of Non-Stoichiometric UO_2," *ibid.,* 9 (1963), 121.

49. Fryxwell, R. E. and Chandler, B. A., "Creep Strength, Expansion, and Elastic Moduli of Sintered BeO as a Function of Grain Size, Porosity, and Grain Orientation," *J. Am. Ceram. Soc.,* 47 (1964), 283.

50. Poteat, L. E. and Yust, C. E., "Creep of Polycrystalline ThO_2," *ibid.,* 49 (1966), 410.

51. Tagai, H. and Zisner, T., "High-Temperature Creep of Polycrystalline Magnesia: I, Effect of Simultaneous Grain Growth," *ibid.,* 51 (1968), 303.

52. Gordon, R. S., Terwilliger, G. R., Bowen, H. K. and Marchant, D. D., *Impurity Effects on the Creep of Polycrystalline Magnesium and Aluminum Oxides at Elevated Temperatures,* AEC AT (11-1)-1591 (December 1967). See also: Terwilliger, G. R., "Creep of Polycrystalline Magnesia," Ph.D. Thesis, University of Utah (1968); and Gordon, R. S., Private communication, September 1968.

53. Fryer, G. M. and Roberts, J. P., "Tensile Creep of Porous Polycrystalline Alumina," *Proc. Brit. Ceram. Soc.,* 6 (1969), 225.

17. Electrical and Magnetic Behavior of Ultrafine-Grain Ceramics

A. J. MOUNTVALA
IIT Research Institute
Chicago, Illinois

ABSTRACT

The electrical and magnetic properties of ceramic materials can be influenced by their microstructural characteristics, including ultrafine-grain size. Salient features of a boundary region which significantly alter or affect the electronic processes in a material are discussed. Electrical characterization of surface characteristics to show the relationship between the dielectric properties of bulk and surface layer in ultrafine particulate $BaTiO_3$ is presented. An electron-beam scanning technique capable of differentiating the electrical characteristics of grain boundaries and adjacent grains has been developed, and preliminary experimental data obtained. The implications of the role of grain boundaries in the electronic behavior of ultrafine-grain ceramics is discussed.

Introduction

It is well known, of course, that the properties of ceramic materials are dependent on microstructure. Considerable work has been done in evaluating the effects of grain size, secondary phases, and composition on the thermal, mechanical, and electrical properties of various ceramic systems. With the recent interest in preparing ultrafine particles and their fabrication into fine-grain sintered ceramics, the effect of decreasing grain size, in the micron and sub-micron region, and the contribution of interfacial or boundary layers to bulk characteristics takes on a new order of importance. Although studies in scattering characteristics of fine particles and the dependence of specific magnetic and ferroelectric properties on microstructure, in particular grain size, have been studied, the precise effects of ultrafine-grain size on the electronic behavior of ceramics have not been thoroughly understood.

As grain size gets smaller and smaller, material properties become increasingly identified with the characteristics of surfaces and boundaries in the material, due to the corresponding increase in the surface-to-volume and grain boundary-to-grain ratios in particulate and sintered materials, respectively. The increasingly

367

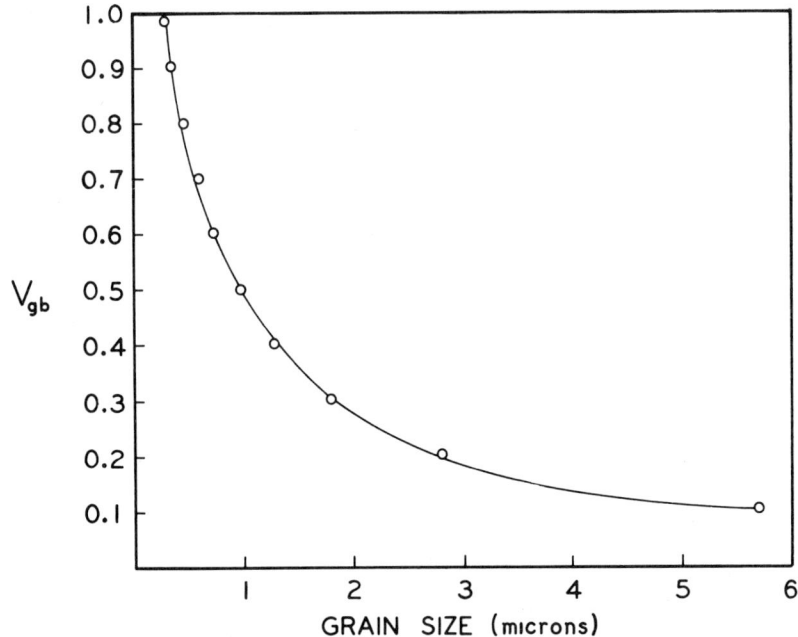

Figure 1. Volume fraction (V_{gb}) of grain boundary as a function of grain size (where: grain layer thickness = 0.1 μ).

important role of the grain boundary in ultrafine-grain ceramics—ultrafine-grain size being arbitrarily defined as less than 2-3 μ—is shown in Figure 1. This shows the relation between the volume fraction of the grain boundary phase, V_{gb}, and grain size in a material, assuming an effective grain boundary layer of 0.1 μ and considering the grains to be spherical. The curve indicates that only after some critical grain size, around 1-2 μ and less, does the value of V_{gb} (and, consequently, the contribution of the grain boundary) increase significantly. Therefore, to understand the effects of ultrafine-grain size on the magnetic and electrical properties of ceramics, it is necessary to study specimens that have average grain sizes of about 2 μ and less, for it is only in this size range that boundary effects become "property-controlling."

Let us now consider the effects of ultrafine-grain size on magnetic and dielectric behavior of specific materials.

Influence of Ultrafine-Grain Size

Magnetic Behavior

The significance of decreasing grain size in determining the magnetic properties of ceramics can be understood by considering the single domain behavior of constituent particles.

With a decrease in particle size the magnetostatic energy drops off as the cube of the particle diameter, D. The domain wall energy decreases as the square of D. Therefore, below some critical diameter, D_{cr}, no magnetic energy is gained by further dividing the particle into Weiss domains. Below D_{cr}, single domain particles would exist in which no stable Bloch walls could be formed.

The critical size, D_{cr}, has been calculated for a number of ferrite materials, on the basis of a simple model [1]. For soft magnetic materials, which have a low magnitude of wall energy and therefore a large wall thickness, D_{cr} is small. For nickel ferrite, a value of 0.08 μ has been calculated [2]. In hard magnetic materials, such as barium ferrite, the wall energy is high and thus the value of D_{cr} is relatively large; for $BaO \cdot 6Fe_2O_3$ it has been calculated to be about 1 μ [3]. Therefore, to understand the magnetic behavior of ultrafine-grain ceramics, one should consider the role of domains and their structure relative to grain diameter.

For instance, the highest coercive force, H_c, is obtained when the energy necessary to change the domain structure is the greatest. In single-domain configurations where domain rotation is predominant, this energy is very high.

The effect of decreasing particle size on the coercive force of iron, $BaO \cdot 6Fe_2O_3$ [4] and MnBi is shown in Figure 2. It is evident that the coercive

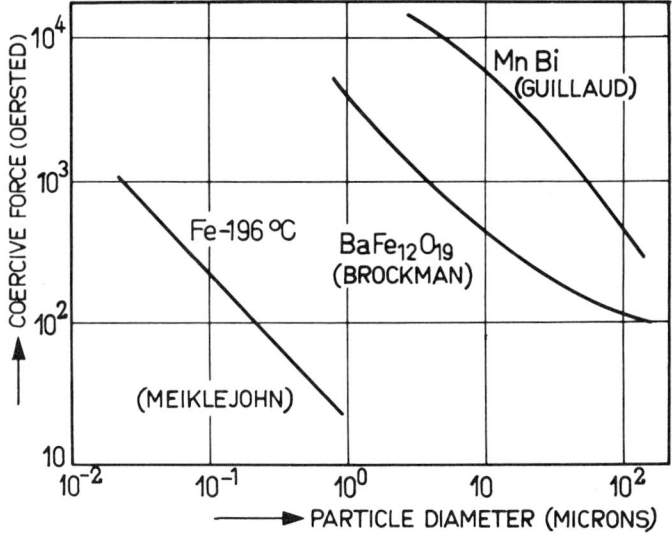

Figure 2. Coercive force H_c as a function of particle size.

force increases with a decrease in particle size. Figure 3 shows the data for MnBi [5] in greater detail. The coercive field increases significantly especially for grain sizes less than 5 μ.

Grain size also has a significant effect on the initial permeability of ferrites. Figure 4 shows the initial permeability as a function of grain size for manganese-

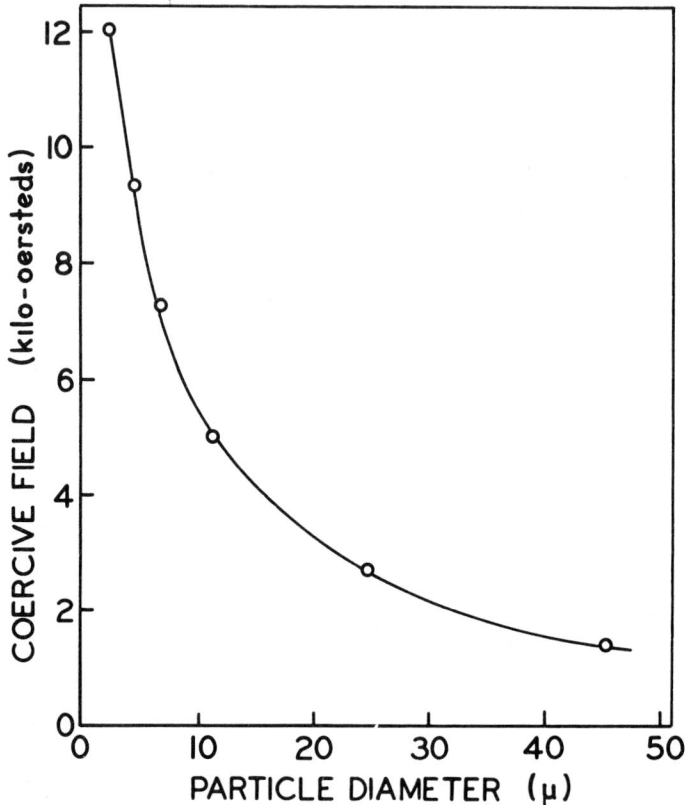

Figure 3. Effect of particle size on coercive force of MnBi (after Guillaud [5]).

zinc ferrites [6]. There are two inflections in the curve. The lower inflection, at about 5 μ, has been attributed to a change from rotational permeability in small grains (less than 5 μ) to wall displacements for larger grains. However, recent data published by Paulus [7], indicates that a substantial increase in density takes place (in the MnZn ferrite system) in the same grain size range. Thus the decreased permeability for fine-grain material may also be due to increased porosity. Also, the chemical composition in this system is very sensitive to the oxygen content during thermal processing, and this must also be borne in mind when considering the behavior of ultrafine-grain ferrite materials.

The role of surface layers in controlling the bulk properties of ultrafine particles is illustrated by studies on the saturation magnetization, I_s, of barium ferrite powders. The values of I_s decrease with decreasing grain size, as shown in Table I [8]. No change in saturation was found for grain sizes greater than 0.5 μ. The decrease in I_s for ultrafine-grain material has been explained on the basis of a surface layer of properties different from that of the bulk, and having a depth,

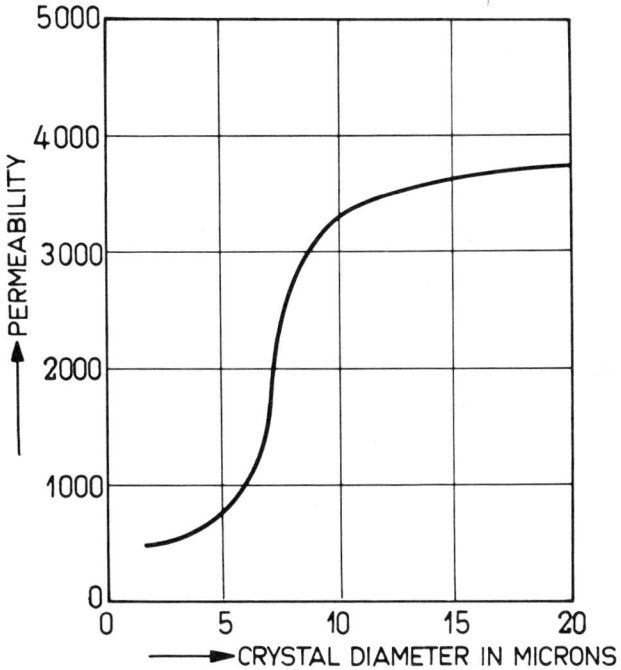

Figure 4. Effect of Grain Size on Permeability μ_o of Mn–Zn Ferrites (after Guillaud [6]).

r_o. We then obtain the following equation:

$$\left(\frac{I_s}{I_{s,o}}\right)^{1/3} = 1 - \frac{2r_o}{D} \tag{1}$$

where

      $I_{s,o}$ is the saturation magnetization of single crystal

      I_s is the saturation magnetization of grain having diameter D

TABLE I

Saturation Magnetization of Ultrafine-
Grain Barium Ferrite Powders

D (μ)	I_s (gauss)
>0.50	~150
0.50	150
0.43	134
0.42	105
0.38	90

The experimental results in Table I give an r_o value of about 0.1 μ. It should be noted that the depth of 0.1 μ for the surface layer should not be considered as typical for all barium ferrites, as this value would also depend to some extent on the method of preparation. However, the value of 0.1 μ is in good agreement with that found for magnetite powders.

Ferroelectric Behavior

A significant amount of work has been done in recent years to establish the fact that dielectric properties of ferroelectrics and piezoelectrics, such as $BaTiO_3$ and lead–zirconate titanates, are strongly dependent on microstructure, in particular on fine grain size [9-13].

The effect of ultrafine grain on dielectric properties of ceramics, is evidenced by the results reported for $BaTiO_3$ [11]. Figure 5 shows the effect of grain size on the relative dielectric constant, K', of hot pressed $BaTiO_3$. The results indicate that K' increases as the grain size decreases from about 7 μ and reaches a maximum value of about 5500, at an average grain size of about 1 μ. The increase in dielectric constant with decreasing grain size is not thoroughly understood. It has been proposed [14] that it is due to an increase in the internal stress of the material brought about by decreased domain twinning in the smaller grains. The work done at the U.S. Army Electronics Command, Ft. Monmouth [11], shows that in their $BaTiO_3$ material, the dielectric constant begins falling off again below 1 μ; other investigators have refuted this experimental observation. This effect of a maxima in K' at 1 μ cannot be explained by the phenomenological theory of Buessem et al. [14], because the stresses in a single domain particle (assuming a critical diameter of 1 μ for domains) should not depend on its absolute size. The fact that below 1 μ grain size the volume frac-

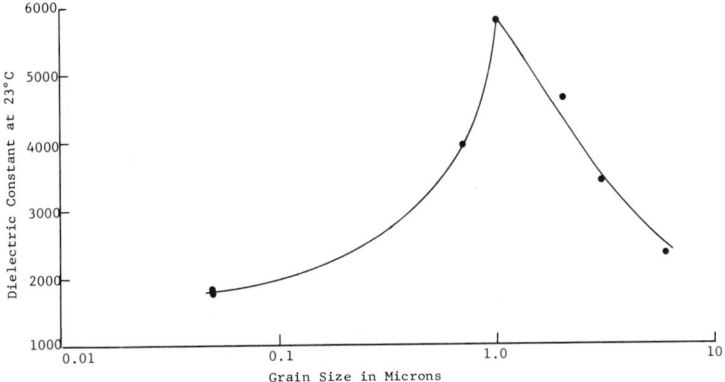

Figure 5. Dielectric constant vs. grain size for $BaTiO_3$ (after Brandmayr [11]).

tion of the grain boundary phase (see Figure 1) becomes greater than 0.5, and would therefore predominate over bulk properties, should also be considered. In order to understand the dielectric behavior of sub-micron grain sized sintered BaTiO₃ material, the effects of porosity, chemical bonding at boundaries, and deviations from stoichiometry must all be considered.

The effects of composition and microstructure on the electrical properties of lead–zirconate titanate ceramics has been investigated by Haertling [12, 15] at Sandia Corporation. Haertling found that significant changes in the electrical properties of hot pressed PZT (Zr/Ti ratio 65/35) containing 2 atomic % Bi ("65/35–2Bi") could be realized by change in bulk density and grain size.

Figure 6 shows the effect of bulk density and grain size on the squareness ratio, R_s, for "65/35-2Bi." These properties are interdependent, as density is a

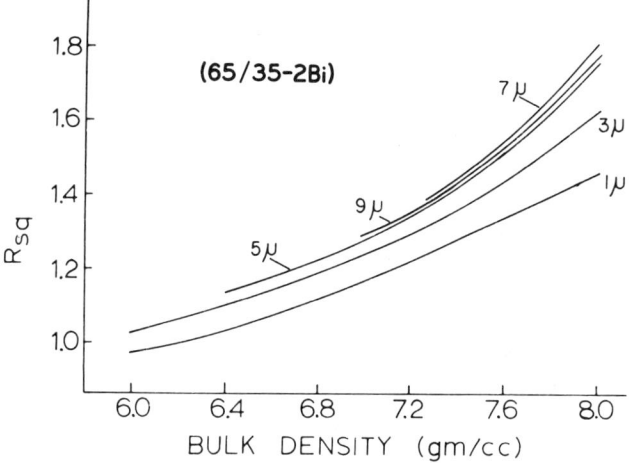

Figure 6. Effect of bulk density and grain size on squareness ratio for lead-zirconate titanate (after G. H. Haertling [15]).

critical consideration in ultrafine-grain ceramics. The R_s ratio increases with increasing density and grain size, although the effect of grain size only becomes noticeable at higher densities.

The remanent polarization, P_r, and coercive force, E_c, behavior of ultrafine-grain piezoelectric material is shown in Figure 7. The P_r sharply decreases, while the E_c increases with decreasing grain size. The data indicate that values of P_r and E_c are not simple functions of grain size but seem to be dependent on two or more exponential terms.

In contrast to the squareness ratio, the coercive force is highly dependent on grain size rather than bulk density, as shown in Figure 8. The E_c increased from 7.9 kv/cm to 12.2 kv/cm as grain size decreased from 10 μ to 1 μ. The increase

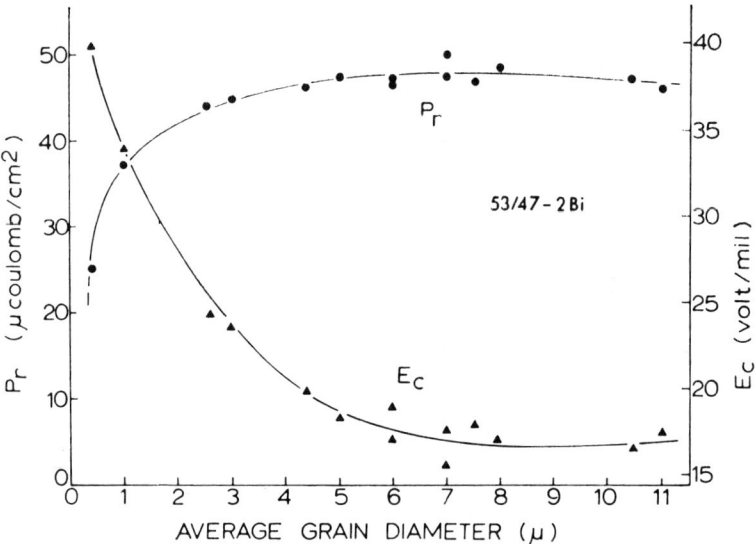

Figure 7. Remanent polarization and coercive field as functions of grain size for lead-zirconate titanate (after G. H. Haertling [12]).

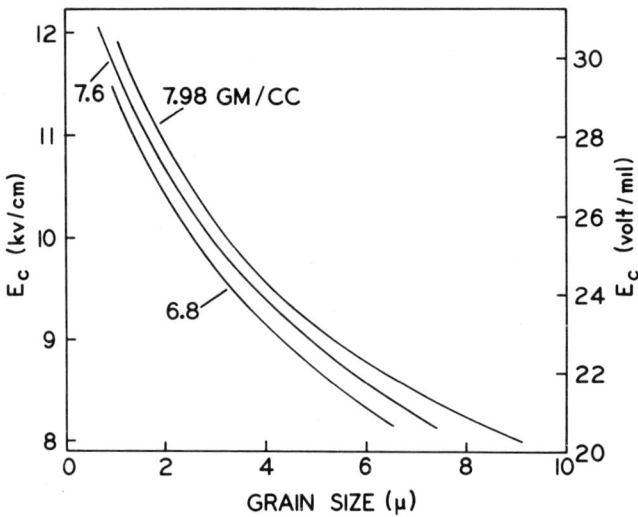

Figure 8. Variation of coercive field with grain size for lead-zirconate titanate (65/35-2 Bi) (after G. H. Haertling [15]).

in E_c is somewhat more significant for grain sizes of 3 μ and less. This increase in switching field is probably due to an increase in internal stresses with decreasing grain size. A decrease in grain size results in a fewer number of domain walls, which also leads to lower mechanical losses and hence higher Q_m, as shown in Figure 9. (Mechanical Q was also observed to have a very slight dependence on density.)

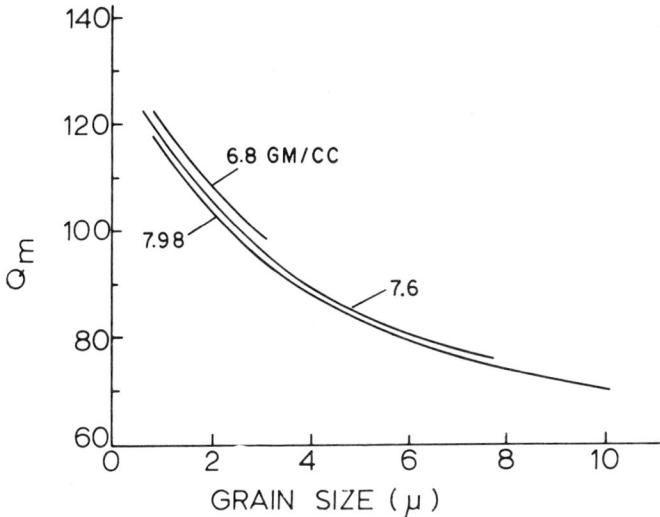

Figure 9. Variation of mechanical Q with grain size for lead-zirconate titanate (65/35-2Bi) (after G. H. Haertling [15]).

As in BaTiO₃, a decrease in grain size was found to increase the relative dielectric constant of material "65/35-2Bi." The ϵ/ϵ_o increased from 450 to 750 for a change of grain size from 8-1 μ, at a density of 7.98 g/cm³ as shown in Figure 10. The observed decrease in ϵ/ϵ_o with increasing porosity is well established, and needs no explanation.

Recent studies [16] on the optical properties of "65/35-2Bi" ferroelectrics have shown sharp and significant differences in the properties of ultrafine-grain materials as compared to those with larger grain sizes. With grain sizes larger than 2 μ, the material behaves as a light scatterer, and scatters light preferentially along the electric dipole direction. Material with grain sizes smaller than 2 μ displays characteristics of optical retarders, and does not preferentially scatter light along a given dipole direction.

The distinctly different and unique optical scattering behavior of material with grain size of about 1 μ and less may be due to a relative absence of domains. The photomicrographs of the large grained material show the presence of 180° domains; this was not the case with the 1 μ grain sized material.

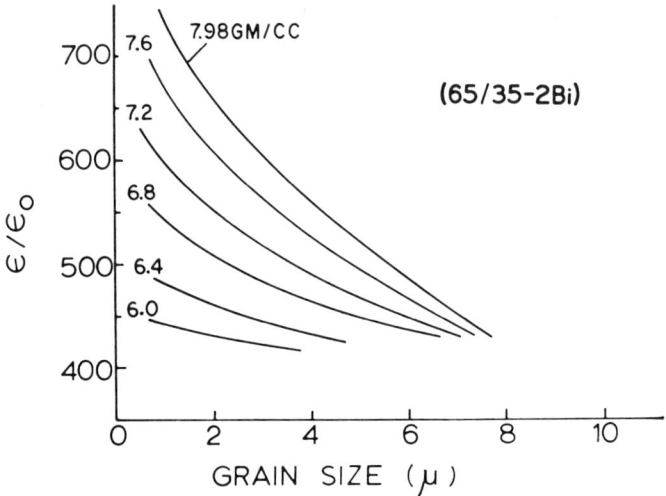

Figure 10. Variation of relative dielectric constant with grain size for lead-zirconate titanate (after G. H. Haertling [15]).

Other Effects

In the preceding section, we have seen how the electrical behavior of ultrafine-grain ferromagnetics and ferroelectrics is significantly different from that of coarse grained materials. In both ferromagnetic and ferroelectric materials, the role and interaction of "domains" have been evoked to explain the differences in properties. It is pertinent, at this time, to ask whether differences in electrical behavior due to ultrafine-grain sizes should exist in other ceramic materials as well.

In recent years, there has been a trend toward finer grained microstructures, along with higher purities, higher densities, and fewer defects in many electronic ceramics. This has resulted in marked improvements in both electrical and physical properties, but there are few direct comparisons which enable one to assign these improvements to any one single factor, such as grain size. For example, fine-grain alumina substrates are of interest because of their smooth surface texture, not because of any improvement in electrical properties, such as voltage breakdown strength. However, there is evidence that in materials other than ferromagnetics and ferroelectrics, ultrafine-grain size does effect specific electronic behavior.

Barta and co-workers [17] investigated the influence of purity and grain size on the resistivity of alumina bodies between $400°C$ and $700°C$, and reported that fine-grain sintered material had a higher resistivity than coarse grained material.

The effect of microstructure on the ionic conductivity of $Zr_{0.84}Ca_{0.16}O_{1.84}$

ceramic specimens (at temperatures less than 800°C) has indicated [18] that specimens of smaller grain size exhibited higher electrical conductivity, and suggests that, in this particular ceramic system, grain boundaries may have a higher conductivity than the bulk.

It can be argued that the effect of grain boundaries is related to the mean free path of the ions or electrons between collisions, which is of the order of interatomic distances for ionic conduction and is usually less than 100–150 Å for electronic conductivity. This implies that except for extremely fine grain sizes—less than 0.1 μ—the effects of boundary scattering are small compared to the lattice scattering. Consequently, decreasing grain size in homogeneous, uniform-composition materials should have little effect.

However, real ceramic materials are not necessarily uniform in composition in the interfacial and boundary regions, and substantial changes do take place due to impurity concentration and deviations from stoichiometry in grain boundaries.

The processing conditions are particularly significant in ultrafine-grain oxide semiconductors, in which the equilibrium of a non-stoichiometric composition with the atmosphere determines its electronic properties. The effect of such thermal conditioning can sometimes be rather unexpected in the case of ultrafine-grain ceramic materials, as indicated by some recent work done in our laboratories [19] on alumina. Figure 11 shows the current–voltage characteristics at 1300°C in a nitrogen atmosphere of polycrystalline alumina (average grain size 1 μ). These I-V traces which have been obtained on bulk specimens quenched from 1575°C, show negative conductance and hysteresis which have an unusual similarity with tunneling phenomena associated with thin films of Al_2O_3 and BeO. No such I-V traces were obtained with coarse grained alumina.

A somewhat speculative explanation for these results is that in the fine-grain material, with an increased number of grain boundaries, the thermal treatment at 1575°C results in an increased concentration of negative space charge, which probably decreases the electric field value beyond which the current is reduced with increasing field. At high local fields, tunneling through negative potential walls may occur.

However, these results do indicate that in ultrafine-grain material, the increased contribution of grain boundaries can introduce anomalous effects due to generation and interaction of defects. Therefore, particularly in ultrafine-grain materials, grain size and stoichiometry effects should not be considered as independent of each other, and it is sometimes difficult to evaluate the importance of grain size studies if information on stoichiometry is neglected.

In considering the behavior of ultrafine size in particulate materials, parameters other than the absolute size or the surface-to-volume ratio should also be considered. Weyl has indicated [20] that the differences in characteristics due to particle size are greater for oxides than for corresponding sulfides or tellurides.

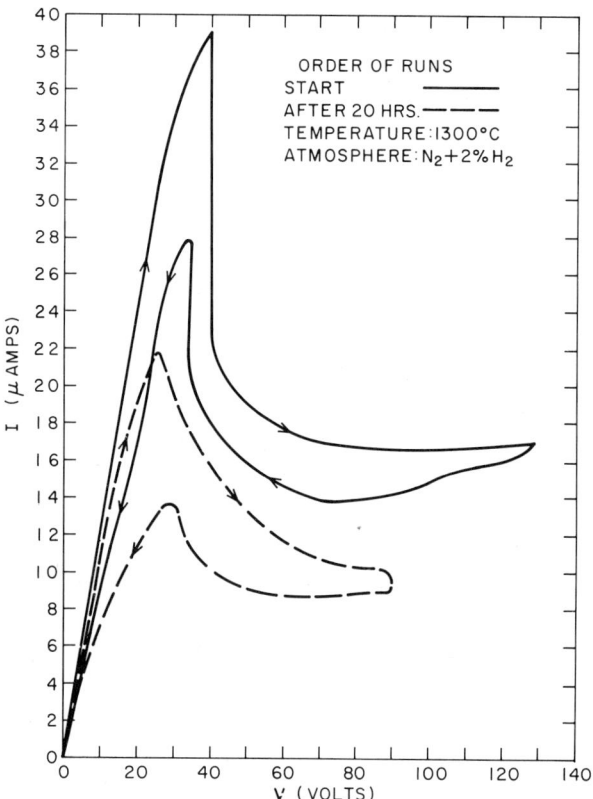

Figure 11. Current voltage characteristics of fine-grained alumina (average grain size 1 μ).

It also seems that a decrease in particle size causes lesser behavior differences, with increasing anion to cation ratios. Borides and carbides are more susceptible than fluorides to fine particle size effects. The reasons for these differences seem to be associated with the ability of some materials to form low energy surfaces. Only high energy surfaces cause a disproportionation of binding forces that affect the material to a considerable depth. It is because of this that marked particle size effects are invariably associated with oxide materials rather than their sulfides.

Electrical Characterization of Ultrafine-Grain Ceramics

If surfaces and grain boundaries are important in affecting the bulk properties of ultrafine-grain ceramics, how best can we characterize them? An equivalent circuit approach has been used to characterize electroceramics. For dielectric materials, the circuit would consist of a capacitance and parallel resistance representing each of the components, namely surfaces, grain boundaries, and bulk. A typical equivalent circuit of a fine-grain ceramic is shown below, where subscripts 'v,' 'b,' and 's' represent the capacitance, C, and resistance, R, of bulk grains, grain boundaries, and surfaces, respectively.

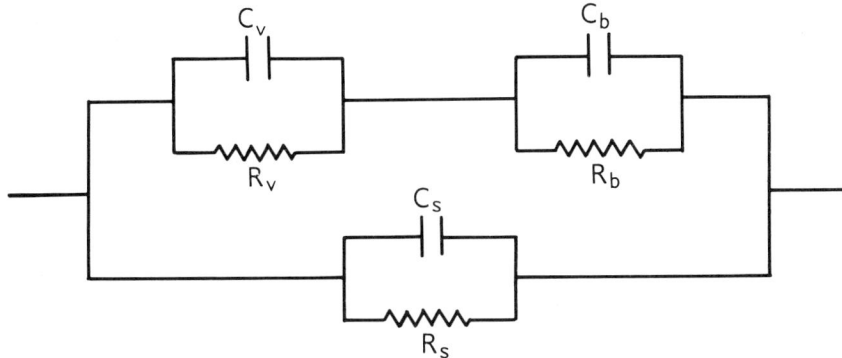

In attempting to characterize ultrafine-grain ceramics for electronic applications, attempts should be made to answer the following questions:

1. What are the properties of the surface defect layer of a particle and how do they change with particle size and oxygen partial pressure?

2. What are the electrical properties of the boundary phase in sintered ceramics and their relationship to grain size and processing environment?

3. What is the correlation between the properties of the defect surface layer of the powder particles and the properties of the boundary phase in the sintered ceramic?

In the following sections, work being done at IITRI under the sponsorship of NASA, Langley Research Center [21], on the role of surfaces and grain boundaries in fine-grain particulate and sintered $BaTiO_3$ materials is presented.

Surface Characteristics of $BaTiO_3$ Particulates

Surface Barrier-Layer Model

The presence of surface layers that alter the properties of barium titanate ($BaTiO_3$) fine particles has been postulated by Kanzig [22, 23], Chynoweth [24], and other investigators. Electron diffraction results by Kanzig and co-workers [22], indicated that deviations from the normal ferroelectric behavior of very small $BaTiO_3$ particles were mainly due to a difference between the properties of the bulk of the particles and those of a surface layer about 100 Å thick. There is experimental evidence [25] that the surface layer has a higher resistivity than the bulk of the particle. It has also been postulated [23] that this surface is of the nature of a Schottky-type depletion layer. The presence of surface layers has been involved in explaining the dependence of switching parameters on thickness and other pyroelectric effects in $BaTiO_3$ single crystals. Surface layers on single crystals, however, may well be of a different type.

This section describes a defect-layer model for the surface of $BaTiO_3$ particles [21] that qualitatively explains their experimental dielectric properties. Figure 12 schematically shows a model for a high-resistivity surface defect layer on a $BaTiO_3$ particle. A $BaTiO_3$ particle probably has a layer of adsorbed oxygen around its outer surface, the strength of the bond of which depends on the energy of the available surface states. We have frequently observed that if $BaTiO_3$ powders are exposed to a low oxygen partial pressure, a wide adsorption band in the near-infra-red region having free carrier characteristics may result.

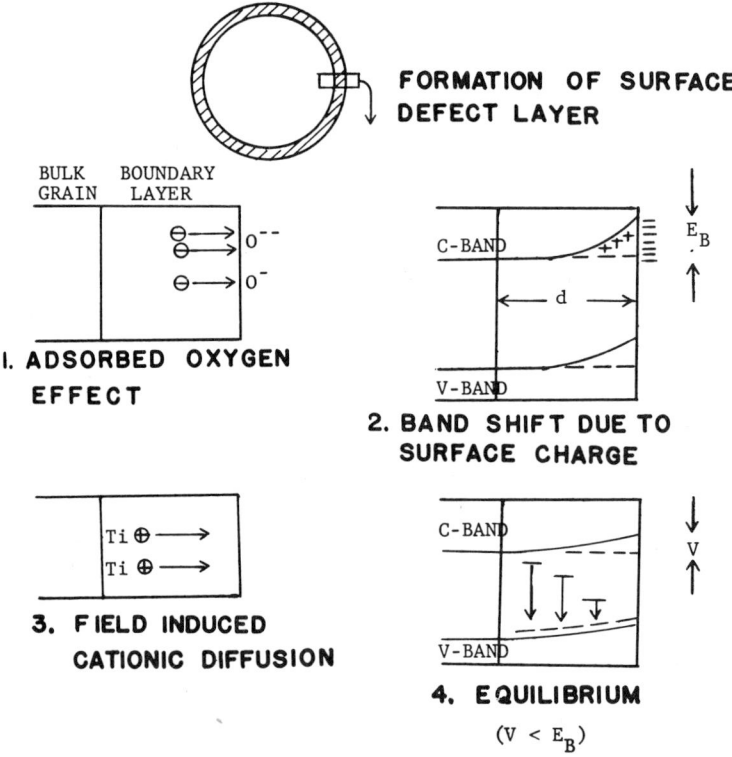

Figure 12. Barrier-layer model for $BaTiO_3$ particle.

The adsorbed oxygen atoms accept electrons from the particle to form the bond and hence represent a negative surface charge. The thickness of the depletion layer, d, depends on the donor concentration, among other factors, and may be considered to be of the order of 10^{-5} cm for an assumed barrier height, E_B, of 1 ev. The resulting high electric field ($\sim 10^5$ v/cm) across the depletion layer can induce cationic migration to the surface, especially since the titanium ion is the most mobile cation in the $BaTiO_3$ structure. The higher mobility of Ti ions is inferred from the crystal chemistry and structure of $BaTiO_3$ by the existence of pronounced ionic polarization effects when $BaTiO_3$ is subjected to dc voltages, and from the molecular quantitative theory of Kinase and Takahashi [26]. The unusual and strong shift of the optical adsorption edge in a high electric field, reported [27] for $BaTiO_3$, is also presumably due to the movement of Ti ions. The adsorption edge is shifted towards higher energies, contrary to the Franz-Keldysh effect.

In our model it is assumed that the titanium ions combine with adsorbed oxygen and form oxygen-deficient TiO_x. Some inconclusive experimental evidence [28] for the presence of TiO_x-type material at the surface has been obtained by electron diffraction of flame-sprayed single-crystal $BaTiO_3$ grains. The cationic migration to the surface ceases whenever sufficient surface charge has been neutralized, and the field is thus decreased. The essential feature of the proposed model is the generation of states near the top of the highest filled band by means of cationic migration to the surface. The resistivity of the depletion layer is increased in two ways: by the adsorption of oxygen, and by the trapping of intrinsic or injected electrons at acceptor states.

Stoichiometric Effects

The importance of stoichiometry on the electrical behavior of ultrafine-particle materials is shown by work done at IITRI [21] on the effect of water vapor on micron and sub-micron sized $BaTiO_3$ powders.

Experiments have been carried out to determine the effects of water vapor on $BaTiO_3$ powders having comparable particle size ($\sim 1~\mu$) and surface areas, but differing BaO/TiO_2 ratios. Figure 13 shows the effect of stoichiometry on dielectric loss due to water vapor. The two powders have nearly identical values of σ_{ac} at a water partial pressure (P_{H_2O}) of 7×10^{-6} atm and in vacuum ($\sim 10^{-6}$ torr). However, with increased values of P_{H_2O} the dielectric loss in the material with excess TiO_2 is enhanced as compared with the stoichiometric material.

On the basis of this experimental evidence, it has been hypothesized that the differences in the dielectric loss behavior, as a function of very slight deviations from stoichiometry, are due to different manners in which water interacts with

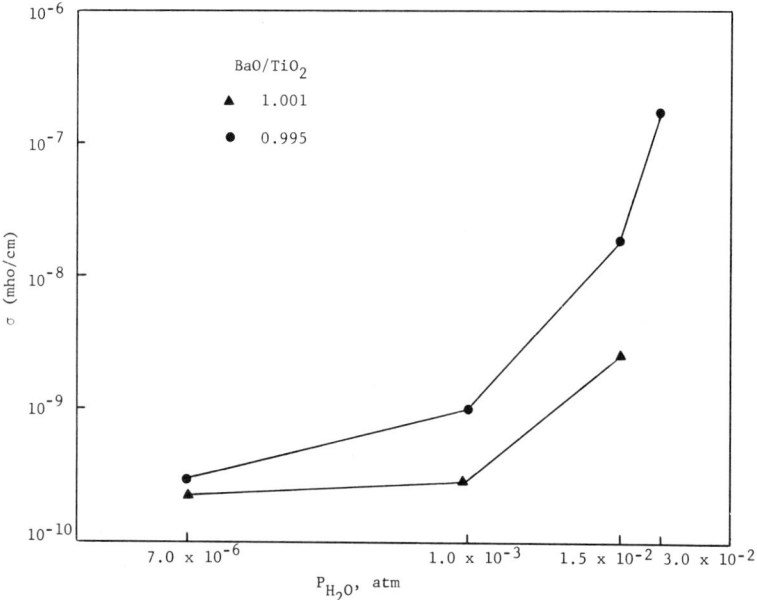

Figure 13. Effect of stoichiometry on dielectric loss due to water vapor on $BaTiO_3$ powders (measured at 1 kHz).

stoichiometric and off-stoichiometric "$BaTiO_3$" surfaces. It is conceivable that in barium titanate material with excess Ti, the injected protons H^{+1} from the adsorbed water, more easily reduces the Ti^{+4} to Ti^{+3}, thus causing an increase in the loss. The exact nature of the adsorbed species and the defects created therein are being verified by a combination of techniques such as infra-red adsorption spectra and n.m.r.

However, these results do indicate that the surfaces of barium titanate powders can be characterized by evaluating the changes of ac conductivity due to controlled partial pressure of water. Further, it may be possible to monitor very small deviations in stoichiometry (BaO/TiO_2 ratio) particularly in ultrafine-particle materials from significant changes in the ac conductivity, $\Delta\sigma$, as a function of P_{H_2O}.

Use of Electron-Beam Scanning Technique for Studying Electrical Behavior of Grain Boundaries

The effects of grain size and the role of grain boundaries in the electrical behavior of ceramics have been studied by carrying out measurements on polycrystalline specimens as a function of grain size, and on bicrystals. Although studies on bulk bicrystals can be useful in explaining the role of grain boundaries in the over-all conduction process, they do not tell us anything about the characteristics of the grain boundary itself. This can only be determined by studying the intrinsic behavior of grain boundaries. There is substantial evidence that the influence of grain boundaries, in typical ceramic systems, extends over a range up to an order of microns. Because of this, and of the inadequacies of the bulk-bicrystal studies, an electron-beam scanning (EBS) technique has been developed for the direct measurements of dielectric properties of the grain boundary region.

The EBS system, developed at IITRI, was first constructed in 1965 [29]. In this technique, one surface of the dielectric material is charged with a scanning electron beam in vacuum. The charge measurement is obtained in the form of a video signal output, which is a measure of the spatially resolved conductivity of the sample.

The mechanical arrangements for dielectric measurements using this technique are straightforward. Figure 14 is a schematic diagram of the EBS system. The

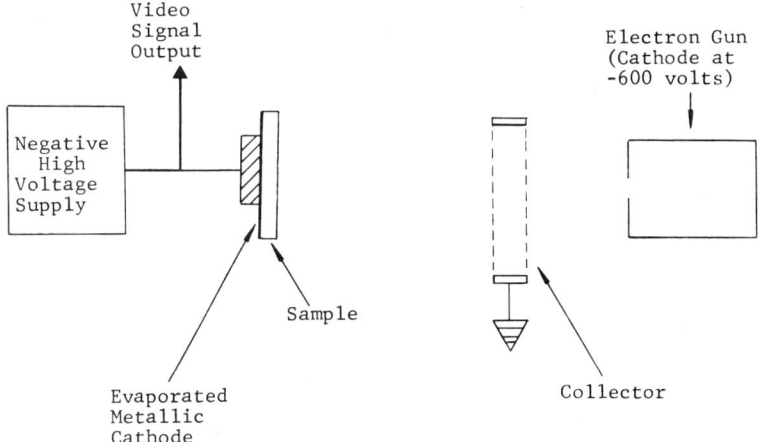

Figure 14. Schematic diagram of the electron-beam scanning system.

sample is backed on one side by a metallic electrode, while the other side is irradiated by an electron beam produced by a conventional electron gun. A collector grid or a large-diameter collector ring is placed close to the sample. The collector electrode may be at ground potential, and the backing electrode is connected to a variable high-voltage supply, initially at ground potential. If the electron accelerating potential is of the order of several hundred volts and if the secondary electron emission coefficient, δ_{eff}, of the insulator is greater than unity at this voltage, then initially the potential of each spot on the insulator which is bombarded by the primary beam will tend to change from zero to a positive value. The sample surface potential will come to an equilibrium value a few volts above ground, when $\delta_{eff} = 1$. The sample is then no longer being charged. If the backing electrode is then slowly raised to a high potential of either polarity while the electron beam continues to scan the sample in a regular pattern, each spot on the front surface of the sample will be brought back to the same equilibrium potential whenever the beam strikes it. The charge deposited on the sample will be opposite in polarity to that of the high-voltage backing electrode and of a magnitude sufficient to charge the sample capacitance to the potential difference between its surfaces, and also to compensate for the conduction which has occurred between its surfaces. The determination of dielectric properties depends on knowing the voltage drop across the sample and the charge deposited on the sample by the electron beam.

The usefulness of the EBS technique in studying the electrical behavior of grain boundaries can be illustrated by describing some work recently initiated in our laboratories, on the properties of grain boundaries in potassium chloride and barium titanate [30].

A conductivity profile was carried out on a KCl bicrystal (twist 12°). Figures 15 and 16 are single line scans across the grain boundary corresponding to scan

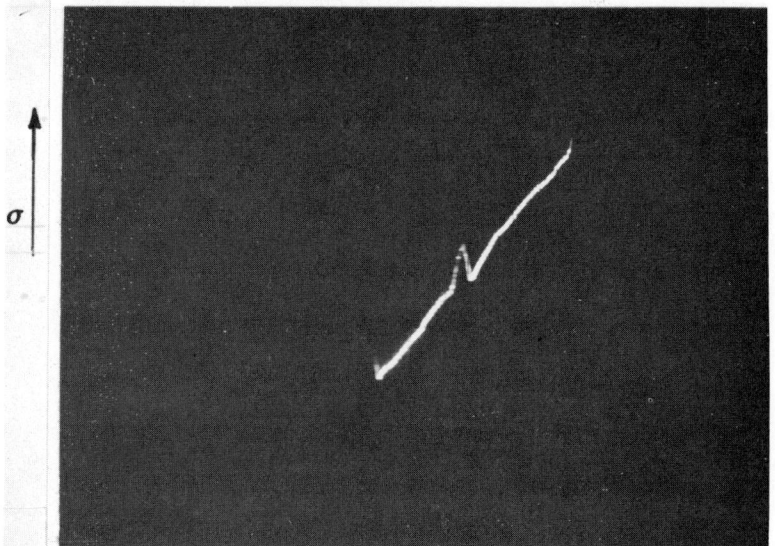

Figure 15. Single line conductivity scan across grain boundary of KCl specimen (scan delay: 2 min).

delays of 2 and 3 minutes, respectively. The beam current was ostensibly the same in both photographs, so that the relative heights of the peaks across the grain boundary region are most likely an indication of the surface potentials of the specimen which are related to the local bulk conductivity—assuming the local

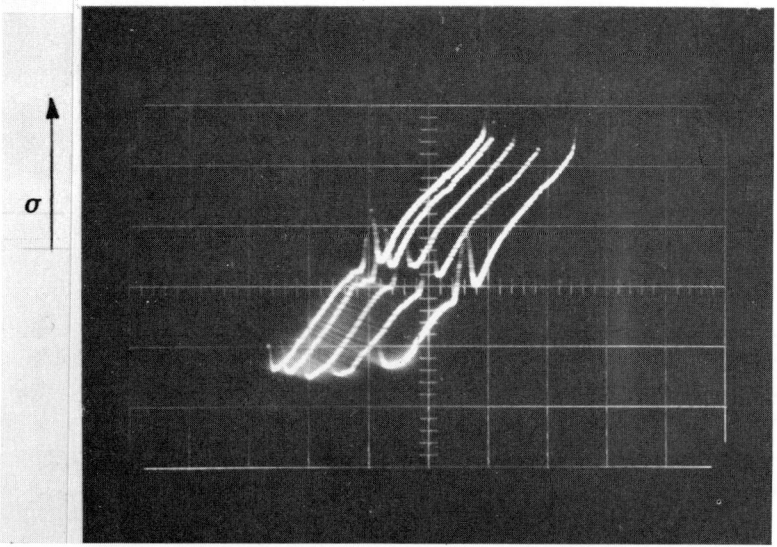

Figure 16. Single line conductivity scan across grain boundary of KCl specimen (scan delay of trace at left: 3 min).

transverse surface resistivity to be relatively high. From Figures 15 and 16, it is evident that an increase in delay time increases the intensity of the conductivity peak in the direction of the higher differential conductivity. This tends to confirm the fact that the conductivity profile peak, at the grain boundary, is a true conductivity effect rather than being due to secondary emission effects.

The present results indicate that twist grain boundaries in KCl have a higher electrical conductivity than the grains. Experiments are in progress to obtain a semi-quantitative measure of the ratios of the conductivities (or conductances) of the grain boundary region to that of the grains, for KCl boundaries as a function of tilt and twist misorientations [30]. Preliminary data have also been obtained on obtaining permittivity profiles on multicrystalline $BaTiO_3$ specimens.

The Electronic "Character" of Grain Boundaries

In order to understand the nature of differences in electrical behavior that could occur in ultrafine-grain ceramics, it is necessary to consider the "electronic characteristics" of boundary (and interfacial) regions. A speculative examination of some of these characteristics is presented in this section.

The grain boundary region in ceramic materials can be considered in a number of different ways. Grain boundaries have often been referred to as "amorphous" in nature. A better way of characterization would be by defining the grain boundary region as having "short-range order" or a high degree of disorder.

The grain boundary region consists of an array of dislocations. In addition, in presently available polycrystalline materials, the impurities, vacancies, and adsorbed gases further contribute to the imperfections of the grain boundaries. Dislocations can have pronounced effects on the electrical properties of materials. Associated with dislocations are energy states lying in the energy gap. The majority of the work on the electrical effects of dislocations has been done on semiconductor materials such as Ge, Si, and InSb. The observations of Gallagher [31] and Pearson et al. [32] on lightly deformed Ge show that the dislocation energy levels are acceptor types; that is, the introduction of dislocations results in a decrease in the number of conduction electrons. The dislocations also scatter electrons and so reduce the mobility. Another effect associated with dislocations in Ge and Si is the formation of an atmosphere of impurity atoms along the dislocation line. This further enhances the recombination by providing localized impurity-trapping levels.

Although the majority of the theoretical and experimental work pertaining to the electrical properties of dislocations has been on broad-band semiconductor materials, some evidence is available on the effects of dislocations on oxide-type materials. Chang [33] has observed that during "transient" creep of Al_2O_3, the electrical resistivity decreases. This has been attributed to the creation of new dislocations carrying charges opposite in sign to that of the free carriers in the bulk of the material. The observed increase in electrical resistivity during "steady state" creep has been attributed to the creation of carrier-trapping point defects in the wake of moving dislocations. Recent experimental data on polycrystalline Al_2O_3 in our own laboratories [19] tend to indicate that charged dislocations at grain boundaries can act as traps for electrons.

Therefore, it is possible that in fine-grain ceramics the presence of an array of dislocations in the grain boundary region acts as a barrier to electronic transport. The dislocation interaction of the grain boundary region would be further dependent on the size of the adjoining grains, and would therefore exert a different "trapping" effect for large and small grain sizes.

In addition to dislocations in the grain boundary region of ultrafine-grain ceramics, impurity ions, vacancies, and adsorbed gases can all interact and contribute towards the creation and interaction of charge carriers in the material. All these imperfections must be considered as traps at grain boundaries, forming potential barriers through which electrons, taking part in the conduction process, have to tunnel.

It is not yet clear how the grain boundary region acquires electrical characteristics different from that of the bulk, or the mechanism whereby it is responsible for creating barriers for electron transport between adjacent grains. However, several possibilities do exist.

For instance, if dislocations accept electrons, the grain boundary charges up negatively and presents a barrier to current flow in n-type materials; if they donate electrons, the grain boundary charges up positively and presents a barrier to current flow in p-type materials.

Deviations from stoichiometry in ceramics such as $BaTiO_3$ further complicate the situation. If the deviation from stoichiometry, for instance, resulted in more anion (oxygen) vacancies rather than cation vacancies, and if dislocations are regarded as a source of vacancies, then they would acquire (in equilibrium) more negatively charged jogs than positive. The dislocation line would then become negative with respect to the bulk and would be surrounded by a positive space charge of anion vacancies. The negative charge of a grain boundary can come from the donors whose electrons are captured by boundary states. As a result, grain boundary charges leave ionized donors behind. Little is also known about the effects of anion impurities in the boundary region of fine-grain ceramics.

Although it is not possible at this time to define the predominant mechanism that characterizes the grain boundary region, it seems that this region can be electronically different from the adjoining grains. The manner in which this defect region interacts with the bulk charge transport processes would depend on the dimensions of the adjoining grains. The effective thickness of the boundary region may be of the order of 0.1 μ, typical for a Schottky-type barrier or space-charge dominated region. For ceramics with grain sizes less than 1 μ, the electrical properties would then be significantly altered by the characteristics of the grain boundary region, and would be different from those of comparatively larger grain sizes ($\geqslant 1 \mu$). Undoubtedly, much remains to be done to unambiguously characterize the properties of the grain boundary region.

· Summary

The magnetic and dielectric behavior of ceramic materials, such as magnetic ferrites and ferroelectric titanates, are susceptible to ultrafine-grain size effects. The significance of decreasing size of particles and grains has been considered from their increased surface-to-volume and grain boundary-to-grain ratios. However, from an atomistic point of view, other parameters such as anion-to-cation ratios should also be considered.

The dielectric behavior of ultrafine $BaTiO_3$ particulates has been explained on the basis of a surface barrier-layer model. The essential features of the proposed model are that cation vacancies created by the process of cationic migration to the surface represent acceptor states which may trap electrons from energy levels above the trap state. The number of electrons available for conduction would be drastically reduced by this means, and the resistivity of the depletion layer would probably rise by several orders of magnitude.

Experimental results have been presented which indicate that the surfaces of ultrafine-particulate barium titanate can be characterized by evaluating the changes in the dielectric loss phenomena. Small changes in stoichiometry of the material (BaO/TiO_2 ratio) can be correlated with a marked change in the ac conductivity as a function of partial pressure of water.

A new electron-beam scanning (EBS) technique for the direct evaluation of the electrical characteristics of the grain boundary region has been developed. EBS profiles indicate that the differences in conductivities between the grain boundary region and the adjoining grains, in potassium chloride, are dependent on boundary misorientation.

The electronic character of grain boundaries and the role of grain boundaries in the electronic behavior of ultrafine-size ceramics have been discussed.

References

1. Kittel, C., "Phys. Theory of Ferromagnetic Domains," *Rev. Mod. Phys.*, 21 (1949), 541–83.
2. Blum, S. L., "Microstructure and Properties of Ferrites," *J. Am. Ceram. Soc.*, 41 (1958), 489–93.
3. Went, J. J., Rathenau, G. W., Gorter, E. W. and Van Oosterhout, G. W., "Ferroxdure, a Class of New Permanent Magnet Materials," *Philips Tech. Rev.*, 13 (1951–52), 194–208.
4. Brockman, F. G., Beck, P. W. and Staneck Jr., W. G., "Research Investigations of Magnetic Materials, Permanent Ceramic Type," Eighth Progress Rept., Contract No. DA–36–039 sc–42503, Philips Research Laboratories, New York.
5. Guillaud, C., "Propriétés d'Aimant Des Poudres Ferromagnetiques cas Particulier de l'Alliage MnBi," *J. Rech. C.N.R.S.*, 2 (1949), 267–78.
6. Guillaud, C., "The Properties of Manganese–Zinc Ferrites and the Physical Processes Governing Them," *Proc. Inst. Electr. Engrs.*, 104B (1957), 165–73.
7. Paulus, M., "Influence des pores et des inclusiens sur la croissance des cristaus de ferrite," *Phys. Stat. Sol.*, 2 (1962), 1325–41.
8. Torkar, K. and Fredriksen, O., "The Effect of Grain Size on Saturation Magnetization of Barium Ferrite Powders," *Powder Metallurgy*, 4 (1959), 105–107.
9. Kniepkamp, H. and Hewang, W., "Über Depolarisationseffekte in polykristallinem $BaTiO_3$," *Z. Angew Physik*, 6 (1954), 385–90.
10. Jonker, G. H. and Noorlander, W., *Science of Ceramics*, Vol. I, Academic Press, New York (1962), 225.
11. Brandmayr, R. J., "Properties of Ultrafine Grained $BaTiO_3$ Super-Pressed at Low Temperatures," Tech. Report ECOM–2719 (August 1966).
12. Haertling, G. H., "Hot-Pressed Lead–Zirconate Lead–Titanate Ceramics Containing Bismuth," *Bull. Am. Ceram. Soc.*, 43 (1964), 875–79.
13. Mountvala, A. J., "Hot-Pressing Piezoelectric and Ferroelectric Materials," *Bull. Am. Ceram. Soc.*, 42 (1963), 120–21.
14. Buessem, W. R., Cross, L. E. and Goswami, A. K., "Phenomenological Theory of High Permittivity in Fine-Grained Barium Titanate," *J. Am. Ceram. Soc.*, 49 (1966), 33–36; and "Effect of Two-Dimensional Pressure on the Permittivity of Fine- and Coarse-Grained Barium Titanate," *ibid*, 49 (1966), 36–39.

15. Haertling, G. H., "Improved Ceramics for Piezoelectric Devices," presented at Western Electric Show and Convention, Los Angeles, California (August 23–26, 1966).
16. Land, C. E., "Ferroelectric Ceramic Electro-optic Storage and Display Devices," presented at IEEE International Electron Devices Meeting, Washington, D.C. (October 18–20, 1967).
17. Barta, R., *et al.*, "Electrical Resistance of Sintered Alumina," *Silikaty*, 1 (1957), 77–83.
18. Tien, T. Y., "Grain Boundary Conductivity of $Zr_{0.84}Ca_{0.16}O_{1.84}$ Ceramics," *J. Appl. Phys.*, 35 (1964), 122–24.
19. Mountvala, A. J., "Effect of Thermal Conditioning on I–V Characteristics of Polycrystalline Al_2O_3," to be published.
20. Weyl, W. A., Private communications.
21. Mountvala, A. J., "Characterization of Ceramic Materials for Microelectronic Applications," NASA Report No. CR-6605 (March 1968).
22. Anliker, M., Burger, H. R. and Kanzig, W., "Das Verhalten von Kolloidalen Seignettealektrika III, Barium Titanat $BaTiO_3$," *Helvetica Physica Acta*, 27 (1954), 99.
23. Kanzig, W., "Space Charge Layer Near the Surface of a Ferroelectric," *Phys. Rev.*, 98 (1959), 549.
24. Chynoweth, A. G., "Surface Space-Charge Layers in Barium Titanate," *Phys. Rev.*, 102 (1956), 705–14.
25. Gerthsen, P. and Hardtl, K., "Eine Methode zum direkten Nachweis von Leitfähigkeitsinhomogenitäten an Korngrenzen," *Z. Naturforsch*, 18a (1963), 423.
26. Kinase, W. and Takahashi, H., "Theory of Spontaneous Deformation of Barium Titanate," *J. Phys. Soc. Japan,* 10 (1955), 942–52.
27. Gaehwiller, Ch., "Absorptionkantenvehschiebung an $BaTiO_3$ in einem elektrischen Feld," *Helvetica Physica Acta*, 38 (1965), 361.
28. Ulrich, D. R., Private communications.
29. Seiwatz, H., "A Nondestructive Electron Beam Scanning System for Measuring Dielectric Properties Including Breakdown," Proc. of the Fifty-First Meeting of the Juniper Committee, Sandia Corporation, May 1966.
30. Mountvala, A. J. and Marks, H. R., "Dielectric Properties of Ceramic Grain Boundaries by Electron Beam Scanning Technique," to be published.
31. Gallagher, C. S., "Plastic Deformation of Germanium and Silicon," *Phys. Rev.*, 88 (1952), 721.
32. Pearson, G. L., Read, W. T. and Morin, F. J., "Dislocations in Plastically Deformed Germanium," *Phys. Rev.*, 93 (1954), 666.
33. Chang, R., "Electrical Resistivity Changes of Al_2O_3 Crystals During Creep," *J. Appl. Phys.*, 34 (1963), 1564.

Index